U0211622

影印版说明

MOMENTUM PRESS 出版的 *Plastics Technology Handbook*（2 卷）是介绍塑料知识与技术的大型综合性手册，内容涵盖了从高分子基本原理，到塑料的合成、种类、性能、配料、加工、制品，以及模具、二次加工等各个方面。通过阅读、学习本手册，无论是专业人员还是非专业人员，都会很快熟悉和掌握塑料制品的设计和制造方法。可以说一册在手，别无他求。

原版 2 卷影印时分为 11 册，第 1 卷分为 ：

塑料基础知识 · 塑料性能

塑料制品生产

注射成型

挤压成型

吹塑成型

热成型

发泡成型 · 压延成型

第 2 卷分为 ：

涂层 · 浇注成型 · 反应注射成型 · 旋转成型

压缩成型 · 增强塑料 · 其他工艺

模具

辅机与二次加工设备

唐纳德 V · 罗萨多，波士顿大学化学学士学位，美国东北大学 MBA 学位，马萨诸塞大学洛厄尔分校工程塑料和加州大学工商管理博士学位（伯克利）。著有诸多论文及著作，包括《塑料简明百科全书》、《注塑手册（第三版）》以及塑料产品材料和工艺选择手册等。活跃于塑料界几十年，现任著名的 Plasti Source Inc. 公司总裁，并是美国塑料工业协会（SPI）、美国塑料学会（PIA）和 SAMPE（The Society for the Advancement of Material and Process Engineering）的重要成员。

材料科学与工程图书工作室

联系电话 0451-86412421

0451-86414559

邮　　箱 yh_bj@aliyun.com

xuyaying81823@gmail.com

zhxh6414559@aliyun.com

影印版

PLASTICS TECHNOLOGY HANDBOOK

塑料技术手册

VOLUME 2

MOLD AND DIE TOOLING

模具

EDITED BY

DONALD V. ROSATO

MARLENE G. ROSATO

NICK R. SCHOTT

哈爾濱工業大學出版社

HARBIN INSTITUTE OF TECHNOLOGY PRESS

黑版贸审字08-2014-093号

Donald V.Rosato,Marlene G.Rosato,Nick R.Schott
Plastics Technology Handbook Volume 2
9781606500828

Originally published by Momentum Press, LLC
English reprint rights arranged with Momentum Press, LLC through McGraw-Hill Education (Asia)

图书在版编目（CIP）数据

塑料技术手册. 第2卷. 模具 =Plastics technology handbook volume 2 mold and die tooling : 英文 /（美）罗萨多（Rosato, D. V.）等主编. —影印本. — 哈尔滨 : 哈尔滨工业大学出版社, 2015.6

ISBN 978-7-5603-5049-3

Ⅰ. ①塑… Ⅱ. ①罗… Ⅲ. ①塑料 – 技术手册 – 英文 ②塑料模具 – 技术手册 – 英文 Ⅳ. ①TQ320.6-62

中国版本图书馆CIP数据核字（2014）第280095号

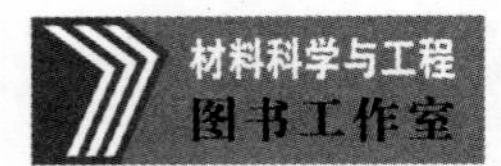

责任编辑 杨 桦 许雅莹 张秀华
出版发行 哈尔滨工业大学出版社
社 址 哈尔滨市南岗区复华四道街10号 邮编 150006
传 真 0451-86414749
网 址 http://hitpress.hit.edu.cn
印 刷 哈尔滨市石桥印务有限公司
开 本 787mm × 960mm 1/16 印张25.5
版 次 2015 年 6 月 第 1 版 2015 年 6 月第 1 次印刷
书 号 ISBN 978-7-5603-5049-3
定 价 130.00元

（如因印刷质量问题影响阅读，我社负责调换）

PLASTICS TECHNOLOGY HANDBOOK

VOLUME 2

EDITED BY

Donald V. Rosato, PhD, MBA, MS, BS, PE
PlastiSource Inc.
Society of Plastics Engineers
Plastics Pioneers Association
UMASS Lowell Plastics Advisory Board

Marlene G. Rosato, BASc (ChE), P Eng
Gander International Inc.
Canadian Society of Chemical Engineers
Association of Professional Engineers of Ontario
Product Development and Management Association

Nick R. Schott, PhD, MS, BS (ChE), PE
UMASS Lowell Professor of Plastics Engineering Emeritus & Plastics Department Head Retired
Plastics Institute of America
Secretary & Director for Educational and Research Programs

Momentum Press, LLC, New York

CONTENTS

FIGURES

Tables

ABBREVIATIONS

AA acrylic acid
AAE American Association of Engineers
AAES American Association of Engineering Societies
ABR polyacrylate
ABS acrylontrile-butadiene-styrene
AC alternating current
ACS American Chemical Society
ACTC Advanced Composite Technology Consortium
ad adhesive
ADC allyl diglycol carbonate (also CR-39)
AFCMA Aluminum Foil Container Manufacturers' Association
AFMA American Furniture Manufacturers' Association
AFML Air Force Material Laboratory
AFPA American Forest and Paper Association
AFPR Association of Foam Packaging Recyclers
AGMA American Gear Manufacturers' Association
AIAA American Institute of Aeronautics and Astronauts
AIChE American Institute of Chemical Engineers
AIMCAL Association of Industrial Metallizers, Coaters, and Laminators
AISI American Iron and Steel Institute
AMBA American Mold Builders Association
AMC alkyd molding compound
AN acrylonitrile
ANSI American National Standards Institute
ANTEC Annual Technical Conference (of the Society of the Plastic Engineers)
APC American Plastics Council
APET amorphous polyethylene terephthalate
APF Association of Plastics Fabricators
API American Paper Institute
APME Association of Plastics Manufacturers in Europe
APPR Association of Post-Consumer Plastics Recyclers
AQL acceptable quality level
AR aramid fiber; aspect ratio
ARP advanced reinforced plastic
ASA acrylonitrile-styrene-acrylate
ASCII american standard code for information exchange
ASM American Society for Metals

ASME American Society of Mechanical Engineers
ASNDT American Society for Non-Destructive Testing
ASQC American Society for Quality Control
ASTM American Society for Testing Materials
atm atmosphere
bbl barrel
BFRL Building and Fire Research Laboratory
Bhn Brinell hardness number
BM blow molding
BMC bulk molding compound
BO biaxially oriented
BOPP biaxially oriented polypropylene
BR polybutadiene
Btu British thermal unit
buna polybutadiene
butyl butyl rubber
CA cellulose acetate
CAB cellulose acetate butyrate
$CaCO_3$ calcium carbonate (lime)
CAD computer-aided design
CAE computer-aided engineering
CAM computer-aided manufacturing
CAMPUS computer-aided material preselection by uniform standards
CAN cellulose acetate nitrate
CAP cellulose acetate propionate
CAS Chemical Abstract Service (a division of the American Chemical Society)
CAT computer-aided testing
CBA chemical blowing agent
CCA cellular cellulose acetate
CCV Chrysler composites vehicle
CEM Consorzio Export Mouldex (Italian)
CFA Composites Fabricators Association
CFC chlorofluorocarbon
CFE polychlorotrifluoroethylene
CIM ceramic injection molding; computer integrated manufacturing
CLTE coefficient of linear thermal expansion
CM compression molding
CMA Chemical Manufacturers' Association
CMRA Chemical Marketing Research Association
CN cellulose nitrate (celluloid)
CNC computer numerically controlled
CP Canadian Plastics
CPE chlorinated polyethylene
CPET crystallized polyethylene terephthalate
CPI Canadian Plastics Institute
cpm cycles/minute
CPVC chlorinated polyvinyl chloride
CR chloroprene rubber; compression ratio
CR-39 allyl diglycol carbonate
CRP carbon reinforced plastics
CRT cathode ray tube
CSM chlorosulfonyl polyethylene
CTFE chlorotrifluorethylene
DAP diallyl phthalate
dB decibel
DC direct current
DEHP diethylhexyl phthalate
den denier
DGA differential gravimetric analysis
DINP diisononyl phthalate
DMA dynamic mechanical analysis
DMC dough molding compound
DN *Design News* publication
DOE Design of Experments
DSC differential scanning calorimeter
DSD Duales System Deutschland (German Recycling System)
DSQ German Society for Quality
DTA differential thermal analysis
DTGA differential thermogravimetric analysis
DTMA dynamic thermomechanical analysis
DTUL deflection temperature under load
DV devolatilization
DVR design value resource; dimensional velocity research; Druckverformungsrest (German

compression set); dynamic value research; dynamic velocity ratio
E modulus of elasticity; Young's modulus
EBM extrusion blow molding
E_c modulus, creep (apparent)
EC ethyl cellulose
ECTFE polyethylene-chlorotrifluoroethylene
EDM electrical discharge machining
E/E electronic/electrical
EEC European Economic Community
EI modulus × moment of inertia (equals stiffness)
EMI electromagnetic interference
EO ethylene oxide (also EtO)
EOT ethylene ether polysulfide
EP ethylene-propylene
EPA Environmental Protection Agency
EPDM ethylene-propylene diene monomer
EPM ethylene-propylene fluorinated
EPP expandable polypropylene
EPR ethylene-propylene rubber
EPS expandable polystyrene
E_r modulus, relaxation
E_s modulus, secant
ESC environmental stress cracking
ESCR environmental stress cracking resistance
ESD electrostatic safe discharge
ET ethylene polysulfide
ETFE ethylene terafluoroethylene
ETO ethylene oxide
EU entropy unit; European Union
EUPC European Association of Plastics Converters
EUPE European Union of Packaging and Environment
EUROMAP Eu^ropean Committee of Machine Manufacturers for the Rubber and Plastics Industries (Zurich, Switzerland)
EVA ethylene-vinyl acetate
E/VAC ethylene/vinyl acetate copolymer
EVAL ethylene-vinyl alcohol copolymer (tradename for EVOH)
EVE ethylene-vinyl ether
EVOH ethylene-vinyl alcohol copolymer (or EVAL)
EX extrusion
F coefficient of friction; Farad; force
FALLO follow all opportunities
FDA Food and Drug Administration
FEA finite element analysis
FEP fluorinated ethylene-propylene
FFS form, fill, and seal
FLC fuzzy logic control
FMCT fusible metal core technology
FPC flexible printed circuit
fpm feet per minute
FRCA Fire Retardant Chemicals Association
FRP fiber reinforced plastic
FRTP fiber reinforced thermoplastic
FRTS fiber reinforced thermoset
FS fluorosilicone
FTIR Fourier transformation infrared
FV frictional force × velocity
G gravity; shear modulus (modulus of rigidity); torsional modulus
GAIM gas-assisted injection molding
gal gallon
GB gigabyte (billion bytes)
GD&T geometric dimensioning and tolerancing
GDP gross domestic product
GFRP glass fiber reinforced plastic
GMP good manufacturing practice
GNP gross national product
GP general purpose
GPa giga-Pascal
GPC gel permeation chromatography
gpd grams per denier
gpm gallons per minute
GPPS general purpose polystyrene
GRP glass reinforced plastic
GR-S polybutadiene-styrene
GSC gas solid chromatography

H hysteresis; hydrogen
HA hydroxyapatite
HAF high-abrasion furnace
HB Brinell hardness number
HCFC hydrochlorofluorocarbon
HCl hydrogen chloride
HDPE high-density polyethylene (also PE-HD)
HDT heat deflection temperature
HIPS high-impact polystyrene
HMC high-strength molding compound
HMW-HDPE high molecular weight–high density polyethylene
H-P Hagen-Poiseuille
HPLC high-pressure liquid chromatography
HPM hot pressure molding
HTS high-temperature superconductor
Hz Hertz (cycles)
I integral; moment of inertia
IB isobutylene
IBC internal bubble cooling
IBM injection blow molding; International Business Machines
IC *Industrial Computing* publication
ICM injection-compression molding
ID internal diameter
IEC International Electrochemical Commission
IEEE Institute of Electrical and Electronics Engineers
IGA isothermal gravimetric analysis
IGC inverse gas chromatography
IIE Institute of Industrial Engineers
IM injection molding
IMM injection molding machine
IMPS impact polystyrene
I/O input/output
ipm inch per minute
ips inch per second
IR synthetic polyisoprene (synthetic natural rubber)
ISA Instrumentation, Systems, and Automation
ISO International Standardization Organization or International Organization for Standardization
IT information technology
IUPAC International Union of Pure and Applied Chemistry
IV intrinsic viscosity
IVD in vitro diagnostic
J joule
JIS Japanese Industrial Standard
JIT just-in-time
JIT just-in-tolerance
J_p polar moment of inertia
JSR Japanese SBR
JSW Japan Steel Works
JUSE Japanese Union of Science and Engineering
JWTE Japan Weathering Test Center
K bulk modulus of elasticity; coefficient of thermal conductivity; Kelvin; Kunststoffe (plastic in German)
kb kilobyte (1000 bytes)
kc kilocycle
kg kilogram
KISS keep it short and simple
Km kilometer
kPa kilo-Pascal
ksi thousand pounds per square inch (psi $\times 10^3$)
lbf pound-force
LC liquid chromatography
LCP liquid crystal polymer
L/D length-to-diameter (ratio)
LDPE low-density polyethylene (PE-LD)
LIM liquid impingement molding; liquid injection molding
LLDPE linear low-density polyethylene (also PE-LLD)
LMDPE linear medium density polyethylene
LOX liquid oxygen
LPM low-pressure molding
m matrix; metallocene (catalyst); meter

mμ micromillimeter; millicron; 0.000001 mm
μm micrometer
MA maleic anhydride
MAD mean absolute deviation; molding area diagram
Mb bending moment
MBTS benzothiazyl disulfide
MD machine direction; mean deviation
MD&DI Medical Device and Diagnostic Industry
MDI methane diisocyanate
MDPE medium density polyethylene
Me metallocene catalyst
MF melamine formaldehyde
MFI melt flow index
mHDPE metallocene high-density polyethylene
MI melt index
MIM metal powder injection molding
MIPS medium impact polystyrene
MIT Massachusetts Institute of Technology
mLLDPE metallocene catalyst linear low-density polyethylene
MMP multimaterial molding or multimaterial multiprocess
MPa mega-Pascal
MRPMA Malaysian Rubber Products Manufacturers' Association
Msi million pounds per square inch (psi $\times 10^6$)
MSW municipal solid waste
MVD molding volume diagram
MVT moisture vapor transmission
MW molecular weight
MWD molecular weight distribution
MWR molding with rotation
N Newton (force)
NACE National Association of Corrosion Engineers
NACO National Association of CAD/CAM Operation
NAGS North America Geosynthetics Society
NASA National Aeronautics Space Administration
NBR butadiene acrylontrile
NBS National Bureau of Standards (since 1980 renamed the National Institute Standards and Technology or NIST)
NC numerical control
NCP National Certification in Plastics
NDE nondestructive evaluation
NDI nondestructive inspection
NDT nondestructive testing
NEAT nothing else added to it
NEMA National Electrical Manufacturers' Association
NEN Dutch standard
NFPA National Fire Protection Association
NISO National Information Standards Organization
NIST National Institute of Standards and Technology
nm nanometer
NOS not otherwise specified
NPCM National Plastics Center and Museum
NPE National Plastics Exhibition
NPFC National Publications and Forms Center (US government)
NR natural rubber (polyisoprene)
NSC National Safety Council
NTMA National Tool and Machining Association
NWPCA National Wooden Pallet and Container Association
OD outside diameter
OEM original equipment manufacturer
OPET oriented polyethylene terephthalate
OPS oriented polystyrene
OSHA Occupational Safety and Health Administration
P load; poise; pressure
Pa Pascal
PA polyamide (nylon)
PAI polyamide-imide
PAN polyacrylonitrile

PB polybutylene
PBA physical blowing agent
PBNA phenyl-β-naphthylamine
PBT polybutylene terephthalate
PC permeability coefficient; personal computer; plastic composite; plastic compounding; plastic-concrete; polycarbonate; printed circuit; process control; programmable circuit; programmable controller
PCB printed circuit board
pcf pounds per cubic foot
PCFC polychlorofluorocarbon
PDFM Plastics Distributors and Fabricators Magazine
PE plastic engineer; polyethylene (UK polythene); professional engineer
PEEK polyetheretherketone
PEI polyetherimide
PEK polyetherketone
PEN polyethylene naphthalate
PES polyether sulfone
PET polyethylene terephthalate
PETG polyethylene terephthalate glycol
PEX polyethylene crosslinked pipe
PF phenol formaldehyde
PFA perfluoroalkoxy (copolymer of tetrafluoroethylene and perfluorovinylethers)
PFBA polyperfluorobutyl acrylate
phr parts per hundred of rubber
PI polyimide
PIA Plastics Institute of America
PID proportional-integral-differential
PIM powder injection molding
PLASTEC Plastics Technical Evaluation Center (US Army)
PLC programmable logic controller
PMMA Plastics Molders and Manufacturers' Association (of SME); polymethyl methacrylate (acrylic)
PMMI Packaging Machinery Manufacturers' Institute
PO polyolefin
POE polyolefin elastomer
POM polyoxymethylene or polyacetal (acetal)
PP polypropylene
PPA polyphthalamide
ppb parts per billion
PPC polypropylene chlorinated
PPE polyphenylene ether
pph parts per hundred
ppm parts per million
PPO polyphenylene oxide
PPS polyphenylene sulfide
PPSF polyphenylsulfone
PPSU polyphenylene sulphone
PS polystyrene
PSB polystyrene butadiene rubber (GR-S, SBR)
PS-F polystyrene-foam
psf pounds per square foot
PSF polysulphone
psi pounds per square inch
psia pounds per square inch, absolute
psid pounds per square inch, differential
psig pounds per square inch, gauge (above atmospheric pressure)
PSU polysulfone
PTFE polytetrafluoroethylene (or TFE)
PUR polyurethane (also PU, UP)
P-V pressure-volume (also PV)
PVA polyvinyl alcohol
PVAC polyvinyl acetate
PVB polyvinyl butyral
PVC polyvinyl chloride
PVD physical vapor deposition
PVDA polyvinylidene acetate
PVdC polyvinylidene chloride
PVDF polyvinylidene fluoride
PVF polyvinyl fluoride
PVP polyvinyl pyrrolidone

PVT pressure-volume-temperature (also P-V-T or pvT)
PW *Plastics World* magazine
QA quality assurance
QC quality control
QMC quick mold change
QPL qualified products list
QSR quality system regulation
R Reynolds number; Rockwell (hardness)
rad Quantity of ionizing radiation that results in the absorption of 100 ergs of energy per gram of irradiated material.
radome radar dome
RAPRA Rubber and Plastics Research Association
RC Rockwell C (R_c)
RFI radio frequency interference
RH relative humidity
RIM reaction injection molding
RM rotational molding
RMA Rubber Manufacturers' Association
RMS root mean square
ROI return on investment
RP rapid prototyping; reinforced plastic
RPA Rapid Prototyping Association (of SME)
rpm revolutions per minute
RRIM reinforced reaction injection molding
RT rapid tooling; room temperature
RTM resin transfer molding
RTP reinforced thermoplastic
RTS reinforced thermoset
RTV room temperature vulcanization
RV recreational vehicle
Rx radiation curing
SAE Society of Automotive Engineers
SAMPE Society for the Advancement of Material and Process Engineering
SAN styrene acrylonitrile
SBR styrene-butadiene rubber
SCT soluble core technology
SDM standard deviation measurement
SES Standards Engineering Society
SF safety factor; short fiber; structural foam
s.g. specific gravity
SI International System of Units
SIC Standard Industrial Classification
SMC sheet molding compound
SMCAA Sheet Molding Compound Automotive Alliance
SME Society of Manufacturing Engineers
S-N stress-number of cycles
SN synthetic natural rubber
SNMP simple network management protocol
SPC statistical process control
SPE Society of the Plastics Engineers
SPI Society of the Plastics Industry
sPS syndiotactic polystyrene
sp. vol. specific volume
SRI Standards Research Institute (ASTM)
S-S stress-strain
STP Special Technical Publication (ASTM); standard temperature and pressure
t thickness
T temperature; time; torque (or T_t)
TAC triallylcyanurate
T/C thermocouple
TCM technical cost modeling
TD transverse direction
TDI toluene diisocyanate
TF thermoforming
TFS thermoform-fill-seal
T_g glass transition temperature
TGA thermogravimetric analysis
TGI thermogravimetric index
TIR tooling indicator runout
T-LCP thermotropic liquid crystal polymer
TMA thermomechanical analysis; Tooling and Manufacturing Association (formerly TDI); Toy Manufacturers of America
torr mm mercury (mmHg); unit of pressure equal to 1/760th of an atmosphere

TP thermoplastic
TPE thermoplastic elastomer
TPO thermoplastic olefin
TPU thermoplastic polyurethane
TPV thermoplastic vulcanizate
T_s tensile strength; thermoset
TS twin screw
TSC thermal stress cracking
TSE thermoset elastomer
TX thixotropic
TXM thixotropic metal slurry molding
UA urea, unsaturated
UD unidirectional
UF urea formaldehyde
UHMWPE ultra-high molecular weight polyethylene (also PE-UHMW)
UL Underwriters Laboratories
UP unsaturated polyester (also TS polyester)
UPVC unplasticized polyvinyl chloride
UR urethane (also PUR, PU)
URP unreinforced plastic
UV ultraviolet
UVCA ultra-violet-light-curable-cyanoacrylate
V vacuum; velocity; volt
VA value analysis
VCM vinyl chloride monomer
VLDPE very low-density polyethylene
VOC volatile organic compound
vol% percentage by volume
w width
W watt
W/D weight-to-displacement volume (boat hull)
WIT water-assist injection molding technology
WMMA Wood Machinery Manufacturers of America
WP&RT World Plastics and Rubber Technology magazine
WPC wood-plastic composite
wt% percentage by weight
WVT water vapor transmission
XL cross-linked
XLPE cross-linked polyethylene
XPS expandable polystyrene
YPE yield point elongation
Z-twist twisting fiber direction

Acknowledgments

Undertaking the development through to the completion of the *Plastics Technology Handbook* required the assistance of key individuals and groups. The indispensable guidance and professionalism of our publisher, Joel Stein, and his team at Momentum Press was critical throughout this enormous project. The coeditors, Nick R. Schott, Professor Emeritus of the University of Massachusetts Lowell Plastics Engineering Department, and Marlene G. Rosato, President of Gander International Inc., were instrumental to the data, information, and analysis coordination of the eighteen chapters of the handbook. A special thank you is graciously extended to Napoleao Neto of Alphagraphics for the organization and layout of the numerous figure and table graphics central to the core handbook theme. Finally, a great debt is owed to the extensive technology resources of the Plastics Institute of America at the University of Massachusetts Lowell and its Executive Director, Professor Aldo M. Crugnola.

Dr. Donald V. Rosato, Coeditor and President, PlastiSource, Inc.

Preface

This book, as a two-volume set, offers a simplified, practical, and innovative approach to understanding the design and manufacture of products in the world of plastics. Its unique review will expand and enhance your knowledge of plastic technology by defining and focusing on past, current, and future technical trends. Plastics behavior is presented to enhance one's capability when fabricating products to meet performance requirements, reduce costs, and generally be profitable. Important aspects are also presented to help the reader gain understanding of the advantages of different materials and product shapes. The information provided is concise and comprehensive.

Prepared with the plastics technologist in mind, this book will be useful to many others. The practical and scientific information contained in this book is of value to both the novice, including trainees and students, and the most experienced fabricators, designers, and engineering personnel wishing to extend their knowledge and capability in plastics manufacturing including related parameters that influence the behavior and characteristics of plastics. The toolmaker (who makes molds, dies, etc.), fabricator, designer, plant manager, material supplier, equipment supplier, testing and quality control personnel, cost estimator, accountant, sales and marketing personnel, new venture type, buyer, vendor, educator/trainer, workshop leader, librarian, industry information provider, lawyer, and consultant can all benefit from this book. The intent is to provide a review of the many aspects of plastics that range from the elementary to the practical to the advanced and more theoretical approaches. People with different interests can focus on and interrelate across subjects in order to expand their knowledge within the world of plastics.

Over 20000 subjects covering useful pertinent information are reviewed in different chapters contained in the two volumes of this book, as summarized in the expanded table of contents and index. Subjects include reviews on materials, processes, product designs, and so on. From a pragmatic standpoint, any theoretical aspect that is presented has been prepared so that the practical person will understand it and put it to use. The theorist in turn will gain an insight into the practical limitations that exist in plastics as they exist in other materials such as steel, wood, and so on. There is no material that is "perfect." The two volumes of this book together contain 1800-plus figures and 1400-plus tables providing extensive details to supplement the different subjects.

In working with any material (plastics, metal, wood, etc.), it is important to know its behavior in order to maximize product performance relative to cost and efficiency. Examples of different plastic materials and associated products are reviewed with their behavior patterns. Applications span toys, medical devices, cars, boats, underwater devices, containers, springs, pipes, buildings, aircraft, and spacecraft. The reader's product to be designed or fabricated, or both, can be related directly or indirectly to products reviewed in this book. Important are behaviors associated with and interrelated with the many different plastics materials (thermoplastics [TPs], thermosets [TSs], elastomers, reinforced plastics) and the many fabricating processes (extrusion, injection molding, blow molding, forming, foaming, reaction injection molding, and rotational molding). They are presented so that the technical or nontechnical reader can readily understand the interrelationships of materials to processes.

This book has been prepared with the awareness that its usefulness will depend on its simplicity and its ability to provide essential information. An endless amount of data exists worldwide for the many plastic materials, which total about 35000 different types. Unfortunately, as with other materials, a single plastic material that will meet all performance requirements does not exist. However, more so than with any other materials, there is a plastic that can be used to meet practically any product requirement. Examples are provided of different plastic products relative to critical factors ranging from meeting performance requirements in different environments to reducing costs and targeting for zero defects. These reviews span products that are small to large and of shapes that are simple to complex. The data included provide examples that span what is commercially available. For instance, static physical properties (tensile, flexural, etc.), dynamic physical properties (creep, fatigue, impact, etc.), chemical properties, and so on, can range from near zero to extremely high values, with some having the highest of any material. These plastics can be applied in different environments ranging from below and on the earth's surface to outer space.

Pitfalls to be avoided are reviewed in this book. When qualified people recognize the potential problems, these problems can be designed around or eliminated so that they do not affect the product's performance. In this way, costly pitfalls that result in poor product performance or failure can be reduced or eliminated. Potential problems or failures are reviewed, with solutions also presented. This failure-and-solution review will enhance the intuitive skills of people new to plastics as well as those who are already working in plastics. Plastic materials have been produced worldwide over many years for use in the design and fabrication of all kinds of plastic products. To profitably and successfully meet high-quality, consistency, and long-life standards, all that is needed is to understand the behavior of plastics and to apply these behaviors properly.

Patents or trademarks may cover certain of the materials, products, or processes presented. They are discussed for information purposes only and no authorization to use these patents or trademarks is given or implied. Likewise, the use of general descriptive names, proprietary names, trade names, commercial designations, and so on does not in any way imply that they may be used freely. While the information presented represents useful information that can be studied or analyzed and is believed to be true and accurate, neither the authors, contributors, reviewers, nor the publisher can accept any legal responsibility for any errors, omissions, inaccuracies, or other factors. Information is provided without warranty of any kind. No representation as to accuracy, usability, or results should be inferred.

Preparation for this book drew on information from participating industry personnel, global industry and trade associations, and the authors' worldwide personal, industrial, and teaching experiences.

DON & MARLENE ROSATO AND NICK SCHOTT, 2011

About the Authors

Dr. Donald V. Rosato, president of PlastiSource Inc., a prototype manufacturing, technology development, and marketing advisory firm in Massachusetts, United States, is internationally recognized as a leader in plastics technology, business, and marketing. He has extensive technical, marketing, and plastics industry business experience ranging from laboratory testing to production to marketing, having worked for Northrop Grumman, Owens-Illinois, DuPont/Conoco, Hoechst Celanese/Ticona, and Borg Warner/G.E. Plastics. He has developed numerous polymer-related patents and is a participating member of many trade and industry groups. Relying on his unrivaled knowledge of the industry and high-level international contacts, Dr. Rosato is also uniquely positioned to provide an expert, inside view of a range of advanced plastics materials, processes, and applications through a series of seminars and webinars. Among his many accolades, Dr. Rosato has been named Engineer of the Year by the Society of Plastics Engineers. Dr. Rosato has written extensively, authoring or editing numerous papers, including articles published in the *Encyclopedia of Polymer Science and Engineering*, and major books, including the *Concise Encyclopedia of Plastics*, *Injection Molding Handbook 3rd ed.*, *Plastic Product Material and Process Selection Handbook*, *Designing with Plastics and Advanced Composites*, and *Plastics Institute of America Plastics Engineering, Manufacturing, and Data Handbook*. Dr. Rosato holds a BS in chemistry from Boston College, an MBA from Northeastern University, an MS in plastics engineering from the University of Massachusetts Lowell, and a PhD in business administration from the University of California, Berkeley.

Marlene G. Rosato, with stints in France, China, and South Korea, has comprehensive international plastics and elastomer business experience in technical support, plant start-up and troubleshooting, manufacturing and engineering management, and business development and strategic planning with Bayer/Polysar and DuPont. She also does extensive international technical, manufacturing, and management consulting as president of Gander International Inc. She also has an extensive writing background authoring or editing numerous papers and major books, including the *Concise Encyclopedia of Plastics*, *Injection Molding Handbook 3rd ed.*, and the *Plastics Institute of America Plastics Engineering, Manufacturing and Data Handbook*. A senior member of the Canadian Society of Chemical Engineering and the Association of Professional Engineers of Canada, Ms. Rosato is a licensed professional engineer of Ontario, Canada. She received a Bachelor of Applied Science in chemical engineering from the University of British Columbia with continuing education at McGill University in Quebec, Queens University and the University of Western Ontario, both in Ontario, and also has extensive executive management training.

Emeritus Professor Nick Schott, a long-time member of the world-renowned University of Massachusetts Lowell Plastics Engineering Department faculty, served as its department head for a quarter of a century. Additionally, he founded the Institute for Plastics Innovation, a research consortium affiliated with the university that conducts research related to plastics manufacturing, with a current emphasis on bioplastics, and served as its director from 1989 to 1994. Dr. Schott has received numerous plastics industry accolades from the SPE, SPI, PPA, PIA, as well as other global industry associations and is renowned for the depth of his plastics technology experience, particularly in processing-related areas. Moreover, he is a quite prolific and requested industry presenter, author, patent holder, and product/process developer. In addition, he has extensive and continuing academic responsibilities at the undergraduate to postdoctoral levels. Among America's internationally recognized plastics professors, Dr. Nick R. Schott most certainly heads everyone's list not only within the 2500 plus global UMASS Lowell Plastics Engineering alumni family, which he has helped grow, but also in broad global plastics and industrial circles. Professor Schott holds a BS in chemical engineering from UC Berkeley, and an MS and PhD from the University of Arizona.

CHAPTER 17

MOLD AND DIE TOOLING

OVERVIEW

When processing plastics, some type of tooling is required. Tools include molds, dies, mandrels, jigs, fixtures, punch dies, perforated forms, and so on. (Tables 17.1 to 17.5; 1, 355). These tools fabricate or shape products. This chapter reviews injection molds and extrusion dies. They fit into the overall flow chart in fabricating plastic products (Table 3.8). Molds and dies for the other processes are discussed in chapters 6 to 16; this chapter includes information applicable to the other molds and dies. The terms for tools are virtually synonymous in the sense that they have some type of female or negative cavity into or through which a molten plastic moves, usually under heat and pressure; they are also used in secondary operations such as cutting dies, stamping sheet dies, and so on.

These tables provide the two most commonly used standards—the American Iron and Steel Institute (AISI) and the German Werkstoff material numbers—and their mean (average) chemical compositions. Note that chemical compositions will always differ from one book to another and from one manufacturer's tool stock list to another. For example, it will be very unlikely that the P20 steel being used will exactly match the chemical composition in these tables. However, they will be close.

Mold and die tools are used in processing many different materials (Table 17.6); many of them have common assembly and operating parts (preengineered since the 1940s), and the aim is to have the tool's opening or cavity designed to form desired final shapes and sizes. They can be composed of many moving parts that require high-quality metals and precision machining (443). For example, with certain processes to capitalize on advantages, molds may incorporate many cavities, adding further to their complexity.

Types of tool	types of material
Anvils	W1
Axes	1045
Blades	
flying shear	D2
granulator	D2
hot shearing	H13
rotary shear	O1
shear	A2, S1, L6
shear, for hard thin materials	D3
shear, heavy duty, cold	D3
shear, thick materials	1.2767
Blocks, die, cold, high pressing stress	6F5
Broaches	O1
Bushings	O1
Cams	O1
Centers, lathe	O1, L3
Chasers	1.2419
Chisels	
hand	S1
pneumatic	S1
Chucks	
collet	W1
split	S4
Collets	O1
Cores	
hammer	W2
hot working	1.2567
Cutters	
circular, for cold rolled strip	D2
milling	O1
strip splitting	O1
Dies	
wire drawing	D2
blanking	D2, D3, S1
coining	A2, O1, S1
cold working	W2
cupping	D2
cutting	W110
cutting, hot	S1
deep drawing, for sheet metal	D2
die casting	H13
die forging, small to medium work	L6, S4
die-casting, to be hobbed	P4
embossing	6F5, 1.2762
embossing, hollow	W110
embossing, large	W110
extrusion, cold working	D2
extrusion, for aluminum and copper	H13, H21
extrusion, for rubber bonded synthetics	H13, H21
extrusion, hot working	H19
forging, cold working	A2, H13
forging, hot working (non-excessive temps.)	S1
forging, hot working	H13, H21
forging, hot working, heavy duty	1.2744
forming, for sheet metal	D2

Table 17.1 Types of tools and materials

gravity, for aluminum die casting	H13, H21
gripper	H13
hammer, drop forge	L6
heading	D2 A, L6
heading, cold working	A2, L6, H13
hobbing	D2, L6
hot extrusion	H21
hot working for non-ferrous metals	1.2567, H21
leather	W1
lower, hot working	H21
medium run	O1
molding part	L6
molding, for abrasive powder (ceramic)	D3
molding, synthetic plastic	P4, 1.2738
nail making, cold working	S1
plastic, corrosive	420
pressing, chemically agressive compounds	1.2316
pressing, component	1.2744, 1.2766, H12
pressure, for aluminum and zinc die cast	H13, H21
punching	W108, W110, A2, D2
resin (artificial), highly stressed	P6
resin, corrosive	420
snap	W108, S1
split hot heading	H13
swaging, hot working (non-excessive temps.)	S1
thread rolling	A2, D2, D3
threading	L3
trimming	A2, D2
Files	1.2008
Gages	
master	D3
plug	D2, O1
ring	D2, O1
Hammers	W1,1045
Hobs	
cutting	O1
master, for cold hobbing	D2, D3, L6, M2, T1
Jaws, chuck	O1
Knives	
cloth cutting	O1
paper cutting	O1
resin bonded material cutting	A2
wood cutting	A2
Liners, mold, for bricks and tiles	D3
Mandrels	
for aluminum tube drawing	H13
for copper tube drawing	H13
for steel tube drawing	S1
Molds	
aluminum or zinc die cast	H13, H21
chemically aggressive material	420, 420mod.
complicated/intricate	A2, D2, O1
compression	H13
compression, for lead, zinc and tin alloys	420, 420mod.

Table 17.1 Types of tools and materials *(continued)*

compressive, high	EN30B
cutt	EN30B, 410
cutt, high wear resisting	D2
die casting, heavy duty	420, 420mod.
glass	420
long production run	A2, D2, O1, 420mod.
plastic cut	D2
plastic injection, very high polish	420mod.
plastic injection, very large	P20
presser casting, for light metals	H11
pressure casting	1.2567
pressure casting, brass	H19
rubber seal	D2
transfer	H13
Pins	
core, for molds	D2
ejector	H13, L1
Plungers	H13
Presses, metal extruding, non-ferrous metals	H12
Punches	
blanking	D3
cold working	L2
cutting, complicated	1.2127
drawing	O1
engraving	S1
head	S1
high silicon and transformer material	D2, D3
hot working	H13, S1
stainless steel sheet and plate	D2
steel sheet and plate	D2, D3
tableting, for abrasive and corrosive powers	D3
Reamers	L3, D3, D2, O1
Rings, hammer roll	1.2766
Rollers	
sheet metal forming	D2
spinning	A2
Rolls	
cold, small	1.2057
tube expanding	O1
Saddles	
hammer	W1, S4
pressing	S4
Saws	
circular, woodworking	1.1830
frame, woodworking	1.1830
metal cutting	1.2442
Scythe	W110
Shears	1045
Sleeves	
abrasive and corrosive powder tabletting	D3
drill	1.2378

Table 17.1 Types of tools and materials *(continued)*

Stamps, minting	W2
Strainers	W108
Taps	
cold	O1, L2
special	D2, M2
Tools	
bending	1.2767
blanking, hot	S1
blanking, nut	S1
burnishing	1.2008
countersinking	L2
cutting, cold	W108
cutting, dynamo and transformer sheets	1.2378
cutting, heavy duty, cold	D3
cutting, medium duty, cold	A2, S1
cutting, medium temperature	S4
cutting, precision, cold	1.2008
cutting, thread machining	O1
drawing, cold	1.2057
embossing	W110, 1.2767
embossing, high stress	1.2762
heading, cold	W2
impact, cold	W108
knurling	A2,1.2057
medium temp. piercing	S1
metal extrusion press	H11
metal extrusion	L6
piercing, hot	H13
pneumatic	S1
pre-forming	S1
press, extrusion, heavy duty	H10A
press, for fine and medium work	D2, O1
press, heavy duty	D2, D3
pressing, hot, complexed engravings	1.2767
pressing, tube	L6
processing, synthetic plastics	P20
punching, heavy duty, cold	D3
punching, hot	S1
punching, medium duty, cold	A2
scraping	L2
sone working, hard	W110
stone working, medium hard	W1
trimming, cold	S4
trimming, hot	H13, M2, S1, S4
trimming, medium temp.	S1, M1, M2

Table 17.1 Types of tools and materials *(continued)*

AISI	Werkstoff	C %	Mn %	Co %	Cr %	Mo %	Ni %	V %	W %
A2	1.2363	1.00	0.60	–	5.00	1.15	–	–	–
A3	–	1.25	0.30	–	5.00	1.00	–	1.00	–
A4	–	1.00	2.00	–	1.00	1.00	–	–	–
A5	–	1.00	3.00	–	1.00	1.15	–	–	–
A6	–	0.70	2.00	–	1.00	1.25	–	–	–
A7	–	2.25	0.30	–	5.25	1.00	–	4.75	–
A8	–	0.55	0.30	–	5.00	1.25	–	–	1.25
A9	–	0.50	0.30	–	5.00	1.40	1.40	1.00	–
A10	–	1.35	1.80	–	–	1.50	1.80	–	–
A11	–	2.45	0.50	–	5.15	1.30	–	9.75	0.50
D1	–	1.00	0.40	–	12.0	1.00	–	–	–
D2	1.2379	1.50	0.40	–	12.0	0.95	–	–	–
D2A	1.2601	1.75	0.40	–	12.0	0.80	–	–	–
D3	1.2080	2.20	0.35	–	12.0	–	–	–	–
D4	–	2.25	0.30	–	12.0	1.00	–	–	–
D5	–	1.50	0.40	3.00	12.0	0.95	–	–	–
D6	–	2.25	0.40	–	12.0	–	–	–	1.00
D7	–	2.30	0.40	–	12.5	0.95	–	4.10	–
F1	–	1.10	0.50	–	–	–	–	–	1.50
F2	–	1.30	0.50	–	0.30	–	–	–	3.75
F3	–	1.25	0.50	–	0.75	–	–	–	3.75
H10	1.2365	0.40	0.30	–	3.25	2.50	–	0.40	–
H10A	1.2885	0.32	0.30	3.00	3.00	2.80	–	0.50	–
H11	1.2343	0.35	0.30	–	5.10	1.50	–	0.40	–
H12	1.2606	0.35	0.30	–	5.10	1.50	–	0.30	1.35
H13	1.2344	0.35	0.30	–	5.10	1.50	–	1.00	–
H14	–	0.40	0.30	–	5.00	–	–	–	5.00
H15	–	0.40	0.30	–	5.00	5.00	–	–	–
H16	–	0.55	0.30	–	7.00	–	–	–	7.00
H19	–	0.40	0.30	4.25	4.25	–	–	2.00	4.25
H20	–	0.35	0.30	–	2.00	–	–	–	9.00
H21	1.2581	0.35	0.30	–	3.40	–	–	0.40	9.40
H21A	–	0.30	0.30	–	2.75	0.60	2.25	0.50	9.25
H22	–	0.35	0.30	–	2.00	–	–	–	11.0
H23	–	0.30	0.30	–	12.0	–	–	–	12.0
H24	–	0.45	0.30	–	3.00	–	–	–	12.0
H25	–	0.25	0.30	–	4.00	–	–	–	15.0
H26	–	0.50	0.30	–	4.00	–	–	1.00	18.0
H41	1.3346	0.65	0.30	–	3.75	8.00	–	1.15	1.80
H42	–	0.60	0.30	–	4.00	5.00	–	2.00	6.00
H43	–	0.58	0.30	–	4.10	8.00	–	2.00	–
H224	1.2713	0.50	0.85	–	0.90	0.30	1.50	–	–
H225	1.2713	0.50	0.85	–	0.90	0.30	1.50	–	–
L1	–	1.00	0.30	–	1.25	–	–	–	–
L2	1.2210	0.80	0.30	–	1.00	–	–	0.20	–
L3	1.2067	1.00	0.40	–	1.50	–	–	0.20	–
L4	–	1.00	0.60	–	1.50	–	–	0.20	–
L5	–	1.00	1.00	–	1.00	0.25	–	–	–
L6	1.2713	0.70	0.70	–	0.75	–	1.50	–	–
L7	1.2303	1.00	0.35	–	1.50	0.40	–	–	–
M1	1.3346	0.80	0.30	–	4.10	8.00	–	1.10	1.50
M2	1.3343	0.95	0.30	–	4.10	5.00	–	1.90	6.10

Table 17.2 American Iron and Steel Institute (AISI) and some BS numbers without their "B" prefix (BH10A/H10A) with comparable Werkstoff numbers and their mean (average) chemical compositions

M3	1.3342	1.10	0.30	–	4.10	5.50	–	2.75	6.10
M4	–	1.30	0.30	–	4.40	5.00	–	4.20	5.90
M6	–	0.80	0.30	12.0	4.00	5.00	–	1.50	4.00
M7	1.3348	1.00	0.30	–	4.00	8.75	–	2.00	1.75
M10	–	0.95	0.30	–	4.00	8.00	–	2.00	–
M15	–	1.50	0.30	5.00	4.00	3.50	–	5.00	6.50
M30	–	0.80	0.30	5.00	4.00	8.00	–	1.25	2.00
M33	1.3249	0.90	0.30	8.00	4.00	9.50	–	1.15	1.50
M34	1.3249	0.90	0.30	8.00	4.00	8.00	–	2.00	2.00
M35	–	0.80	0.30	5.00	4.00	5.00	–	2.00	6.00
M36	–	0.80	0.30	8.00	4.00	5.00	–	2.00	6.00
M41	1.3246	1.10	0.30	5.00	4.25	3.75	–	2.00	6.75
M42	1.3247	1.10	0.30	8.00	3.75	9.50	–	1.15	1.50
M43	–	1.20	0.30	8.25	3.75	8.00	–	1.60	2.75
M44	–	1.15	0.30	12.0	4.25	6.25	–	2.00	5.25
M46	–	1.25	0.30	8.25	4.00	8.25	–	3.20	2.00
M47	–	1.10	0.30	5.00	3.75	9.50	–	1.25	1.50
M48	–	1.50	0.30	9.00	3.90	5.10	–	3.00	10.0
M50	1.3551	0.81	0.30	0.25	4.00	4.25	0.10	1.00	0.25
M52	–	0.90	0.25	–	4.00	4.50	–	1.90	1.25
M61	–	1.80	0.35	–	3.90	6.40	–	4.90	12.5
M62	–	1.30	0.30	–	3.90	10.5	–	2.00	6.25
O1	1.2510	0.90	1.15	–	0.50	–	–	–	0.50
O2	1.2842	0.90	1.60	–	–	–	–	–	–
O6	–	1.45	0.65	–	–	0.25	–	–	–
O7	–	1.20	0.30	–	0.75	–	–	–	1.75
P1	–	0.10	0.30	–	–	–	–	–	–
P2	–	0.07	0.30	–	2.00	0.20	0.50	–	–
P3	–	0.10	0.30	–	0.60	–	1.25	–	–
P4	1.2341	0.07	0.30	–	5.00	0.75	–	–	–
P5	–	0.10	0.30	–	2.25	–	–	–	–
P6	1.2735	0.10	0.30	–	1.50	–	3.50	–	–
P20	1.2330	0.35	0.30	–	1.70	0.40	–	–	–
P21	–	0.20	0.30	–	–	–	4.00	–	–
P30	1.2766	0.30	0.55	–	1.20	0.30	4.10	–	–
S1	1.2542	0.50	0.30	–	1.50	–	–	0.20	2.00
S2	–	0.50	0.40	–	–	0.50	–	–	–
S3	–	0.50	0.40	–	0.75	–	–	–	1.00
S4	–	0.55	0.80	–	0.35	–	–	0.35	–
S5	–	0.55	0.75	–	–	0.40	–	–	–
S6	–	0.45	1.40	–	1.50	0.40	–	–	–
S7	–	0.50	0.30	–	3.25	1.40	–	–	–
T1	1.3355	0.70	0.30	–	4.10	–	–	1.10	18.0
T2	–	0.80	0.30	–	4.10	0.85	–	2.10	18.2
T3	–	1.05	0.30	–	4.10	–	–	3.00	18.0
T4	1.3255	0.90	0.30	5.00	4.10	0.85	–	1.00	18.0
T5	1.3265	0.80	0.30	8.00	4.10	0.85	–	2.10	18.2
T6	–	0.80	0.30	12.0	4.50	–	–	1.50	20.0
T7	–	0.75	0.30	–	4.10	–	–	2.10	14.0
T8	–	0.80	0.30	5.00	4.10	0.85	–	2.10	14.0
T9	–	1.20	0.30	–	4.00	–	–	4.00	12.0

Table 17.2 American Iron and Steel Institute (AISI) and some BS numbers without their "B" prefix (BH10A/H10A) with comparable Werkstoff numbers and their mean (average) chemical composition *(continued)*

T15	1.3202	1.50	0.30	5.00	4.00	–	–	5.00	12.0
T20	–	0.80	0.30	0.50	4.60	0.80	–	1.50	21.8
T21	–	0.65	0.30	1.00	3.80	0.70	0.40	0.50	14.0
T42	1.3207	1.30	0.30	9.50	4.10	3.10	0.40	3.00	9.00
W1	–	1.00	0.30	–	–	–	–	–	–
W2	–	1.00	0.30	–	–	–	–	0.25	–
W3	–	1.00	0.30	–	–	–	–	0.50	–
W4	–	1.00	0.30	–	0.25	–	–	–	–
W5	–	1.00	0.30	–	0.50	–	–	–	–
W6	–	1.00	0.30	–	0.25	–	–	0.25	–
W7	–	1.00	0.30	–	0.50	–	–	0.20	–
W108	1.1525	0.80	0.30	–	0.18*	–	–	–	–
W109	–	0.90	0.30	–	0.18*	–	–	–	–
W110	1.1545	1.05	0.30	–	0.18*	–	–	–	–
W112	1.1663	1.20	0.30	–	0.18*	–	–	–	–
W209	–	0.90	0.30	–	0.18*	–	–	0.25	–
W210	1.2833	1.05	0.30	–	0.18*	–	–	0.25	–
W310	–	1.05	0.30	–	0.18*	–	–	0.45	–
6G	–	0.55	0.80	–	1.00	0.45	–	0.10	–
6F2	–	0.55	0.75	–	1.00	0.35	1.00	0.10	–
6F3	–	0.55	0.60	–	1.00	0.75	1.80	0.10	–
6F4	–	0.20	0.70	–	–	3.35	3.00	–	–
6F5	–	0.55	1.00	–	0.50	0.50	2.70	0.10	–
6F6	–	0.50	–	–	1.50	0.20	–	–	–
6F7	–	0.40	0.35	–	1.50	0.75	4.25	–	–
6H1	–	0.55	–	–	4.00	0.45	–	0.85	–
6H2	–	0.55	0.45	–	5.00	1.50	–	1.00	–
410	1.4006	0.11	0.80	–	12.5	–	–	–	–
416	1.4005	0.15	1.20	–	13.0	0.50	–	–	–
420	1.4021	0.20	0.80	–	13.0	–	–	–	–
431	1.4057	0.19	0.80	–	16.0	–	1.60	–	–
440A	–	0.70	0.80	–	17.0	0.65	–	–	–
440B	–	0.85	0.80	–	17.0	0.65	–	–	–
440C	–	1.10	0.80	–	17.0	0.65	–	–	–

*Estimated chrome content.

Table 17.2 American Iron and Steel Institute (AISI) and some BS numbers without their "B" prefix (BH10A/H10A) with comparable Werkstoff numbers and their mean (average) chemical composition *(continued)*

Werkstoff	AISI/SAE	C %	Mn %	Co %	Cr %	Mo %	Ni %	V %	W %
1.1525	W108								
1.1545	W110								
1.1625	W1								
1.1645	W1								
1.1663	W112								
1.1673	–	1.40	0.50	–	–	–	–	–	–
1.1730	1045	0.45	0.70	–	–	–	–	–	–
1.1740	–	0.60	–	–	–	–	–	–	–
1.1750	W1								
1.1820	–	0.55	0.70	–	–	–	–	–	–
1.1830	–	0.86	–	–	–	–	–	–	–
1.2002	–	1.20	0.40	–	0.40	–	–	–	–
1.2008	–	1.40	0.50	–	0.35	–	–	–	–
1.2057	–	1.00	0.50	–	1.10	–	0.20	–	–
1.2067	L3								
1.2080	D3								
1.2082	420								
1.2083	420	0.42	–	–	13.0	–	–	–	–
1.2162	–	0.21	1.25	–	1.20	–	–	–	–
1.2127	–	1.00	1.05	–	1.05	–	–	–	–
1.2210	L2								
1.2241	L2								
1.2303	L7								
1.2311	P20								
1.2312	P20								
1.2316	–	0.36	–	–	16.0	1.20	–	–	–
1.2330	P20								
1.2332	4142								
1.2341	P4								
1.2343	H11								
1.2344	H13								
1.2361	440B								
1.2363	A2								
1.2365	H10								
1.2367	–	0.37	–	–	5.00	3.00	–	0.60	–
1.2378	–	2.20	0.30	–	13.0	–	–	–	0.80
1.2379	D2								
1.2419	–	1.15	0.95	–	1.10	–	–	–	1.20
1.2436	–	2.20	0.30	–	12.0	–	–	–	–
1.2442	–	1.05	–	–	0.50	–	–	–	2.10
1.2510	O1								
1.2515	–	1.05	0.50	–	0.35	–	–	0.20	0.75
1.2542	S1								
1.2547	S1								
1.2550	S1								
1.2562	–	1.25	0.50	–	0.35	–	–	0.20	3.50
1.2567	–	0.32	–	–	2.40	–	–	0.60	4.30
1.2581	H21								
1.2601	D2A								
1.2606	H12								

Table 17.3 Werkstoff numbers with comparable AISI numbers or a near-matching chemical composition

1.2631	–	0.53	–	–	8.30	1.20	–	–	1.20
1.2663	A2								
1.2678	H19								
1.2711	6F2								
1.2713	L6								
1.2714	L6								
1.2718	6F5								
1.2721	L6								
1.2735	P6								
1.2738	–	0.40	1.45	–	1.95	0.20	1.05	–	–
1.2744	–	0.57	–	–	1.10	0.80	1.70	0.10	–
1.2762	–	0.70	–	–	1.50	0.70	0.50	–	0.30
1.2764	–	0.19	0.40	–	1.25	0.20	4.00	–	–
1.2766	–	0.32	–	–	1.20	0.20	4.10	–	–
1.2767	–	0.45	0.40	–	1.35	0.25	4.00	–	–
1.2770	–	0.85	–	–	–	–	0.80	0.10	–
1.2787	431ss								
1.2826	S4								
1.2833	W2								
1.2842	O2								
1.2885	H10A								
1.3202	T15								
1.3207		1.30	0.30	9.50	4.10	3.10	–	3.10	9.00
1.3243	–	0.90	0.30	5.00	4.10	5.00	–	2.00	6.20
1.3246	M41								
1.3247	M42								
1.3249	M33/34								
1.3255	T4								
1.3265	T5								
1.3342	M3								
1.3343	M2								
1.3344	M3								
1.3346	H41/M1	0.82	0.40	–	3.80	8.60	–	1.15	1.75
1.3348	M7								
1.3355	T1								
1.3505	–	1.00	0.50	–	1.45	–	–	–	–
1.3551	M50								
1.3554LW	M2								
1.4005	–	0.12	–	–	12.5	–	–	–	–
1.4006	410								
1.4014LW	420								
1.4021	420								
1.4028	–	0.33	–	–	13.5	–	–	–	–
1.4034	–	0.46	–	–	13.0	–	–	–	–
1.4057	431								
1.4110	–	0.60	–	–	14.1	0.60	–	0.10	–
1.4112	440B								
1.4140	–	0.43	–	–	13.2	–	1.00	–	–
1.4528	–	1.07	–	1.50	17.0	1.10	–	0.10	–
1.5864	–	0.32	–	–	1.20	0.20	4.10	–	–
1.6582	4340								
1.6747	–	0.30	–	–	1.25	0.30	4.25	–	–

Table 17.3 Werkstoff numbers with comparable AISI numbers or a near-matching chemical composition *(continued)*

Symbol	Element	Melting point °C	Properties and uses
Al	Aluminum	660	Widely used light metal
Ar	Argon		An inert gas used in TIG welding
Be	Beryllium	1,285	A light metal used to toughen copper
C	Carbon		An essential element in steel, especially in hardenable steels
Cr	Chromium	1,900	A corrosion resistant material which increases hardenability and resistance to wear. An essential element in stainless steel and heat resistant steels
Co	Cobalt	1,495	Used mainly in high-speed steels and permanent magnets
Cu	Copper	1,083	A metal of high electrical and heat conductivity and alloyed with other metals to give brasses and bronzes
Fe	Iron	1,536	A fairly soft white metal when pure. A major element in steel
He	Helium		A light gas generally used together with argon gas to weld heavier sections of aluminum and copper
Mg	Magnesium	651	Used as an alloy in aluminum to increase its work hardening ability and corrosion resistance to sea water
Mn	Manganese	1,260	Used in steel making as a deoxidant, it also increases the hardenability and tensile strenath but decreases ductility. Also added to non-heat treatable aluminum to improve its mechanical properties
Mo	Molybdenum	2,620	Increases hardenability in steels. Used in high speed steels and also used in stainless to increase resistance to corrosion
Ni	Nickel	1,458	A widely used metal in steels, coppers and aluminums to improve toughness
Pb	Lead	327	A heavy metal used as an alloy to improve machinability in many metals. Higher levels (1% plus) may cause weldability problems
Si	Silicon	1,427	Generally used as a powerful deoxidizer in steels. It also increases strength and ductility in aluminums and if combined with magnesium in aluminum it allows precipitation hardening
W	Tungsten	3,410	Generally the main constituent in high-speed steel. Also used as the electrode in TIG welding
V	Vanadium	1,720	Added to steels to increase hardenability and also to give a greater resistance to shock loading
Z	Zinc	419	Widely used for galvanizing mild steel and when alloyed with copper makes brass. Also used as a basis for some die-casting alloys and when added to aluminum drastically increases strength and allows precipitation hardening

Table 17.4 Elements and their symbols

Metals	Examples
Carbon steels	1010, 1020, 1040, 1050, 1090, B1112
Tool Steels	01, A2, D2, S1, H13, M2
Cast irons	Class 20, Class 35, Ductile 60-45-10
Copper alloys	ETP copper CDA 110, DHP copper CDA 122, Tin bronze CDA 905, Be copper CDA 172, Yellow brass CDA 360, Phosphor bronze CDA 521
Nickel alloys	Monel, Hastelloys, Ni, Cr, B alloys, Pure nickel
Alloy steels	4140, 4340, 4620, 9310
Special steels	Nitralloy, Marage 200-350, Weathering
Stainless steels	303, 304 (CF-8, 316 (CF-8M), 420, 440C, 17-4 PH (CB-7Cu)
Aluminum alloys	3003, 5052, 6061, 7075, 355, 380
Zinc	ASTM B86-48, No XX5
Titanium	Pure Ti (Gr 1-3), Ti-6A1-4V
Magnesium	AZ51, AZ63

Table 17.5 Examples of different metals used in tools

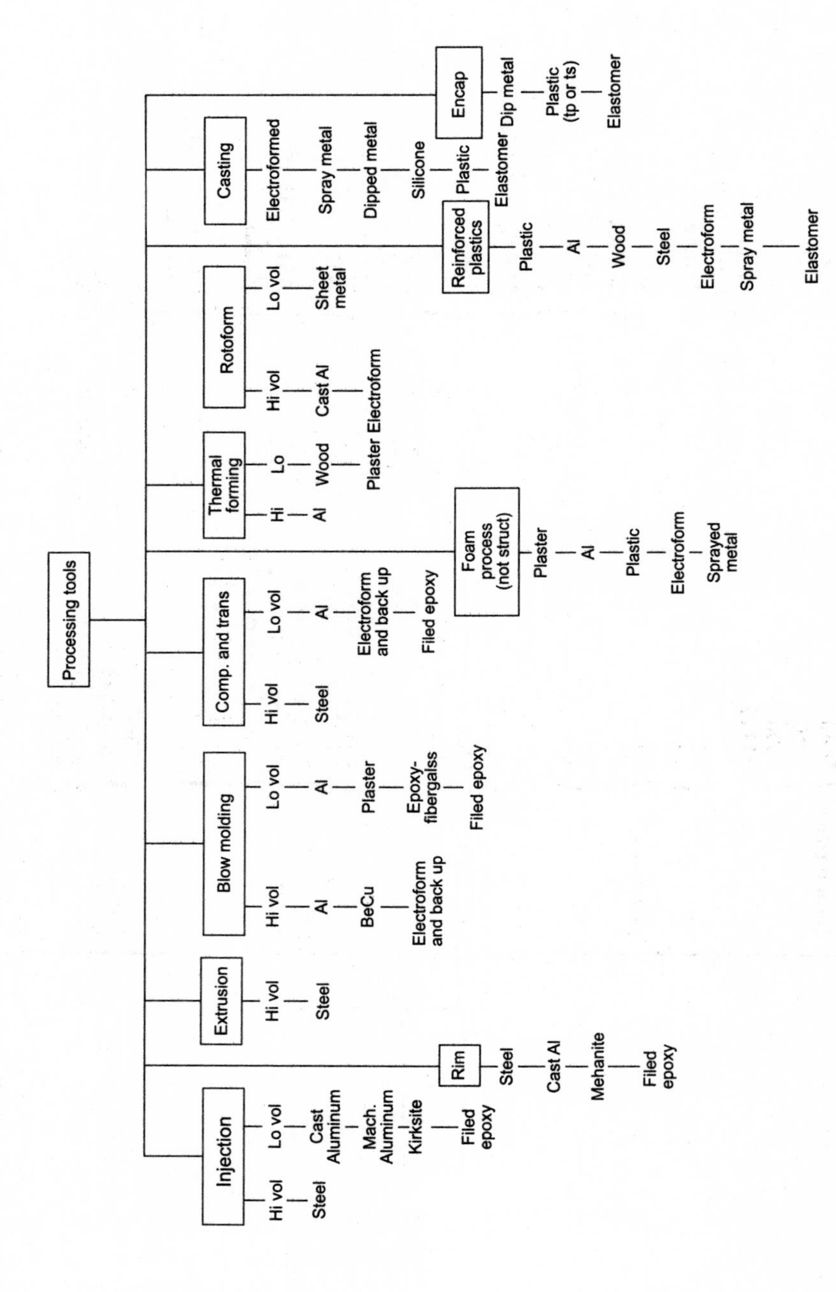

Table 17.6 Examples of mold and die tools for different fabricating processes

Tools of all types can represent up to one-third of the company's manufacturing investment. The total tool activity is summarized in Figure 17.1 (432). Metals, specifically steels, are the most common materials of construction for the rigid parts of tools. Some mold and die tools cost more than the primary processing machinery with the usual approaching half the cost of the primary machine. About 5% to 15% of tool cost is for the materials used in its manufacture, design about 5% to 10%, tool building hours about 50% to 70%, and profit at about 5% to 15%. Table 17.7 provides a cost comparison of molds in terms of the properties of plastic.

As shown in Table 17.6, various materials are used for construction. Examples of different types include wood and plastic for thermoforms and reinforced plastics (RPs; chapters 7 and 15).

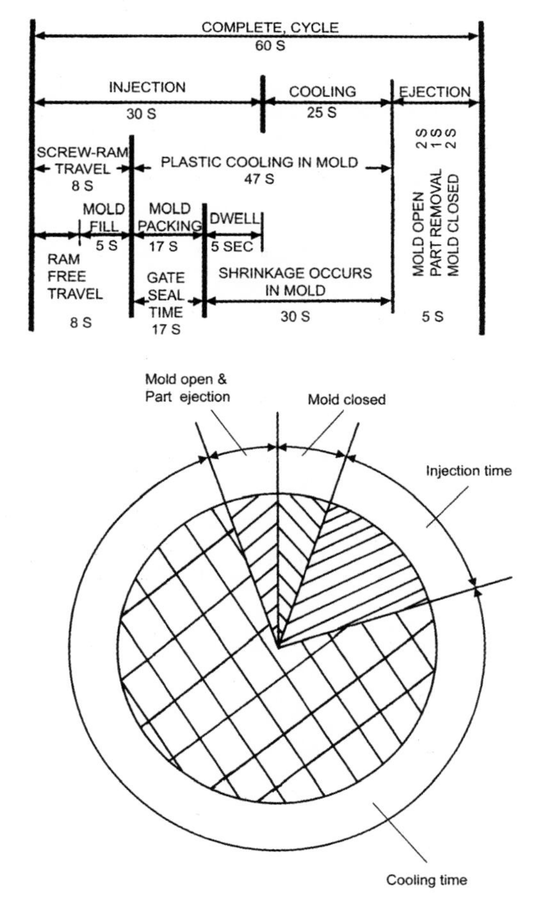

Figure 17.1 Flow chart for typical tool activity.

Molding		Materials	Flexural modulus × 10^3 kg/mm^3	Impact strength	Heat resistance (HDT 18.6)°C	Paintability by baking (140–150°C)	Weight ratio[1] (Equiflextural modulus)	Moldability	Cost[2] (Mold cost)[3]
Compression molding	Hot press	Polyester + GF (SMC)	1	O ~ Δ	> 200	◎	0.65	O ~ Δ	O (O)
		Polyester + GF (BMC)	1.1	↑	↑	◎	0.6	↑	↑
		Polyester + GF (High-strength SMC)	1.6 ~ 4.2	◎	↑	—	0.4 ~ 0.5	↑	O (O)
	Cold press	Polyester + GF (Resin injection)	0.8	O	150 ~ 200	Δ ~ X	0.62	Δ	◎ (◎)
		Polyester + GF (Hand lay-up)	—	—	—	—	—	—	—
	Stamping	PP + GF or sawdust, (AZDEL, etc.)	0.6	◎	160	—	0.5	◎	Δ ~ O (O)
		Nylon + GFTF (STX, etc.)	0.8	◎	215	O	↑	◎	Δ (O)
Filament winding		Epoxy + CF (CFRP)	15	O	> 200	—	0.2	Δ	Δ ~ x (O)
Injection molding		PP + GF, talc (EPDM) AS + GF	0.6 ~ 0.4	↑	120 ~ 105	—	0.5	O	◎ (—)
		PBT or nylon + GF	1.2 ~ 1.4	↑	205 ~ 215	◎ ~ O	0.5	↑	Δ (O)
		Foamed styrene or ABS (+ GF)	2.4 ~ 2.5	O ~ Δ	80 (100)	—	0.4 ~ 0.6	↑	O (O)
	RLM	Urethane + GF (RRIM)	0.1 ~ 0.2	◎	O ~ Δ	Δ ~ X	—	O ~ Δ	Δ (◎ ~ O)

Note: 1. Ratio based on sheet metal weight as 1
2. Relative comparison for 400–500 kg
3. Mold cost for sheet metal
Symbols: ◎ Excellent; O Good; Δ Fair; × No good

Table 17.7 Examples of cost comparison of molds in terms of the properties of plastic

The RP tools listed in Table 17.8 also include vacuum and pressure flexible bags. Tool operations vary from fabricating solid products to making foamed products, such as using a steam chest (Fig. 17.2) to produce expandable polystyrene (EPS) foams (chapter 8).

The proper choice of materials of construction for openings or cavities is paramount to achieving quality, performance, and longevity of tools. Desirable properties are good machinability of

Property	Natural rubber (molded)	Natural rubber (latex)	Nitrile	Neoprene	Butyl	Polyrethane	Silicone (RTV)	PVC
Tensile strength (kgf/cm^2)	210	210	105	140	140	280	70	140–210
(MN/m^2)	20.5	20.5	10.5	13.5	13.5	27.5	7	13.5 – 20.5
(lbf/in^2)	3000	3000	1500	2000	2000	4000	1000	2000 – 3000
Hardness range, Shore A	30–90	40	40–95	40–95	40–75	75–90	40–85	–
Tear resistance	Very good	Very good	Fair	Good	Good	Excellent	Poor	Fair to good
Abrasion resistance	Excellent	Good	Good	Good	Good	Excellent	Poor	Fair
Resilience	Excellent	Excellent	Fair	Good	Bad	Good	Excellent	Bad
Compression set	Good	Fair	Good	Good	Fair	Poor	Fair	Poor

Table 17.8 Typical properties of various RP mold bag materials

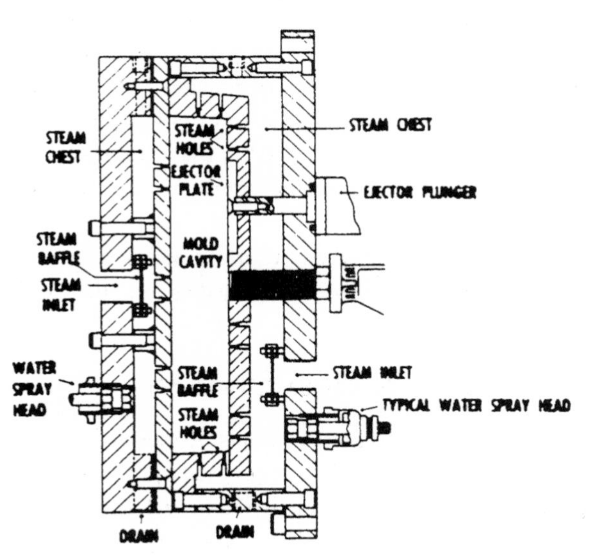

Figure 17.2 Example of a steam chest mold for producing expandable polystyrene (EPS) foams.

component metal parts, material that will accept the desired finish (polish, for example), ability to transfer heat rapidly and evenly, capability of sustained production without constant maintenance, and so on. (Tables 17.9 to 17.12). As the technology of tool enhancements continues to evolve, tool manufacturers have increasingly turned to such enhancements in the hope of gaining advantages.

There are now a wide variety of enhancement methods and suppliers, each making its own claims on the benefits of its products. With so many suppliers offering so many products, the decision on which technology to try can be time consuming. There are toolmakers that do not have the resources to devote to a detailed study of all of these options. In many cases they treat tools with methods that have worked for them in the past, even though the current application may have different demands and newer methods have been developed. What can help is to determine their needed capabilities and features, such as hardness, corrosion resistance, lubricity, thermal conductivity, polishing, coating, and repairing (Tables 17.9, 17.13 to 17.17).

There are many tool metals such as D2 steel that are occasionally used in their natural state (soft) when their carbon content is 1.40 to 1.60 wt%. Tool metals, are generally used in a pretoughened state (not fully hardened). For an unsophisticated hardness test, use a file across a discreet corner of your tool and compare your finding with Table 17.18 (355).

By increasing hardness, longer tool life can often be achieved. Increased wear properties are especially critical when fabricating with abrasive glass- and mineral-reinforced plastics. This is important in high-volume applications and high-wear surfaces, such as mold gates inserts and die orifices. Some plastic materials release corrosive chemicals as a natural by-product during fabrication. For example, hydrochloric (HCl) acid is released during the tooling of polyvinyl chloride (PVC). These chemicals can cause pitting and erosion of untreated tools surfaces. Untreated surfaces may rust and oxidize from water in the plastic and humidity and other contaminants in the air.

Polishing and coating tools help to meet product surface requirements. Improved release characteristics of fabricated products are a common advantage of tool coatings and surface treatments. This can be critical in applications with long cores, low draft angles, or plastics that tend to stick on hot steel in hard-to-cool areas. Coatings developed to meet this need may contain polytetrafluoroethylene (PTFE; Teflon). Metals such as chrome, tungsten, or electroless nickel provide inherent lubricity.

The following tables are guides pertaining to mold purchasing and follow-ups. The approach outline can be used in developing guides to purchasing dies and other tools (Tables 17.19 to 17.23).

MATERIAL OF CONSTRUCTION

In addition to the basic processing machinery, molds or dies are required in almost all processes used for the manufacture of the many different products produced worldwide. These parts may be of a simple design made from wood, as usually used in RP bag molding (chapter 15). For more sophisticated processes such as injection molding (IM), extrusion, and blow molding (chapters 4 to 6), the mold may be composed of many parts requiring high-quality metals and precision

AISI Designation Description	Typical Rc	Hardening Temp (°F)	Tempering Temp (°F)	Heat Treatability	Compressive Strength	Corrosion Resistance	Wear Resistance	Toughness	Machinability	Polishability	Weldability	Thermal Conductivity
4140	30-36	1500	1200	10	4	1	2	8	6	5	4	5
P20	30-36	1600	1100	10	4	2	2	9	6	8	4	5
414SS	30-34	1550	750	10	4	7	3	9	4	9	4	2
420SS	35-40	1885	1050	10	4	6	3	9	4	9	4	2
P5	59-61	1575	450	6	6	2	8	6	10	7	9	3
P6	58-60	1475	425	6	6	3	8	7	10	7	8	3
O1	58-62	1475	475	7	9	1	8	3	8	8	2	5
O6	58-60	1475	500	6	8	1	8	4	10	5	2	5
H13	50-52	1875	1000	6	7	3	6	7	9	5	2	4
S7	54-56	1725	550	8	8	3	7	5	9	8	3	4
A2	56-58	1750	1000	9	9	3	9	3	8	7	2	4
A6	56-58	1600	450	7	8	2	8	4	10	7	4	5
A10	58-60	1475		7	9	2	9	5	8	6	2	5
D2	56-58	1850	950	9	8	4	10	3	4	6	1	2
420SS	50-52	1885	480	8	6	7	6	6	7	10	6	2
440SS	56-58	1900	425	7	8	8	8	3	6	9	4	2
M2	60-62	2225	1125	8	10	3	10	2	4	6	2	3
ASP23	61-63			8	10	4	10	5	4	7	2	3
BECU (2%)	36-42	625	NR	7	2	6	1	1	10	9	7	9
BECU (2%)	26-30			7	2	6	1	1	10	9	7	9.5
BECU (0.5%)	20-24	900	NR	7	1	7	1	1	10	9	9	10

Table 17.9 Examples of the properties of different tool materials

	Weldability (Repair)	Polishability	Thermal Fatigue	Heat Transfer	Abrasion Resistance	Corrosion Resistance	Toughness	Machinability	Comments
P-20	2	2	2	2	3	3	4	4	Large cavities, cores - eliminate heat treat process and associated warpage and cracking.
H-13	3	2	4	2	3	5	4	4	Thermal fatigue resistance, polishability. Mostly chosen for zinc, aluminum die casting.
420 S.S.	3	2	5	2	1	2	2	4	Corrosion resistance (poor thermal conductivity).
P-6	4	4	2	2	3	3	4	4	Easily machinable. Welds, repairs well. Low carbon steel - not dimensionally stable in heat treatment.
O-1	3	2	2	2	3	2	3	3	Oil hardening - pins, small inserts, etc.
S-7	3	5	3	3	3	3	2	4	Shock resisting steel - long cores - where subjected to mechanical loads (slides, lifters)
A-2	3	4	4	4	3	2	4	1	Good abrasion resistance, polishability Air hardening - heat treat stable.
D-2	1	4	4	5	3	1	4	1	Extreme abrasion resistance. Used as gate inserts, etc. for filled resin applications.
Aluminum	5	1	3	1	5	1	1	5	Prototype, short run and structural foam molding.

Table 17.10 Guide to different tool materials, where 5 is best

Element	Effect	Element	Effect
Aluminum (Al)	Combines with nickel and titanium to form an intermetallic compound, which precipitates on aging and provides strength and hardness.	Nickel (Ni)	Usually added to improve hardenability of low-allow steels. In maraging steels, nickel combines with aluminum and titanium to form an intermetallic compound that increases hardness and strength on aging. Large amounts of nickel also assist in corrosion resistance.
Carbon (C)	Very influential in controlling hardness, depth of hardness, and strength.		
Chromium (Cr)	A carbide-forming element that contributes strongly to hardenability and abrasion and wear resistance. Additional amounts of chromium, greater than are needed for carbide formation, remain in solution and enhance corrosion resistance.	Silicon (Si)	Principal function is as a deoxidizing agent during melting. In higher quantities, it retards tempering, thus allowing higher tempering and operating temperatures.
Cobalt (Co)	An element added to the maraging steels to improve strength.	Titanium (Ti)	Found in maraging steels, where it acts as a potent strengthener by combining with nickel and/or aluminum to form an intermetallic compound, which precipitates on aging.
Manganese (Mn)	Combines with free sulfur to form discrete sulfide inclusions and improve hot workability. It is also a deoxidizing agent. In larger quantities, it increases hardenability by decreasing the required quenching rate. It is the principal element used to obtain quenching by air cooling, which minimizes distortion.	Tungsten (W)	Increases hardness, strength , and toughness.
Molybdenum (Mo)	Promotes hardenability in mold steels. The elevated tempering requirement increases the steel's strength at higher operating temperatures and provides more complete relief of residual stresses for greater dimensional stability.	Vanadium (V)	A strong carbide-forming element, which is usually added to control grain size and increase wear resistance.

Table 17.11 Examples of improving/changing properties of tool materials via alloying

Alloy Designation	Cost Index	Tensile Strength 0.2% Yield	Working Hardness	Coeff. of Thermal Expansion	Thermal Conductivity
	(AISI 4140=1)	KSI	Rockwell C	10^{-6}/°F	BTU/FT²/HR/°F
AISI 4140	1.0	100	27-30	12.7	19
AISI P 20	1.3	120	28-35	7.1	17
AISI H 13	3.5	180	40-45	7.1	17
UNS S42000	2.5	200	28-30	6.5	15
PH 15.5	4.5	175	38-40	6.2	12
MAR 18(300)	7.5	290	48-56	5.6	17

Table 17.12 Example of costs and properties of tool materials, including alloys

Tool Material	Thermoplastics unfilled	Thermoplastics glass filled: Low pressure thermosets SMC, BMC	High pressure thermosets phenolics, ureas, diallylls, melamines, alkyds	Prototype injection molds,	Structural foams	Casting of liquid resins	Blow molds
Carbides		●	●				
Steel, nitriding		●	●				
Steel, carburizing		●	●				
Steel, water hardening	●	●	●				
Steel, oil hardening	●	●	●				●
Steel, air hardening	●	●	●				●
Nickel, cobalt alloy	●						●
Steel, prehardened 44 Rc	●						●
Beryllium, copper	●						●
Steel prehardened 28 Rc							●
Aluminum bronze					●		●
Steel low alloy & carbon				●	●		●
Kirksite (zinc alloy)				●	●		●
Aluminum, alloy				●	●		●
Brass				●		●	●
Sprayed metal				●		●	
Epoxy, metal filled				●		●	
Silicone, rubber				●		●	

Table 17.13 Hardness of tool materials for a few different plastic materials and processes

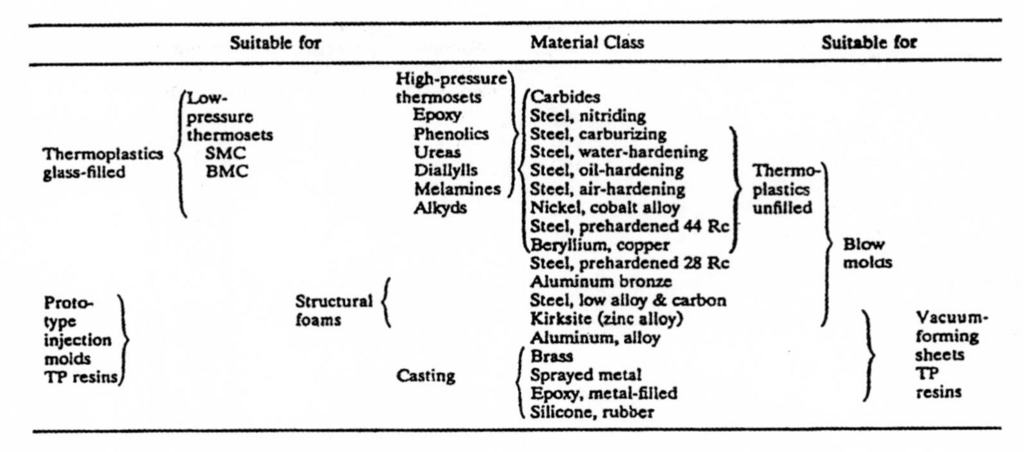

Suitable for			Material Class	Suitable for	
		High-pressure thermosets: Epoxy, Phenolics, Ureas, Diallyls, Melamines, Alkyds	Carbides		
Thermoplastics glass-filled	Low-pressure thermosets: SMC, BMC		Steel, nitriding		
			Steel, carburizing	Thermoplastics unfilled	Blow molds
			Steel, water-hardening		
			Steel, oil-hardening		
			Steel, air-hardening		
			Nickel, cobalt alloy		
			Steel, prehardened 44 Rc		
			Beryllium, copper		
			Steel, prehardened 28 Rc		
	Structural foams		Aluminum bronze		
Prototype injection molds TP resins			Steel, low alloy & carbon		
			Kirksite (zinc alloy)		Vacuum-forming sheets TP resins
			Aluminum, alloy		
		Casting	Brass		
			Sprayed metal		
			Epoxy, metal-filled		
			Silicone, rubber		

Table 17.14 Example of tool materials arranged in order of hardness

machining. To capitalize on its advantages, the mold may incorporate many cavities and dies with openings of different shapes, which adds further to its complexity.

The quality of machine tools used is absolutely critical to the efficiency in manufacturing plastic products. A significant factor is the computer numerical control (CNC) systems for machining all kinds of tools. Although the skill and experience of the toolmaker remain an essential factor, the CNC program gives an operator precise control over all machining operations. Many of these are lengthy and repetitive, which makes them ideal for digital computerization.

Most tools have to be handled very carefully and must be properly maintained to ensure their proper operation. They are generally very expensive and can be highly sophisticated. The major cost is in machine-building labor.

Choosing the proper materials for constructing openings or cavities is paramount to quality, performance, and longevity (i.e., the number or length of products to be processed). Using low-cost materials to meet high performance requirements will compromise integrity. For example, for 90% of molds and dies, the cost of the material of construction can be less than 5% of the total cost; for the rest, the cost of materials can be up to 15% of the total cost. Thus it does not make sense to compromise tool integrity to save a few dollars; it does make sense to use the best material for the application. The cost of labor, particularly for manufacturing the tool, is the bulk of the tool's cost.

They are used in processing many different materials with many of them having common assembly and operating parts with the target to have the tool's opening or cavity designed to form desired final shapes and sizes. For example, with certain processes to capitalize on advantages, molds may incorporate many cavities, adding further to its complexity. Sheet dies incorporate many thickness adjustment controls. Many tools, particularly for IM, have been preengineered as standardized products that can be used to include cavities, different runner systems, cooling lines, unscrewing mechanisms, and so on.

Vickers, HV	Rockwell C	Brinell	Vickers, HV	Rockwell C	Brinell
940	68	–	410	41	388
920	67	–	400	40	379
900	67	–	390	39	369
880	66	767	380	38	360
860	65	757	370	37	350
840	65	745	360	36	341
820	64	733	350	35	331
800	64	722	340	34	322
780	63	710	330	33	313
760	62	698	320	32	303
740	61	684	310	31	294
720	61	680	300	29	284
700	60	656	295	29	280
690	59	647	290	28	275
680	59	638	285	27	270
670	58	630	280	27	265
660	58	620	275	26	261
650	57	611	270	25	256
640	57	601	265	24	252
630	56	591	260	24	252
620	56	582	255	23	243
610	55	573	250	22	238
600	55	564	245	21	233
590	54	554	240	20	228
580	54	545	230	18	219
570	53	535	220	15	209
560	53	525	210	13	200
550	52	517	200	11	190
540	51	507	190	9	181
530	51	497	180	6	171
520	50	488	170	3	162
510	49	479	160	0	152
500	49	471	150	–	143
490	48	460	140	–	133
480	47	452	130	–	124
470	46	442	120	–	114
460	46	433	110	–	105
450	45	425	100	–	95
440	44	415	95	–	90
430	43	405	90	–	86
420	42	397	85	–	81

Table 17.15 Different hardness conversions

METAL	Btu/hr ft °F
ALUMINUM ALLOYS:	
1100	137.9
2024	108.9
6061	99.2
7075	70.1
STAINLESS STEELS:	
303 SS	9.7
410 SS	14.5
414 SS	14.3
420 SS	14.4
440 SS	14.0
TITANIUM ALLOY (Grade 5)	4.5
TITANIUM CARBIDE	15.7
COPPER AND ALLOYS:	
Pure copper	227.6
Free machining (1%Pb)	222.5
Cartridge brass (70%)	70.2
Naval brass	67.7
Manganese bronze	62.9
Phosphor bronze	41.1
Ampco® 18 @ 92 R_b	31.4
Ampco® 21W @ 29 R_c	26.6
Ampco® 22W @ 35 R_c	24.2
Ampco® 940 @ 94 R_b	125.0
Ampco® 97 @ 77 R_b	190.0
BeCu Moldmax® @ 40 R_c	60.5
BeCu Moldmax® @ 30 R_c	75.5
BeCu Protherm® @ 96 R_b	145.6
(BeCu alloys from Brush-Wellman)	
IRON:	
Pure iron	43.0
Cast iron	27.1
Low carbon steel alloys	30.0
High carbon steel alloys	26.1
TOOL STEEL:	
A-2	15.0
A-6	15.0
A-11 (CPM 10V®)	12.4
D-2	11.4
H-13	14.2
O-1	18.5
O-6	18.5
P-20	16.8
S-7	16.5
M-2	12.3
T-15	12.1

Table 17.16 Thermal conductivity of tool materials

TYPE	200 °F	400 °F	800 °F
1020	6.5	6.7	7.1
4140	6.8	7.1	
6150	6.8	7.1	7.4
W1	5.8	6.1	7.3
W2	7.4		8.0
S1	6.9	7.0	7.5
S5	6.4		7.0
S6	6.4		7.0
S7	6.8	7.0	7.4
O1	5.8	5.9	7.1
O2	6.2	7.0	7.7
A2	5.8	5.9	7.2
A6	6.4	6.9	7.5
D2	5.6	5.7	6.6
D3	6.3	6.5	7.2
D4	6.2		6.9
H10	6.1		6.8
H11	6.2	6.9	7.1
H13	5.8	6.4	6.8
H14	6.1		
H19	6.1	6.1	6.7
H21	6.9	7.0	7.2
H22	6.1		6.4
T1	5.3	5.4	6.2
T5	6.2		
T15		5.5	6.1
M1		5.9	6.3
M2	5.6	5.2	6.2
M3			6.4
M4			6.4
M7		5.3	6.4
M10			6.1
L2	7.4		8.0
L6	6.3	7.0	7.0
P2	7.0		7.6
P20	6.5		7.1
SS 303,4	9.6	9.9	
SS 316	8.8	9.0	
SS 414	5.8	6.1	
SS 420	5.7	6.0	
SS 440	5.7		
Ti alpha alloy	4.6		4.8
Ti alpha-beta	5.0	5.1	5.2
Ti beta alloy	5.2	5.4	5.6
Ampco 18, 21, 22	9.0		
Ampco 940	9.7		
FREE MACH Cu	9.8		
BeCu (2%)	9.7		
BeCu (0.5%)	9.8		
6061 Aluminum	13.1		
INCONEL	6.4		

Table 17.17 Thermal-expansion coefficients of tool materials

Findings	HRC	Possible carbon content
Easy to file (soft)	25–30	0.2–0.3%
Hard but possible to file (medium)	40–55	0.3–0.7%
Unable to file (hard)	60–70	0.6–2.5%

HRC = Hardness Rockwell C

Table 17.18 HRC file check

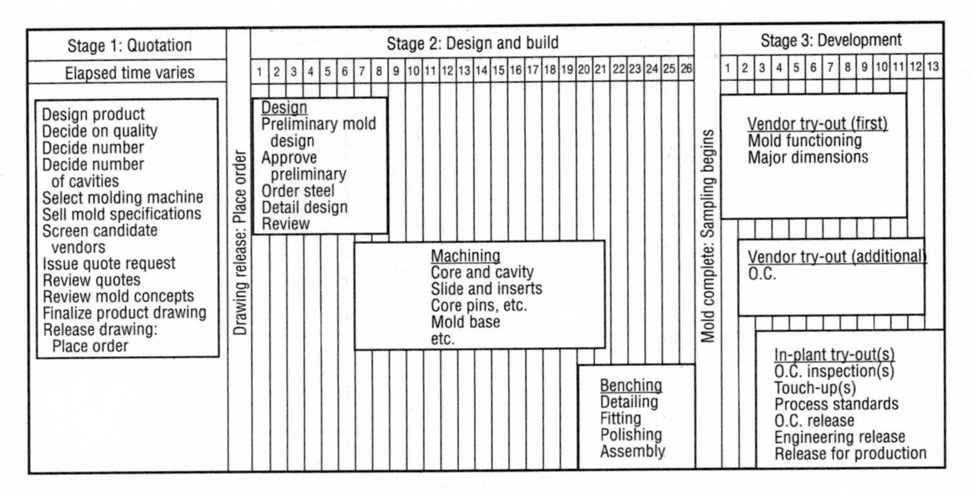

Table 17.19 Example of a schedule, in weeks, for purchasing of a mold

1. Detailed design to be reviewed and approved by customer prior to mold construction.
2. Mold base to be hardened to minimum specified amount.
3. Specify mold surface hardness.
4. All cavities should include engraving to identify molded part cavity ID.
5. Specify ejection system to be used.
6. Mold surfaces which directly form the molded part should have direct water cooling when possible (as opposed to nearby plate cooling which must then cool the insert which cools the molded part.
7. Water lines and fittings should not be located in clamping slots.
8. Mold to include parting line alignment locks in addition to leader pins..
9. Specify polishing and/or coating requirements for cavity (s).
10. Specify that no welding shall take place without notification to and approval by the customer.

Table 17.20 Guide for mold construction

1. ___ Was latest issue part drawing used?
2. ___ Will mold fit press for which intended? Are press ejectors specified?
3. ___ Are daylight and stroke of press sufficient for travel and ejection?
4. ___ Are reverse views correct?
5. ___ Are one guide pin and one return pin offset?
6. ___ Do guide pins enter before any part of mold?
7. ___ Can mold be assembled and disassembled easily?
8. ___ Has allowable draft been indicated?
9. ___ Is plastic material and shrinkage factor specified?
10. ___ Are mold plates heavy enough?
11. ___ Are mold parts to be hardened clearly specified?
12. ___ Are sufficient support pillars located and specified?
13. ___ Are waterlines, steam lines, thermocouple holes, or cartridge holes shown and specified?
14. ___ Does water in/out location clear press tie bars and clamp locations?
15. ___ Is ejector travel sufficient?
16. ___ Are stop buttons under ejector bar specified?
17. ___ Are ejector pins sufficient? Specified?
18. ___ Is the steel type for mold parts specified?
19. ___ Have eyebolt holes been provided?
20. ___ If stripper type, does stripper plate ride on guide pins for full stroke?
21. ___ Do loose mold parts fit one way only? (Make foolproof.)
22. ___ Will molded part stay on ejector side of mold?
23. ___ Can molded part be ejected properly?
24. ___ Have trademarks and cavity numbers been specified?
25. ___ Has engraving been specified?
26. ___ Has mold identification been specified?
27. ___ Has plating or special finish been specified?
28. ___ Is there provision for clamping mold in press?
29. ___ Are runners, gates, and vents shown and specified?
30. ___ Does mold have stock number on mold sections?

Table 17.21 Example of a mold checklist

TO: Date: ___/___/___

JOB NUMBER ______________ YOUR P.O. NUMBER ______________

DESCRIPTION: ______________

____

____

REVIEW PROBLEMS

____

____

____

____

____

SCHEDUAL DELAYS WITH REASONS

____

____

____

____

____

Current Date: ___/___/___

COMMENTS: As of this date, the mold is _____% complete.

____

____

____

____

____

Signed: ______________

Title: ______________

Table 17.22 Example of a mold progress report

COMPANY NAME

Address

MOLD PROGRESS REPORT

Part Name ____________ P.O. Number ____________ Job Number ____________

Customer ____________ Scheduled Delivery: Original __/__/__ Date of Report __/__/__

Attention of: ____________ Current __/__/__ Report by: ____________

	DESIGN	MODEL/HOBS	CAVITIES	CORES/INSERTS	MOLD BASE	POLISH	ASSEMBLY
Estimated Completion	__/__ % complete	__/__ % complete	__/__ % complete	__/__ % complete	__/__ % complete	__/__ % complete	__/__ % complete
Week							
1. __/__							
2. __/__							
3. __/__							
4. __/__							
5. __/__							
6. __/__							
7. __/__							
8. __/__							
9. __/__							
10. __/__							
11. __/__							
12. __/__							
13. __/__							
14. __/__							
15. __/__							
16. __/__							
17. __/__							
18. __/__							
19. __/__							
20. __/__							

Table 17.23 Example of a detailed mold progress report

Material of construction choices range from computer-generated plastic to specialty alloys or even pure carbide tooling with the usual made from steels. Everyone from purchasing agents to shop personnel must consider the ramifications of tool performance requirements. One may consider the softest tool that will do the job because it is usually the least expensive to build, but it also requires special and careful handling and has a limited life.

Different materials of construction principally use different grades of steels; others include types such as aluminum, beryllium copper alloy, and brass, Kirksite, sintered metal, steel-powder-filled epoxy plastic, silicone, plaster of paris, RP, sand, wood, and flexible plastic. The proper choice of materials of construction for their openings or cavities is paramount to quality, performance, and longevity (number or length of products to be processed). Desirable properties are good machinability of component metal parts, material that will accept the desired finish (polished, textured, etc.), ability with most molds or dies to transfer heat rapidly and evenly, capability of sustained production without constant maintenance, and so on.

Different materials of construction are used to manufacture tools that principally use steel. Some of the tool materials incorporate different special metals providing improvements in heat transfer, wear resistance of mating mold halves, and so on. These special metals include beryllium copper alloy, brass, aluminum, Kirksite, and sintered metal (Table 17.11). Table 17.24 provides tool materials in addition to those reviewed in Table 17.1. (Since Table 17.24 is long, it is located at the very end of this chapter.)

STEEL

P20 steel is a popular, high-grade forged tool steel relatively free of defects; it is available in a pre-hardened steel. It can be textured or polished to almost any desired finish and it is a tough material. H-13 is usually the next most popular steel used. Stainless steel, such as 420 SS, is the best choice for optimum polishing and corrosion resistance. Other steels and materials, such as copper alloys with fast cooling, aluminum with low cost and fast cooling, are used to meet specific requirements to meet tool life and cost. The choice of steel is often limited, particularly by the available sizes of blocks or plates that are required for the large tools.

As a tooling guide to life expectancy, consider P-20 steel for long runs (1 million products), QC-7 aluminum for medium runs (250,000 products), sintered metal for short runs (100,000 products), and filled epoxy plastic for relatively shorter runs (50 to 200 products).

The flow surfaces of the tool usually have protective coatings, such as chrome plating, to provide corrosion resistance. With proper chrome-plated surfaces, microcracks that may exist on the steels are usually covered. The exterior of the die is usually flash chrome plated to prevent rusting. Where chemical attack can be a severe problem (processing PVC, etc.), various grades of stainless steels are used with special coatings. Coatings will eventually wear, so it is important that a reliable plater properly recoat the tool; this is usually done by the original tool manufacturer.

The needs of the vast majority of materials, particularly steel, can be satisfied with a relatively small number of these materials. The most widely used steels have been given identifying numbers of the AISI. The properties of the tool material usually are as follows: (1) chemical compositions; (2) wear resistance to provide a long life; (3) toughness to withstand processing and particularly factory handling; (4) high modulus of elasticity so that the die channels do not deform under melt operating pressure and the die's weight; (5) high compression strength, which is very important; (6) high uniform thermal conductivity; (7) machinability so that good surface finishes can be applied, particularly near the die exit; and (8) ability to be repaired.

Important requirements for tools are high compression strength at the processing temperatures of the plastics, wear resistance especially in regard to the increasing use of reinforcing fibers, adequate toughness, possibly corrosion resistance, and good thermal conductivity. In addition, so that the tools may be manufactured economically, good machinability is expected and, in certain cases, also cold hobbing is of less importance since wire and die sink electrical discharge machining (EDM) has taken over most of these applications that require hobbing. Dimensional stability during heat treatment generally is necessary.

Different characteristics and performances identify steels. For example, higher hardness of steel improves wear, dent, and scratch resistance and polishability; but lowers machinability and weldability. High sulfur content degrades the stainless qualities and polishability of the steel. Hardness as a measure of the internal state of stress of the steel has an adverse effect on weldability, fracture toughness, and dimensional stability. Other behaviors and characteristics are shown in Tables 17.25 to 17.27 and Figure 17.3.

Different treatments provide different performances. For example, low-carbon steel undergoing a surface-hardening process is used to resist wear and abrasion. It is used in molds, dies, and other machine parts. The steel is heat-treated in a box packed with carburizing material, such as wood charcoal, and heated to 1090°C (2000°F) for several hours.

An economic advantage of the cast tool is the possibility of casting the cooling core in place, where required, simultaneously with the tool. In tools of small or medium sizes, ducts can be introduced during the casting operation. In large tools it has proved to be advantageous to cast cooling chambers, which are subsequently closed off by plates. But the effectiveness of this cooling is slight, and it must be decided whether this cooling effect is adequate for the fabricated plastic product.

Gray cast iron or castings are used sometimes for large size tools. Cast tools normally consist of steel castings and, moreover, of those grades of steel whose chemical composition largely corresponds to that of forged or rolled grades. Case-hardened steels still have a tough core as well as a wear-resistant surface after any heat treatment. The high surface hardness and modern deoxidation methods offer the best conditions for polishing. In special cases this property can be improved even further with the help of remelting processes (electroslag refining).

Considering the various stresses to which tools are exposed and requirements they are expected to meet, different steel conditions cannot be met with one grade of steel. When heating to the required treatment temperatures and afterward when quenching, there is also the action due to the different wall thicknesses that reach the higher or lower temperatures at different times. The resulting temperature differences necessarily lead to stresses, which can result in the yield point of the tool material being exceeded, particularly at the higher temperatures.

In addition, there is the danger that thermal stresses at sharp-edged transitions, at deeper machining grooves, or at thin sections between the cavity and the cooling core may be relieved not only in the form of a plastic deformation but even as a crack. The aim is to design the tool with as symmetrical a configuration as possible and with a largely uniform distribution of mass in regard to the heat treatment.

AISI NO.	THERM CONDUCT. (Btu/hr ft °F)	HARDNESS (Rc)	TEMPERING TEMP (°F)	DISTORTION DURING HARDENING	COMPRESSIVE STRENGTH	TOUGHNESS	RESISTANCE TO WEAR
A2	15	60/62	400	LOWEST	VERY HIGH	MEDIUM	HIGH
A6	15	58/60	350	LOWEST	HIGH	MEDIUM	MED - HIGH
A10		58/60	400	LOWEST	VERY HIGH	MEDIUM	HIGH
P20	22	28/34	1150	PREHARDENED	MED	VERY HIGH	LOW
420 SS	14.3	50/52	850	LOW	MED - HIGH	MED	MED
440 SS	13.8	56/58	425	LOW	HIGH	LOW	HIGH

Table 17.25 Properties of the more popular tool materials

AISI Type	Crucible steel designations Designation	General Characteristics	Toughness	Dimensional Stability in Heat Treatment	Machinability (Annealed)	Polishability (Heat-treated)	Typical Applications
P-20	CSM-2	Medium carbon (0.30%) and chrome (1.65%). Available prehardened (300 Bn).	10[1]	6	9 (prehardened)	8 (prehardened)	Excellent balance of properties for injection and compression molds of any size.
H-13	NuDie V	Hot-work die steel; 5% chrome. May be hardened to about 50 Rc.	9	8	9	9	Higher hardness than P-20; good toughness and polishability. Used for abrasion resistance in RP molds and high-finish injection molds.
A-2	Air Kool	Cold-work die steel, high carbon (1.0%) 5% chrome. May be hardened to about 60 Rc.	8	9	8	7	High hardness for abrasion-resistant, long-wearing compression and injection molds. Limited to small sizes.
D-2	Airdi 150	Cold-work die steel; high carbon (1.55%) 11.5% chrome. May be hardened to about 60 Rc.	7	9	5	6	Highest abrasion resistance. Difficult to machine. Susceptible to stress cracking. Small molds only.
414	CSM 414	Stainless steel; 12% chrome, 2% Ni, 1% Mn, low carbon (0.03%).	10	10	9	9	"Stainless version" of P-20; similar properties and uses.
420	CSM 420	Stainless steel; 13% chrome, 0.80% Mn, medium carbon (0.30%).	9	10	8	10	"Stainless version" of H-13; similar properties and uses. Very stable in heat treatment, takes high polish.
4145	Holder block	Medium carbon (0.50%) and chrome (0.65%). Available prehardened.	10	10 (prehardened)	10 (prehardened)	6 (prehardened) 7 (fully hardened)	Low-cost steel, for mold bases and large molds. Not suited to high-quality finish.

[1] 10 = best

Table 17.26 Examples of tool steels with applications

Steels	AISI-SAE Designations	Amount of Alloy
Carbon	10XX[1]	Plain with 1.00% maximum Mn
	11XX	Resulfurized
	12XX	Resulfurized and rephosphorized
	15XX	Plan with 1.00 to 1.65% Mn
Manganese	13XX	1.65% Mn
Nickel	23XX	3.50% Ni
	25XX	5.00% Ni
Nickel-chromium	31XX	1.25% Ni; 0.65 to 0.80% Cr
	32XX	1.75 Ni; 1.07% Cr
	33XX	3.50% Ni; 1.50 to 1.57% Cr
	34XX	3.00% Ni; 0.77% Cr
Molybdenum	40XX	0.20 or 0.25% Mo
	44XX	0.40 or 0.52% Mo
Chromium-molybdenum	41XX	0.50, 0.80, or 0.95% Cr; 0.12, 0.20, 0.25, or 0.30% Mo
Nickel-chromium-molybdenum	43XX	1.82% Ni; 0.50 or 0.80% Cr; 0.25% Mo
	47XX	1.05% Ni; 0.45% Cr; 0.20 or 0.35% Mo
	81XX	0.30% Ni; 0.40% Cr; 0.12% Mo
	86XX	0.55% Ni; 0.50% Cr; 0.20% Mo
	97XX	0.55% Ni; 0.20% Cr; 0.20% Mo
	98XX	1.00% Ni; 0.80% Cr; 0.25% Mo
Nickel-molybdenum	46XX	0.85 or 1.82% Ni; 0.20 or 0.25% Mo
	48XX	3.50% Ni; 0.25% Mo
Chromium	50XX	0.27, 0.40, 0.50, or 0.65% Cr
	51XX	0.80, 0.87, 0.92, 0.95, 1.00, or 1.05% Cr
Chromium with 1.00% C min	50XXX	0.50% Cr
	51XXX	1.02% Cr
	52XXX	1.45% Cr
Chromium-vanadium	61XX	0.60, 0.80, or 0.95% Cr; 0.10 or 0.15% V minimum
Tungsten-chromium	72XX	0.75% Cr; 1.75% W
Silicon-manganese	92XX	0.00 or 0.65% Cr; 0.65, 0.82, or 0.85% Mn; 1.40 or 2.00% Si

[1]First two digits denote type of steel; second two digits indicate carbon wt–percentage

Table 17.27 Examples of tool steel alloys (first two digits denote type of steel; second two digits indicate carbon weight percentage)

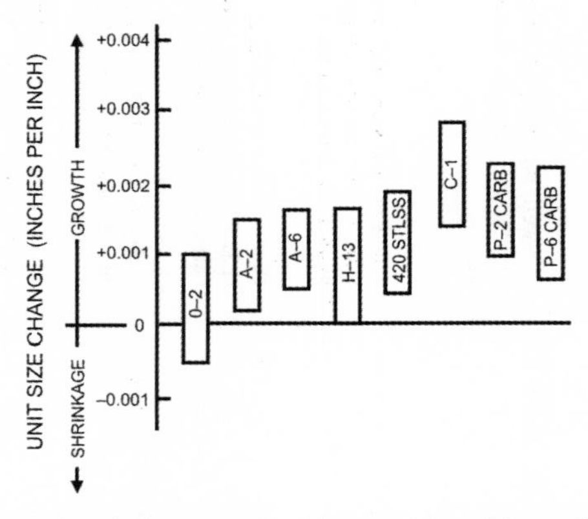

Figure 17.3 Examples of dimensional changes of tool materials subjected to heat treatment.

In machining as well as in noncutting shaping processes, stresses develop chiefly as a result of the solidification of surface layers near the edge. These stresses may already exceed the yield point of the respective material at room temperature and consequently lead to plastic deformations. Since the yield point decreases with increasing temperature, additional stresses can be relieved by plastic deformation during the subsequent heat treatment. In order to avoid unnecessary and expensive remachining, it is advisable to eliminate these stresses by stress-relief annealing (443).

When processing particularly abrasive plastics, the wear can still be too high even when using high-carbon, high-chromium steels. Metallurgical melting cannot produce steels with even higher amounts of carbides. In such cases, hard material alloys, produced by powder metallurgy, are available as a tool material. These alloys contain about 33 wt% of titanium carbide, which offers high wear resistance because of its very high hardness.

Like other tool steels, hard material alloys are supplied in the soft-annealed condition, when they can be machined. After the subsequent heat treatment, which should if possible be carried out in vacuum-hardening furnaces, the hard materials attain a hardness of about 70 HRC. Because of the high carbide content, dimensional changes after the heat treatment are only about half as great as those in steels produced by the metallurgical melting processes.

An efficient tool is not only the selection of the correct steel but also subject to a proper and careful heat treatment. A single error can destroy whatever has been invested up to this time in the cost of the steel and in the machining. Therefore, besides appropriate facilities for the heat treatment, knowledge of a theoretical and practical nature is an absolute requirement for carrying out the heat treatment properly.

The concept of distortion is frequently confused with that of dimensional change. Distortion involves changes in shape that are based on avoidable mistakes in the tool construction, pretreatment, and heat treatment. Dimensional changes are changes in shape that can be attributed to necessary crystal structural transformations and unavoidable thermal stresses. In contrast to distortions, dimensional changes are unavoidable.

Machining rolled or forged steels makes most of the tools. Extensive machining is required, particularly for large tools. When producing large tools, extensive machining work must frequently be carried out in the manufacturing process. For this reason, consideration is given to using blanks with a premachined cavity for constructing large tools. In the past, disadvantages of this technique existed (high allowances, local impurities, shrink holes or subcutaneous blowholes, etc.). Steel foundries are able to produce tools with close tolerances and the smallest allowances for machining.

Types are available that offer a slight allowance for grinding. The tolerances for these tools, whose patterns are made from metal or plastic, are +0.25%, based on the nominal size. With little effort for making the pattern and the tool it is possible to cast tools for processing plastics with only a slight allowance for machining of about 1 to 1.5 mm per surface, depending on the cavity. Another grade are tool blanks with a machining allowance of at least 5 mm per surface. Even this relatively rough finish can be of interest economically for the manufacture of large tools. Before deciding in favor of any of the forms in which the tools are supplied, consideration should be given to the cost of the material and to the machining required for the tool in question. Comparisons should be made of the costs of a cast tool and the costs of the pattern required.

In order to attain a high hardness at the surface, it is necessary to increase the carbon content in the case by adding carburizing agents during the heat treatment. Usually a carburizing depth of about 0.6 to 1 mm is aimed for and, after quenching and tempering, a case hardness of 58 to 62 HRC is achieved. Because of the extensive heat-treatment cycle, certain dimensional changes cannot be avoided; these require an additional expenditure in the finishing.

Because of the low carburizing depth, subsequent changes in the engraving are no longer possible. Should they be necessary, it is recommended that the tool be refinished in all the regions of the engraving area. Those areas of the surface of the tool that have already been subjected to carburizing in the first heat treatment will be supercarburized by the additional heat treatment.

Nitriding steels are preferred for high surface hardnesses. Use is made of aluminum alloyed steels Cr/Al/Mo and Cr/Ai/Ni. Because of the lower toughness of the nitrided layer, the depth of the nitrided case is limited. These steels have a lesser surface hardness after nitriding, but it is possible to increase the nitriding depth. The total service life of the tools of aluminum-free nitriding steels is consequently longer than that of the aluminum alloyed steels. In general, nitriding steels are supplied in the already quenched and tempered condition. The aluminum-alloyed types have a higher core strength than aluminum-free grades.

They are used in special cases for tool construction, such as for tools with very thin sections. But since the core strength of these steels is low, the hot-work steels, which are also known from their use in light-metal die-casting tool construction, are preferred. The high retention of tempering of these hot-work steels, which are normally premachined in the soft-annealed condition, permits tools to be heat-treated to a higher core strength without a decrease in strength in any further nitriding treatment.

Quenched and tempered tool steels are normally supplied and used. Since a heat treatment is not required there are no problems due to unavoidable dimensional changes and associated reworking. The time required for manufacturing a tool is reduced. The availability of standard tools and premachined bars and plates in these steel grades contributes to the reduction in manufacturing time.

In order to improve the machinability of quenched and tempered tool steels, grades with a higher sulfur content of about 0.05% to 0.08% are available. Since the sulfide inclusions in the steel have a low hardness, they impair the polishability only slightly. On photoetching, however, the sulfide can lead to a nonuniform structure. For this reason quenched and tempered tool steels with a low sulfur content are required despite the disadvantages with respect to machinability.

The advantages of using fully hardening tool steels rather than case-hardening steels for the manufacture of tools are primarily that the heat treatment is simpler and that making corrections to the cavity at a later time is possible without a new heat treatment. However, the greater risk of cracking can occur, particularly for tools with a greater cavity depth, because tools from these steels do not have a tough core. Moreover, the tougher steels with a carbon content of about 0.4% do not attain the high surface hardness of about 60 HRC, which is desirable with respect to wear and polish.

There is the low-carbon steel that can under go a surface hardening process. This steel is used to resist wear and abrasion. The steel is heat-treated in a box packed with carburizing material, then allowed to cool slowly.

Maraging steels for tools are special steels that were developed originally as high-strength steels for aviation and space travel. They are suitable for tools with particularly complicated mold cavities or die shapes. These special steels are supplied in the solution-annealed condition. In spite of their higher tensile strength the soft nickel-martensite of these steels can be machined readily. After a subsequent aging treatment at a temperature of 480°C to 490°C (896°F to 914°F), these steels reach a higher tensile strength. When the machining operation is performed prior to the heat treatment, consideration has to be given so that the uniform reduction in volume occurs at about 0.05%. If a plastics with high filler content is to be processed, it may result heavy tool wear.

It is usually recommended that maraging steels be nitrided. This can be combined with the process of precipitation hardening. It is a low treatment temperature that is possible in an appropriate gas atmosphere.

The cost for using stainless steel is not excessive but the benefit toward eliminating corrosion inside water channels and at O-ring face seals, for example, is great if the tool is expected to run a long time. Conversely for a mold, if you have a product that only runs 250,000 pieces and at end of the year the mold will be retired, then 4130/4140 plates (No. 2 steel) will be sufficient. If the molding material is corrosive, such as processing PVC, then stainless steel may be needed for molding surfaces. If the molding plastic has certain fillers then abrasive wear resistance is more of a concern.

ALUMINUM

In the past aluminum (Al) tools were considered "soft" tooling, and never considered for high-volume production. Unfortunately, taking advantage of the many benefits of Al were delayed; they provide quick turnaround times, faster outputs, lower costs, and in many cases, higher quality for their fabricated plastic products. Changes started to occur in 1991 when IBM formally announced that they had completed a five-year study of using aluminum molds for high-volume production (82).

This study demonstrated that, compared to identically designed and built tool steel molds, Al molds cost 50% less, delivered in half the amount of time, and produced higher quality products during cycle times that were reduced by 20% to 50%. With the better thermal conductivity of Al compared to steel, cycle time is reduced. From a handling and operating approach, the more lightweight Al provides about a 3:1 weight advantage. Other advantages include lower cost with machining that is much faster than tool steel. Heat dissipation is very uniform, resulting in products that retain excellent dimensional stability because there are fewer tendencies to distort due to shrink differences. The even heat distortion also reduces physical residual stress (82, 165).

Note that Al may not be the best choice for every application, and someone familiar with the advantages, disadvantages, costs, contacts, and basic concepts involved with using Al molds should study each situation. However, there are few cases that would not benefit from Al molds.

Data were compiled for a number of molds and various product designs, using a variety of plastic molding compounds including glass fiber loadings as high as 40 wt%. Molds were still producing parts after as many as 2 million cycles. Thus the age-old concept that Al molds were not tough enough to withstand the rigors of high-volume, long-term production runs of injection-molded products disappeared. Since that announcement, many other companies have come forward to admit that they too had the same experience. What became obvious was that many molders had been using Al molds for production but were hesitant to let others know, especially their customers, because of the "soft tool" stigma that had been placed on aluminum molds for so long.

Al has become more popular in the production of tools that include injection blow molding, extrusion blow molding, IM, extrusion, vacuum-forming layup molds, rubber molds, shoe molds, load cells, foam molds, prototype tooling, carpet-forming molds, robotics, general tools, jigs, and fixtures. With the automobile industry driving everything as far as large molds are concerned, production expectations have been upgraded dramatically. It is common to see aluminum-plate molds that have run anywhere from 500,000 to millions of products.

Certain grades of Al are capable of being machined properly and withstanding the pressures involved with IM. Al is usually three times the cost of tool steel by weight, but it is ⅓ the weight of tool steel, so the initial cost is the same for both.

Regarding wear and hardening, Al wears better if it is not hardened. This is due to the nature of the material to even out the microscopic, molecular, wavy surface over time. Instead of galling, Al surfaces slide. But many users do utilize some type of hardening, whether through anodizing or with another surface treatment. Anodizing can produce a surface hardened up to an Rc (Rockwell C) of 65. This is accomplished by converting up to 0.002 in of the Al surface to an oxide and adding an additional 0.002 oxide to that surface for a total of 0.004 in hardened layer. However, anodizing will chip away corners that are 90° or less. While this can be repaired by removing the anodized finish and reanodizing, it is not a cost-effective remedy. The anodizing process will form a layer approximately 0.004 in thick. Half of this will be embedded in the aluminum and half will be on top of the surface. This must be taken into consideration when machining cavities.

There are other methods of hardening the surface of Al. One of the most popular is a method listed under many different commercial names that includes a PTFE material in a nickel base. This is applied to the surface and can be stripped and reapplied as necessary.

Al can be difficult to identify. It can look like anything; only its weight will give it away. Al can be plated with hard chromium, which will give it an 80 HRC mirror finish; it can be bright red or green or it can look and feel like cast iron because it can have tool steel inserts fitted. Fortunately most forming and tooling Al is easy to recognize, like wrought tool aluminum (1050A, 6061/HE20, 6063/HE9, 6063A, 6082/HE30, 7020/HEl 7, ALUMEC 79 and 89) and cast aluminum (LM6, LM25, LM5; 355). Tables 17.28 to 17.30 provide some information on Al.

The best grade of Al to use is aircraft quality that is known as a T-6 grade (7075). It does not gum up cutters and grinders. It can be welded, and EDM (plunge and wire) can be used for machining. Al will machine at speeds approximately four times faster than tool steel and two to three times faster than prehardened P-20 steel. The surface can be polished using standard procedures with a

Property	Aluminum		Steel	
	QC-7	7075-T6	P-20	H-13
Typical hardness, Rockwell C	16	14	28 to 32	52 to 54
Typical yield strength, ksi	79 to 74	73 to 48	125 to 135	225
Thermal conductivity, Btu/hr/sq ft/deg F/ft	91	75	20.0	16.3
Average coefficient of thermal expansion, inches/deg F X 10^6	12.8	13.1	7.10	6.10
Density, lb/cu in	0.101	0.101	0.284	0.280

Table 17.28 Property comparison of aluminum and steel

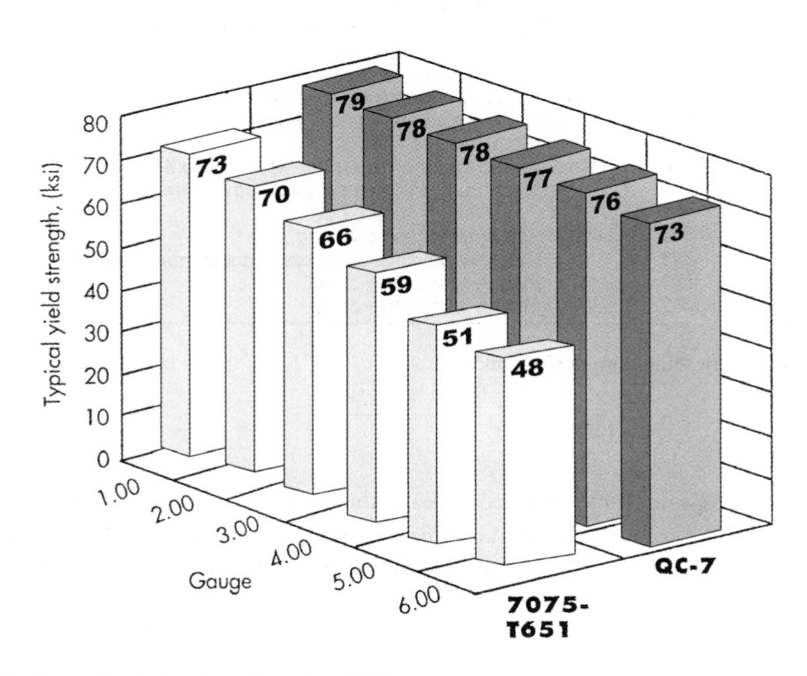

Table 17.29 Strength of aluminum based on thickness

Designation	Description
F	As fabricated, temper from shaping process only.
O	Annealed, recrystallized, softest temper for wrought.
H1	Strain hardened only.
H2	Strain hardened and then partially annealed.
H3	Strain hardened and then stabilized.
W	Solution heat–treated.
T	Thermally treated to produce tempers other than F,O, or H.
T1	Cooled from elevated temperature and then naturally aged.
T2	Cooled from elevated temperature, cold worked, and then naturally aged.
T3	Solution heat–treated, cold worked, and then naturally aged, improved strength.
T4	Solution heat–treated and then naturally aged to a substantially stable condition. Not cold worked.
T5	Artificially aged only. Rapid cool process like casting.
T6	Solution heat–treated and then artificially aged. Not cold worked after heat treated. Common class.
T7	Solution heat–treated and then stabilized, good growth control and residual stress.
T8	Solution heat–treated, cold worked, and then artificially aged. Cold worked to improve strength.
T9	Solution heat–treated, artificially aged, and then cold worked to improve strength.
T10	Cold worked and then artificially aged . Rapid cooling after heat treatment then cold worked for strength.
T51	Stress–relieved by stretching.
T52	Stress–relieved by compressing.
T53	Stress–relieved by combined stretching and compressing.
T42	Wrought only, properties of T4.
T62	Wrought only, properties of T6.

Table 17.30 Wrought aluminum performance

fewer number of steps due to the immediate smoothing ability of the aluminum microstructure. Because it is so light, Al can be machined on smaller equipment and at faster feeds. The more contoured the surfaces, the more savings available using aluminum because of the fast speeds and heavy feeds that can be used.

PREHEATING

Generally, preheats of aluminum and copper (Cu) are designed to minimize cracking, but this is not the reason why Al and Cu are preheated—although even Al and Cu would be less likely to crack when preheated. Pure Al and Cu are both excellent conductors of heat. When they are being reparied, the heat from a heated tungsten arc is very quickly absorbed, giving the welder the impression that not enough amps are being used. The natural reaction to this is to keep turning up the amps until a molten pool is obtained. This is when problems can start. Tungsten can disintegrate, and an inability to feed the filler wire into the pool can develop because the arc is too strong. To counteract this condition, one needs to preheat the base material to reduce the amount of heat loss from the pool. When preheating Al, it is very important to stay below 180°C (350°F), because the Al will start to lose its original properties if it is held at elevated temperatures and the tool aluminum will become soft. Preheats on tool coppers should be limited to 400°C (750°F) on chromium coppers, 350°C (660°F) on nickel coppers, and 215°C (600°F) on beryllium coppers.

To help to keep your preheat to a minimum, consider changing the argon gas to an argon-helium or helium-argon gas. This will raise the temperature of your weld pool without increasing your amps. Mixed gases are very important when welding heavier sections of Al and especially Cu.

ALUMINUM ZINC

Kirksite is a popular alloy of aluminum and zinc to make prototype tools or short production runs. The low melting alloy with a high degree of heat conductivity makes it easy to produce a tool. Pouring temperatures are as low as 800°F as compared to 3000°F for steel and 2000°F for beryllium-copper castings.

COPPER

There are many types of copper alloy used in the manufacture of tools across the whole spectrum of industry. This section reviews types of tool copper used in the manufacture of mold tools. Most tool coppers connected with mold tools tend to be two main types: mold-face coppers and working coppers.

The mold-face coppers (with colors of copper and light copper) will be of low-alloy copper. If the tool is soft it may contain low levels of nickel (2 wt%), chromium (0.5%), and/or beryllium (0.5%). If the tool is hard, it may contain high levels of beryllium (2% to 3%). Copper mold tools of light copper color will be of a higher alloy copper. These tools may contain high levels of nickel (10%) and silicon (3%) and have additions of chromium.

These coppers and light coppers tend to be used in the manufacture of mold cores, cavities, pinchoffs (nips), sprue bushings, hot runner systems, core pins, ejector pins, blow pins, and so on. Some mold-face tool coppers include the trade names AMPCO (83, 91, 95, 97, 940, 945) and UDDEHOLM (Moldmax and Protherm).

The working copper tools with yellow copper color that are used behind the scenes on tools will be a high alloy copper. They are likely to contain 8% to 15% aluminum, 3% to 5% ferrite, and, if they are hard, up to 3% beryllium. Uses for these include slides, bushes, wear plates, gibs, mold-locking devices, sleeve bearings, guide-pin bushings, guide rails, lifter blades, and pin bushings. Trade names include AMPCO (18, 21/21W, 22/22W, 25, and M4; 433).

BERYLLIUM COPPER

A popular copper is BeCu, which must be carefully handled (434). It is used in tools to provide relatively fast heat transfer. There are two basic families of BeCu alloys: those with high heat conductivity and those with high strength. Heat conductivity is about ten times greater than those of stainless steels and tool steels. Conductivity is double that of aluminum, such as alloy 7075, and higher than that of others. They have higher hardness and strength than aluminum. Table 17.31 compares BeCu to other materials, including Ampco Metal's new copper alloy called MoldMATE 90.

When heat-treated, they are the strongest of all copper-based alloys. Copper alloys such as beryllium copper (BeCu) and bronze are sometimes used for the high copper content that yields outstanding thermal conductivity in the tools. They can be used just as inserts in the tool to help direct heat transfers in critical areas of the tools.

COPPER ZINC

Brass, an alloy of copper and zinc, is used in the manufacture of molds and dies. One of its desired properties is good heat transfer (for fast cooling). Care has to be taken with these tools in structural foam molding. Some chemical blowing agents have decomposition properties that corrode the copper alloy.

OTHER ALLOYS

Examples of other alloys include bismuth, lead, tin, cadmium and, where very low melting points are required, indium. Different expansion properties for the different alloys exist. The extreme is with bismuth that expands at a rate of 3.3 vol% when it solidifies. By adjusting the composition, it is possible to obtain alloys with dimensional stability on casting or a significant degree of growth. Lead is added when this growth is more than 25%; this growth persists after solidification and can be detected as late as a month later.

MoldMATE versus other mold materials

Material	Thermal conductivity, Btu/hr/ft/°F	Hardness, Rockwell	Tensile strength, ksi	Yield strength, ksi
MoldMATE 90	90	31 RC	136	126
BeCu (2.0% Be)	60	36 RC	175	155
BeCu (0.5% Be)	125	95 RB	110	90
P20 tool steel	22	30 RC	140	110
Aluminum	80	86 RB	78	74

Source: Ampco Metal Inc.

Table 17.31 Properties of beryllium copper versus other tool materials

These are relatively soft, heavy metals that are brittle to shock yet perform well when subjected to heated plastic melt flow under sustained load. Their mechanical properties improve with age. Conductivity of heat and electricity is rather poor. Being stable metals, they have the great advantage that they can be remelted and repeatedly used. They are normally applied by simple gravity casting, but they also lend themselves to pressure die-casting and can be sprayed. This technique is unrivaled for high-fidelity reproduction and makes possible the application of alloys with low melting points to the pattern surface at well below their casting temperature.

From the tool-making point of view, the use of fusible alloys has two important advantages where exact reproduction is concerned. First, because so little heat is involved, no thermal stresses are left in the tool and the tendency toward warping is negligible. Second, because fusible alloys containing bismuth and have no casting contraction, the sprayed metal shell is dimensionally an exact replica of the pattern. Spraying with alloys is comparable in several respects to electroforming.

METAL SPRAY

Low-cost tooling metal spraying with low melting alloys is another method. It satisfies the demand for quickly made, low-cost molds. Metal spraying is performed in a way similar to paint spraying, with the important difference being that the metal has to be liquefied by heat. Although metal leaves the gun at a temperature in the range of 200°C to 300°C, the very small particles produced in the nozzle cool so rapidly in flight that they solidify almost instantaneously on the pattern. Each tiny droplet is roughly spherical while it is in the air, but on impact it spreads out into a liquid film. Succeeding droplets do the same, and a well-bonded structure can be produced as long as the correct temperature is maintained.

Problems can develop. If the droplets have cooled so much that they are virtually solid by the time they contact the pattern, they neither flatten out nor fuse together and make only point contact with their neighbors. This results in a weak and brittle deposit with a good deal of entrapped air. On the other hand, if the spray is too hot, a puddle of metal forms and is then blown away by the air stream. In practice it is not difficult to work at the right temperature, providing it is understood that the secret of success is to keep a proper distance between the nozzle and the pattern. Adjustment of this distance is the most effective way of controlling temperature.

There is high-temperature and low-temperature metal spraying. In the former, solid metal is fed slowly into an oxyacetylene flame; in the latter, a charge of solid metal is liquefied inside the spray gun by means of electric heating elements that will melt metals up to a maximum melting point of about 200°C. This means that only fusible (low melting point) alloys can be handled.

POROUS METAL

Porous metals are a different material of construction. They are material made up of bonded or fused aggregate (powdered metal, coarse pellets, etc.) such that the resulting mass contains numerous open interstices of regular or irregular size allowing air or liquid to pass through the mass of the mold. They are used in different processes, particularly thermoforming. They are used to help with the removal of air and other gases from the mold cavities, which otherwise cannot be released using more conventional methods such as the mold parting line or knockout pins.

SOFT TOOLING

There is a special material category of so-called soft tooling that covers materials such as elastic or stretchable tools made of rubber or plastic elastomer (silicone, etc.), rather than the usual materials, so that products of complex shapes can be removed without tool side actions. Among other benefits, they can be stretched to remove cured products having undercuts.

In general, soft tooling can be anything other than the usual steels used in the production of tools. It includes materials such as cast or machined aluminum grades, cast plastics (epoxy, silicone, etc.), cast rubbers, and cast zincs. By definition they are the least expensive and more flexible and are usually faster to fabricate; they also have limited production capabilities compared to steel tools. Today's choices range from computer-generated plastic tools to specialty alloys or even pure carbide. However, each of them has limitations in durability and capabilities.

These tools last a relatively limited time if they are properly prepared and maintained. Steel, wear-resistant edge plates can be used to expand life expectancy. Preventative maintenance, such as cleaning, is very important. To clean, use a tool cleaner designed to loosen residue that collects in parting lines and vents. The cleaning fluid should be compatible with the tooling material.

MANUFACTURING

Different conventional metal-cutting methods are used to meet requirements based on the type of material used and the configuration of the tool. For example, there is the process of photochemical machining (PCM) is recognized by the metalworking industry as one of several effective methods for the fabrication of metal parts. The technique—also called photoetching, chemical etching, and chemical blanking—competes with stamping, laser cutting, and EDM. It uses chemicals, rather than mechanical or electrical power or heat, to cut and blank metal.

PCM has several distinct advantages over these other processes. Low tooling costs associated with the photographic process, quick turnaround times, and the intricacy of the designs that can be achieved by the process are some of the advantages, as are high productivity and the ability to manufacture burr- and stress-free parts. Of paramount importance in this process are the cost savings associated with generating prototypes.

The advantages of using fully hardening tool steels rather than case-hardening steels for the manufacture of tools are primarily the simpler heat treatment and the possibility of making corrections to the cavity at a later time without a new heat treatment. However, the greater risk of cracking is a disadvantage, particularly for tools with a greater cavity depth, because tools from these steels do not have a tough core. Moreover, the tougher steels with a carbon content of about 0.4% do not attain the high surface hardness of about 60 HRC, which is desirable with respect to wear and polish.

Sometimes the mechanical action of the tool may require certain steel selections so as to permit steel-on-steel sliding without galling. Tooling surfaces of precision optics will need steel that can be polished to a mirror finish. If the inserts will receive coatings to further enhance performance, then steel characteristics to receive coating or endure a coating process must be considered (application temperature versus tempering temperature). Hot runner mold components often use hot work steel because of its superior properties at elevated temperatures. Very large molds and short-run molds may use prehardened steel (270 to 350 Brinell) to eliminate the need for additional heat treatment.

When tool steels of high hardness are used, they are supplied in the soft annealed condition (hardened mold inserts for cores, cavities, other molding surfaces and gibs, wedge locks, and so on are typically hardened to a range of 48 to 62 Rc). They are then rough machined, stress relieved, and machine finished and go to heat treatment for hardening and tempering to desired hardness. After this heat treatment, the core or cavity typically must then be finish ground and/or polished. In some applications, there will be additional coatings or textures to further treat the tool surfaces.

While steels having higher hardness typically have increased wear resistance, they should be selected when appropriate because those same steels may be more difficult to machine. A D2 or an A2 insert will have higher hardness and excellent wear resistance, but will be much more difficult to grind to its final shape and size than a H13 or S7 insert will be. Inserts such as those with titanium carbide alloy are even more difficult to grind, but they afford outstanding wear resistance.

When processing particularly abrasive plastics, the wear can still be too high even when using high-carbon, high-chromium steels. Metallurgical melting cannot produce steels with even higher amounts of carbides. In such cases, hard-material alloys, produced by powder metallurgy, are available as tool materials. These alloys contain about 33 wt% of titanium carbide, which offers high wear resistance because of its very high hardness.

Like other tool steels, hard-material alloys are supplied in the soft-annealed condition where they can be machined. After the subsequent heat treatment, which should if possible be carried out in vacuum-hardening furnaces, the hard materials attain a hardness of about 70 HRC. Because of the high carbide content, dimensional changes after the heat treatment are only about half as great as those in steels produced by the metallurgical melting processes.

In machining as well as in noncutting shaping processes, stresses develop chiefly as a result of the solidification of surface layers near the edge. These stresses may already exceed the yield point of the respective material at room temperature and consequently lead to metallic plastic deformations. Since the yield point decreases with increasing temperature, additional stresses can be relieved by plastic deformation during the subsequent heat treatment. In order to avoid unnecessary, expensive remachining, it is advisable to eliminate these stresses by stress-relief annealing.

ELECTRIC-DISCHARGE MACHINING

EDM, also called spark erosion, is a method involving electrical discharges between a graphite or copper anode and a cathode of tool steel or other tooling material in a dielectric medium. The discharges are controlled in such a way that erosion of the workpiece takes place, developing the required contours. The positively charged ions strike the cathode so that the temperature in the outermost layer of the steel rises so high as to cause the steel layer to melt or vaporize, forming tiny drops of molten metal that are flushed out as chippings into the dielectric.

EDM is a widely utilized method of cavity and core-stock removal. Electrodes fabricated from materials that are electrically conductive are turned, milled, ground, and developed in a large variety of shapes that duplicate the configuration of the stock to be removed. The electrode materials include graphite, copper, tungsten, copper-tungsten, and other electrically conductive materials. Special forms of EDM can now be used to polish tool cavities, produce undercuts, and make conical holes from cylindrical electrodes.

ELECTROFORMING

This process is used for the production of single or low numbers of cavities, as opposed to others requiring many cavities. The process deposits metal on a master in a plating bath. Many proprietary processes exist. The master can be constructed of such materials as plastic, RP, plaster, or concrete that is coated with silver to provide a conductive coating. The coated master is placed in a plating tank and nickel or nickel-cobalt is deposited to the desired thickness of up to about 0.64 cm (0.25 in). With this method, a hardness of up to 46 Rc is obtainable. The nickel shell is reinforced with different materials (copper, plastic, etc.) to meet different applications. Sufficient thickness of copper allows for machining a flat surface, which enables the cavity to be mounted into a cavity pocket.

SURFACE FINISH

Tooling surfaces such as mold cavities and die openings require meeting certain surface finish (Tables 17.32 and 17.33). Rather than identifying the finish required as dull, vapor-honed satin, shiny, and so on, there are standards such as a diamond polishing compound, SPI (originally SPI/SPE) Mold Standard Finish, and the American Association's standard B46.1 Surface textures (extremely accurate surface measurements; a near-perfect system) are used. ASA B46.1 corresponds to the Canadian standard CSA B 95 and British standard BS 1134.

As shown in Table 17.34, there are different manufacturing methods that identify surface finish. Surface finishes are also used on plasticator screws (and other tools). The processes used for hardening screws and barrels are not necessarily the same as those applicable to tool components (chapter 3).

The diamond polishing compound contains diamond particles within a certain microinch range (Table 17.35). It provides a guide to surface finish.

Grade	Vacuum	Cast Iron Chips	Salt Bath	Protective Gas
STAVAX ESR ORVAR GRANE ARNE RIGOR	400 grain	180—220 grain or rougher	180 grain or rougher	220 to 400 grain
SVERKER 21	400 grain	180 grain or rougher	60 grain	220 to 400 grain

Table 17.32 Various heat treatments versus finish of Uddeholm tool steels

Europe Grain size	USA Grit (Mesh)	FEPA 30GB Grain size	FEPA 31GB Grain size	Abrasive size range µm
46	60			425—355
60	80	50		300—250
90	100	80		180—150
100	120	100		150—125
120	150	120		125—105
180	180	180		90—75
220	240	220	P 240	63—53
			P 320	48—44
400	320		P 400	37—33
600			P 500	32—28
			P 600	27—24
	400		P 800	23—20
			P 1000	19—17
800			P 1200	16—14
1000				10
	600			8
	800			7
	900			6

Table 17.33 Different grain standards used for surface finishes

Process	Roughness Height (µin.)
Flame cutting	250–2,000
Sawing	63–2,000
Drilling	63–500
EDM & CM	32–1,000
Milling	16–1,000
Turning	16–1,000
Boring	16–1,000
Reaming	16–250
Tumbling	2–63
Grinding	2–125
Honing	2–63
Lapping	1–32
Superfinishing	1–32
Polishing	1–32
Sand casting	250–2,000
Ceramic casting	32–250
Investment casting	32–250
Pressure casting	32–125
Forging	63–500
Diecasting	16–250

Table 17.34 Identification of surface finish based on manufacturing process

Diamond Compound	Finish	Particle Size (μin.)	Approximate Mesh	Color
1-8M	Super	0-2	14,000	Ivory
3-7M	Very high	2-4	8,000	Yellow
6-48M	Mirror	4-8	3,000	Orange
9-6M	High	8-12	1,800	Green
15-5M	Fine	12-22	1,200	Blue
30-4H	Lapped	22-36	600	Red

Table 17.35 Diamond-particle compound relates to surface finish

The SPI's standard consists of six steel disks finished to various polish levels and covered with protective plastics caps processed in tools with those finishes. Roughness of the disks range from 0 to 3 μin (highest level of finish) to 15 μin, and all others are coarser. The No. 1 finish is equivalent of 8000 grit diamond (0 to 3 μm range). No. 2 finish is the surface finish obtained when the final phase of polishing is completed using 1200 grit diamond (up to 15 diamond am). No. 3 finish uses a 320 grit abrasive cloth. No. 4 finish uses a 280 grit abrasive stone. No. 5 finish is achieved with 240 grit, dry blast at 100 psi at a distance of 5 in. No. 6 finish is equivalent to 24 grit, dry blast at 100 psi from a distance of 3 in. Figures 17.4 to 17.6 provide an introduction to ASA B46.1 nomenclature and symbols.

POLISHING

When involved in polishing and plating, particularly when in a hurry, decide what is required, and then research your options. More often than not, customers come rushing over or calling on the moldmaker with an urgent polishing and plating job that has to be done yesterday. Taking this action to produce a first-quality product requires certain details (435).

Recognize that a tool shop essentially farms out four things: heat treating, texturing, polishing, and plating. The very first thing that must be determined is the order in which these need to be done. Good planning means good timing. Plan polishing or plating early (Table 17.36). When you are about to start polishing and plating, consider the metals that are in the tool and the properties of the plastics that will be processed in the tools. Consult with your polishing and plating vendor to find out what is recommended for optimal production times. The key to success here is not just optimal production times for your customer; it is also optimal production times for you.

A general requirement for all tools is that they have a high polish where the plastic melt contacts the tools. Other parts of the tools may require a degree of polishing (smooth) permitting parts to fit with precision, eliminating melt leaks in the tools. A large part of tool cost is polishing, which can represent from 5% to 30% of the tool cost.

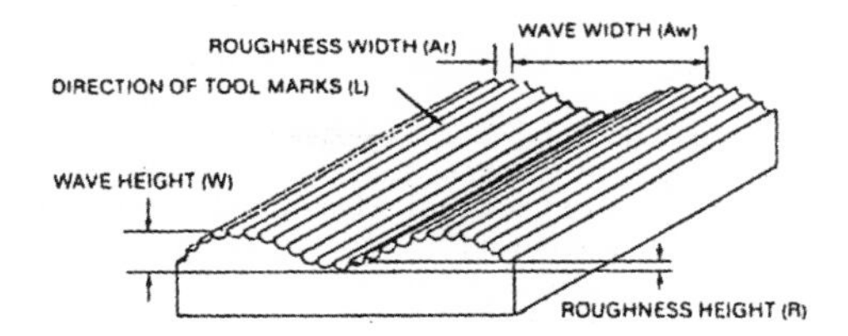

Figure 17.4 Terms identifying tool surface roughness per ASA B46.1 standard.

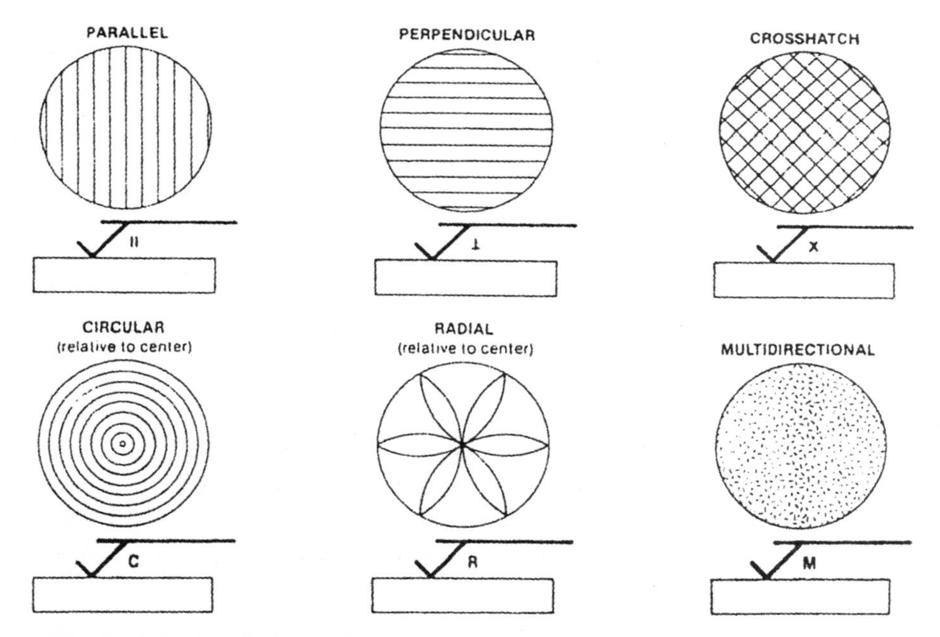

Figure 17.5 Symbols identified on tool per ASA B46.1 standard.

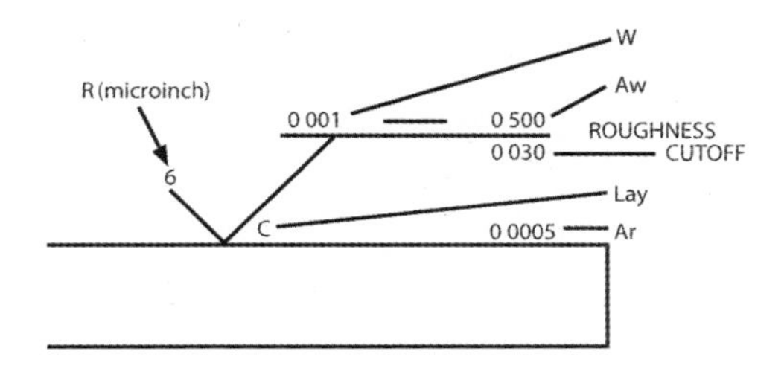

Figure 17.6 Illustrating roughness at a given point on a tool surface per ASA B46.1 standard.

Grinding and Polishing Steps / UHB Grades		Rough Grinding -Grain Size Before Heat Treatment Carried out by machine, stone or emery powder with suitable lubricant*	Fine Grinding -Grain Size (Prepolishing) After Heat Treatment With stones or emery powder and suitable lubricant*	Polishing with Diamond Paste Grain size and polishing tool
GRANE ARNE RIGOR SVERKER 21	55 HRC 58 HRC 60 HRC 60 HRC	180—400 (Wet) 180—400	220 320 400 600 800	14 μ on Hard Wood then 6 μ, 3 μ on Soft Wood then 1 μ on Fibre/Felt (if required)
STAVAX ESR ORVAR	54 HRC 52 HRC	180—400 180—400 180—400 60—400	220 320 400 600	25 μ, 14 μ on Hard Wood then 6 μ, 3 μ on Soft Wood then 1 μ on Fibre/Felt (if required)
STAVAX ESR IMPAX GRANE	300 HB 300 HB 300 HB	180—400 180—400 180—400	220 320 400 600 800	6 μ, 3 μ on Soft Wood then 1 μ on Fibre/Felt (if required)

* e.g. Paraffin

Table 17.36 Polishing sequences

Polishing a tool begins when the designer puts the finishing information on the drawing that includes hardness (Figs. 17.7 and 17.8). Such terms as "mirror finish" and "high polish" are ambiguous. The only meaningful way is to use an accepted standard to describe what has to be polished and to what level. It is also important that the designer specify a level of polish no higher than is actually needed for the job, because going from just one level of average roughness to another greatly increases the cost of the tool. An estimate on the cost of polishing as a function of the roughness of the finish is summarized in Figure 17.9.

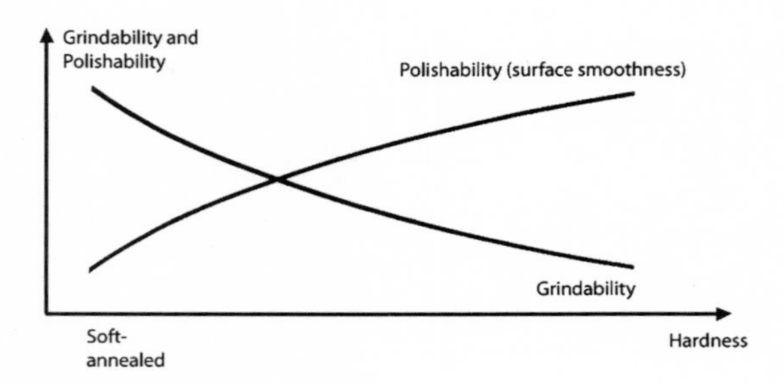

Figure 17.7 Polishability versus hardness.

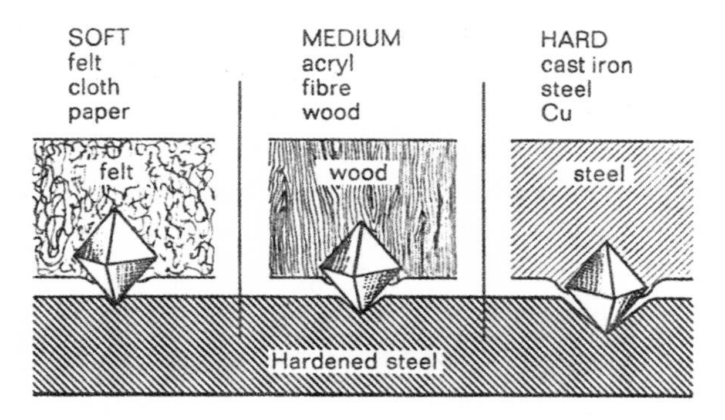

Figure 17.8 Comparison of polishing tool hardness.

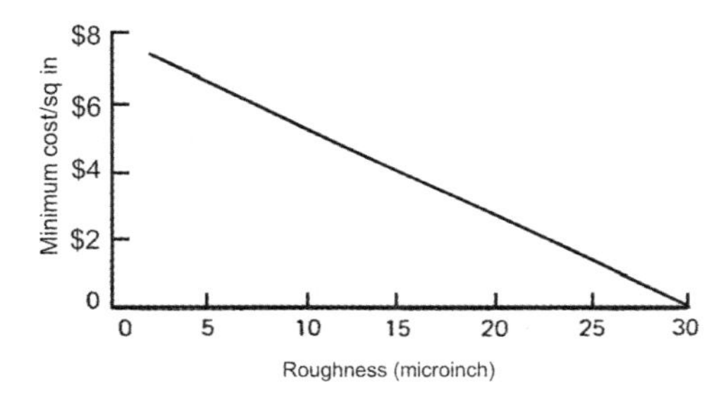

Figure 17.9 Cost of polishing tool steels.

Polishing provides a desirable appearance. However, it is done to obtain a desired surface effect on the fabricated product to facilitate the ejection of the molded product, free melt flow through the die orifice, prepare the tool for another operation such as etching or plating, and so on. If a plastic product is to be plated, it is important to remember that plating does not hide any flaws; it accentuates them. Therefore, on critical plate jobs, the tool polish must be better than for non-plated parts.

Polishing is used to remove the weak top layer of metal. It may be weak from the stresses induced by machining (conventional, EDM, etc.) or from the annealing effect of the heat generated in cutting. When it is not removed, this layer very often breaks down, showing a pitted surface that looks eroded or corroded.

The techniques used to get a good and fast polish is to make sure the part is as smooth as possible before polishing. If EDM is used, the final pass should be made with a new electrode at the lowest amperage. If the part is cast or hobbed, the aim is to have a master finish with half the roughness of the desired tool finish. If machining, the last cut should be made at twice the normal speed, the slowest automatic feed, and a depth of at least 0.0025 cm (0.001 in). A steady stream of dry air must be aimed at the cutting tool to move the chips away from the cutting edge.

Do not use a lubricant during the last machining cut (443). The cutting tool should be freshly sharpened, and the edges honed after sharpening. The clearance angle of the tool is usually 6° to 9°, and if a milling cutter or reamer is used, it should have a minimum of four flutes. Roughness is given as the arithmetical average in microinches. A microinch is one millionth of an in (10^{-6} in) and is the standard term used in the United States. Sometimes written as MU in or μin, it is equal to 0.0254 micron. A micron, one-millionth of a meter (10^{-6} m), is the standard term in countries that use the metric system. It is often written as micrometer or μm and is equal to 39.37 μin.

Orange Peel

Polishing can damage the tool material unless it is properly done. An example of a common defect is orange peel. It is a wavy surface effect that results when the metal is stretched beyond its yield point by overpolishing and takes a permanent set. Further polishing only make matters worse with small particles breaking away from the surface. The harder the steel, the higher the yield point and therefore the less chance of orange peel. Hard carburized or nitrided surfaces are much less prone to this problem. To avoid orange peel, polish the tool by hand. With powered polishing equipment, it is easier to exceed the yield point of the metal. If power polishing is done, use light passes to avoid overstressing.

Orange-peel surfaces usually can be salvaged by the following procedure: Remove the defective surface with a very fine-grit stone; stress-relieve the tool; restone; and diamond-polish. If orange peel reoccurs after this treatment, increase the surface hardness by nitriding with a case depth of no more than 0.005 in and repolish the surface.

Art of Polishing

The usual first phase in tool finishing (called benching) involves the use of both hand tools and power-assisted grinders to prepare the surfaces to be polished. Depending on the last machining operation and roughness of the remaining stock, the toolmaker will select the method of preparing the surface in the quickest manner possible. Typical hand tools include files and diemaker rifflers. The files range from 20 to 80 teeth/in and will assist in removing stock quickly and accurately. Rifflers are available in a wide range of shapes to fit into any conceivable contour found on a tool and have from 20 to 220 teeth/in.

There are a variety of cutting tools, from carbide rotary cutters to abrasive drum, band, cartridge, tapered cone, and disk-to-flap wheels. They are used to reduce the roughness of the surface in the quest for the desired tool surface finish. Power-assisted tools include electric, pneumatic, and ultrasonic machines, which can be equipped with a wide variety of tool holders. Rotary pneumatic and electric grinders, with the tool mounted either directly or in a flexible shaft, are popular with the toolmaker for rapidly removing large amounts of stock. With the grinding or cutting medium mounted in a 90° or straight tool, depending on the surface, the toolmaker will start with the coarser medium and work the surface progressively to the finer medium. The finishing process can best be described as one where each preceding operation imparts finer and finer scratches on the surface. The initial stages are perhaps the most important in the finishing process. Spending too little time in these phases normally will be detrimental to the final surface finish.

The use of abrasive stones, either worked by hand or in conjunction with reciprocating power tools, normally is the next phase in tool finishing. These stones are a combination of grit particles suspended in a bonding agent. For general-purpose stoning, silicon carbide is available in type-A stones for steel hardness under 40 R, and type-B stones for higher hardness. Most stones are available in square, rectangular, triangular, and round shapes in grits of 150, 240, 320, 400, and 600. Other stones consist of aluminum oxide in type E for working EDM surfaces, type M that has oil impregnated for additional lubrication, and type F for added flexibility or reduced breakage of thin stones.

The next phase in benching usually consists of using abrasive sheets or disks to continue smoothing the tool. All the finished work is carried out in the direction of the scratches installed in the line of draw or ejection. Machine, file, stone, or abrasive marks installed perpendicular to the direction of draw are detrimental to the removal of the part from either the cavity or core, and must be avoided. At this point, the desired tool surface finish may have been achieved, or the piece parts may be grit-blasted, and finishing is considered complete. These finishes may be acceptable for nonfunctional core/cavity surfaces in which the part does not pose a problem.

HAND POLISHING

There are people that just polish and there are polishers that can engrave or bench by hand or bench any shape on any tool block. Most engravers used to utilize a collection of Swiss files that they developed throughout the years; different jobs demanded different shapes and so their collections grew. Seeing an engraver with a hundred or more scrapers of different shapes only attested to the person's skill and long years of patient work.

Different tools demand different approaches. There can be a job that has a core and it requires a 0.030 in radius on all tooling corners. The polisher knows that pulling a scraper toward himself will be ten times faster than pushing one away from himself. This action occurs because for some reason the human hand has more natural depth-control pressure when it pulls than when it pushes. Today many polishers simply grind or blitz radiuses.

Carbide blanks have become an extremely useful tool in the benching and engraving tool box; blanks of 3⁄16 in and 1⁄4 in diameter are a perfect fit to most people's hands and the polisher can retain a good feel without the fingers cramping or getting sore. The sense of touch is the secret to being a successful engraver. The overall secret to benching is that the tool is the boss and not the polisher. No contour will come out looking professional with uncontrolled nicks and gouges in the steel. Other approaches include rotating the scraper.

Only experimentation will demonstrate to you how a scraper will cut. This is because each polisher naturally uses a different amount of speed in the stroke, uses a different amount of pressure with the hands, and holds the scraper at a different angle. The true way to learn which combination is best is to try the following: (1) scrape away from your body as fast as you can (exactly like peeling a potato) using a light pressure; (2) look at the size and type of the chips that you are producing; and (3) try to produce an exact pile of chips that are all the same. This is the beginning of control.

There is also the finishing aid involving ultrasonic finishing and polishing systems. The ultrasonic controls deliver strokes that can be adjusted to between 10 and 30 μm at speeds up to 22,000 cycles per second. This action eliminates costly hand finishing and is extremely effective in polishing deep narrow ribs that have been EDM installed

PROTECTIVE COATING/PLATING

OVERVIEW

There is a distinction between platings and coatings. Generally, thin layers of metals applied to the surface of tool components are considered platings. The application of alloys, fluorocarbons, fluoropolymers (such as PTFE; chapter 2) or dry lubricants is considered a coating. With few exceptions, treatments involve processes and chemicals that should not be used anywhere near a fabricating machine (because of corrosiveness), and they are best handled by custom plating and treating shops that specialize in their use.

Tool coatings and platings are typically used to enhance tool performance in one or more of the following areas: wear resistance, corrosion resistance, improved tool release, resizing components, or any of those properties. No single treatment is ideal for solving all these problems. There are treatments that resist the corrosion damage inflicted by chemicals such as hydrochloric acid when processing PVC, formic acid or formaldehyde with acetals, and oxidation caused by the interaction between tools and moisture in the plant atmosphere. Release problems require treatments that decrease friction and increase lubricity in mold cavities.

Tools can be subjected to sweating and moisture condensation, particularly during the summer months. This can lead to corrosion and rust, and in turn, to poor finishes and inferior-quality fabricated products. By keeping the air in the plant or around the tool dry, you can not only eliminate rust but also improve product quality and increase your production rate.

The fabricator must determine which problems could cause the greatest loss of productivity and then select the treatments that will be most effective in solving the problems. The next sections describe examples of common tool coatings.

CHROMIUM

Chromium can be applied at a wide range of thicknesses, which makes it a good choice for resizing and enhancing wear resistance; hardness is 70 Rc and thicknesses from 0.00005 to over 0.030 in can be achieved. Chrome's corrosion resistance is not as good as electroless nickel. Careful location of anodes and electrodes is needed to minimize edge buildup and to achieve even thickness.

NICKEL

Nickel can be applied over a wide range of thicknesses up to 0.100 in. It is very good for resizing. Hardness will only be 40 to 50 Rc. It is sometimes referred to as sulfamate nickel. Anodes and electrodes are required and edges can have excessive buildup unlike the electroless nickel. Coating can be applied as more than needed and machined back to size by grinding or EDM.

ELECTROLESS NICKEL

With its more even deposition, electroless nickel coating is excellent for corrosion resistance; it has better corrosion resistance than chrome because of its reduced porosity. However, it has less wear resistance. Plated hardness is approximately 47 to 48 Rc, but can be heat-treated to 70 Rc. Thickness is best at 0.0001 to 0.0005 up to 0.002 in. Good, even deposition without heavy buildup on corners is possible with electroless nickel. It can also coat deep recesses or coat inside water lines, among other usually difficult areas.

NICKEL-PTFE

Nickel-PTFE has PTFE dispersed as small particles into a nickel coating. The result is a longer retention of the PTFE lubricity properties with durability and possibly improved corrosion resistance. The usual thickness of nickel-PTFE coatings is 0.0002 to 0.003 in. Hardness characteristics are the same as electroless nickel.

TITANIUM NITRIDE

Titanium nitride (TiN) is a hard coating with a gold appearance that is obtained via a physical vapor deposition (PVD) process. The process typically is applied at 500°C (930°F). This may anneal steel hardness if a reduced application temperature is not requested. Best coating is at normal application temperature, which works well with H-13 and steels heat treated and tempered above 500°C (930°F). It is a very hard coating at 82 Rc and 80 millionths of an inch to 0.0002 in thick. There are other PVD nitride variations, but TiN is the most common in use.

DIAMOND BLACK

Diamond BLACK is a ceramic PVD coating, but at a reduced temperature of approximately 120°C (250°F) or less; thus it has no adverse effects on steel hardness. The resulting surface hardness is 94 Rc and the surface thickness is 80 millionths of an inch. It has excellent hardness, lubricity, and wear resistance.

Numerous metals other than these reviewed have been used to coat the components of tools (Tables 17.37 to 17.39). Gold plating can be used to create a protective surface for PVC and some fluorocarbon materials. It prevents tarnishing, discoloration, and oxidation of the tool surface. Gold

Material	*Application method*
Chromium	Plating
Nickel	Plating
Electroless nickel	Solution treatment
Tufram TFE aluminum	Deep anodizing process followed by TFE impregnation; used on aluminum alloys
TFE ceramic	Spray and bake application; used for all die materials that can withstand 250°C bake
Tungsten silicide	Solution treatment; used on steel and ferrous alloys
Tungsten carbide	Explosion impact or flame spray with plasma arc; used for all high-melting metals to improve abrasion resistance
Aluminum oxide	Plasma flame spray; used for extreme abrasion resistance; used on steel dies but usually limited to small dies because of expansion problems; works best on 18-8 stainless
PTFE	Spray and bake application; used for low-friction and low-adhesion application; poor abrasion resistance
Polyimide, aramid	Straight organic coatings with high softening points (450–550°C), which are applied by spray and baked; low friction characteristics against some resins (for example, PVC); moderate abrasion resistance
Filled polyimide, aramid	Aramid and polyimide systems containing TFE and other fluorocarbon resins to improve the friction properties

Table 17.37 Examples of coatings based on material used

will protect the original finish and provide a hydrogen barrier. Fifty millionths of an inch of gold plating (0.00005 in [0.0013 cm]) is adequate for the purpose. Gold can also be used as a primer under polished chrome. Platinum and silver have also been used to plate tools, and they share with gold the notable drawback of high cost.

PROBLEMS

Tool wear cannot be prevented. This wear should be observed, acknowledged during maintenance checkup, and dealt with at intervals in the tool's useful life; otherwise, the tool could be allowed to wear past the point of economical repair. Periodic checks of how platings and coatings are holding up will allow the fabricator to have a tool resurfaced before damage is done to the tool. A poorly finished tool that is being used for the first time has its heat, pressure, and exposure to plastic actually

Process	Material	Applied to	Purpose
Coating by impingement, molecularly bound	Tungsten disulfide	Any metal	Reduce friction and metal-to-metal wear with dry film, nonmigrating
Coating by impingement, organically bound	Graphite	Any metal	Reduce sticking of plastics to mold surface, can migrate
Electrolyte plating	Hard chrome	Steel, nickel, copper alloys	Protect polish, reduce wear and corrosion (except for chlorine or fluorine plastics)
	Gold	Nickel, brass	Corrosion only
	Nickel	Steel and copper alloys	Resist corrosion except sulfur-bearing compounds, improve bond under chrome, build up and repair worn or undersized molds
Electroless plating	Nickel	Steel	Protect nonmolding surfaces from rusting
	Phosphor nickel	Steel and copper	Resist wear and corrosion
Nitriding	Nitrogen gas or ammonia	Certain steel alloys	Improve corrosion resistance, reduce wear and galling, alternative to chrome and nickel plating
Anodizing	Electrolytic oxidizing	Aluminum	Harden surface, improve a wear and corrosion resistance

Table 17.38 Examples of coatings based on process used

reworking its surface. Fragmented metal is pulled out of the metal fissures, and plastic forced into them. While the fissures are plugged with plastic, the fabricator may actually be processing plastic against plastic.

Starting up a tool that has a poor finish can damage the tool that did not experience proper presurfacing. If the tool surface is unsound (no prior treatment was used), a thin layer of metal plating, particularly chrome plating, will not make it correct. A poorly prepared surface makes for poor adhesion between treatment and the base metal. The effectiveness of a surface treatment depends on not only the material being applied, but also the process by which it is applied. For any plating or coating to adhere to the surface of a tool component, it has to bond to the surface. The bonding may be relatively superficial, or a chemical or molecular bond may accomplish it. The nature and strength of the bond directly affect the endurance and wear characteristics of the plating or coating. The experience of the plater is an important factor in applications where cut-and-dried or standard procedures have not been developed.

	Coating	Application Process	Application Temperature	Substrates	Appearance	Hardness	Thickness	Coefficient of Friction
Thin Intermetallic Coatings	*DICRONITE DL-5* Modified tungsten disulfide in lamellar form.	Air-delivered dry metallic lubricant. Bonds instantly by penetrating steel without binders or adhesives.	Ambient.	All materials and platings.	Dark silver-gray or bluish color.	1.0–1.5 Moh's Scale	0.00020 in. or less	0.030 against itself.
	WS2 Modified tungsten disulfide in lamellar form.	Pressurized air. No binders or adhesives.	Ambient.	All stable metal substrates.	Slight blue-gray color. Rhodium finish.	Takes on the hardness of substrate.	0.00002 in.	Dynamic 0.030 Static 0.070
	POLY-OND Nickel phosphorous impregnated with polymers.	Electroless nickel deposition sprayed or dipped in *Teflon*, then baked to cure and set the PTFE into the surface.	700°F (370°C) bake is typical. Depends on target R_c.	Almost any material. Steels, aluminum, brass, bronze, cast iron all acceptable.	Shiny, silver color. May have initial whitish residual appearance.	50 R_c [68–70 R_c if baked at 750°F (400°C) for 1 hr].	0.000050 to 0.0025 in.	0.06 under 200 lb. kinetic load, against itself.
PTFE Infused Into Metallic Coatings	*TFE-LOK* PTFE particles locked into a hard chromium electrodeposited surface.	Chromium is heated to expand pores, which are infiltrated with cryogenically cooled PTFE particles under high pressure.	350°–400°F (175°–205°C)	Steels, stainless steel, aluminum, copper and copper alloys.	Dark to medium gray color. Smooth texture on polished surfaces.	688–70 R_c	0.002–0.005 in. Not used for salvage work.	0.04 estimated. Approaches that of PTFE.
	NEDOX-SF2 Modified electroless nickel infused with PTFE polymer.	Nickel alloy is modified to increase porosity. PTFE is applied, then heat treated for set, cure, and hardness.	750°F (400°C) typical.	All ferrous alloys. Some nonferrous alloys, including aluminum.	Gold color. Grainy texture from lattice structure.	Up to 70 R_c	0.0002–0.003 in. (0.0015 in. maximum preferred)	Dynamic 0.12 Static 0.18
Nickel/Phosphorous/PTFE Codepositions	*NITUFF* *Teflon* coating applied over hardcoat anodizing.	Precision hardcoat anodizing is dipped into *Teflon*. Can be applied only to aluminum.	<200°F (<95°C)	Aluminum. Low copper and silicon content preferred.	Color varies with aluminum alloy, temper, and coating thickness.	>62 R_c	0.001 in. total thickness; half penetrates, half is added.	Depends on surface finish before coating.
	NICKLON 10.5% phosphorous nickel alloy with 25% PTFE suspended in the matrix.	Codeposition of nickel-*Teflon*.	145°F (65°C)	Any metallic substance.	Milky gray appearance. Can be polished to a #2 diamond finish.	48 R_c (70 R_c if heat treated.)	0.0003–0.0005 in.	Static 0.07 Dynamic 0.03
	NICOTEF Submicron PTFE particles suspended in a nickel-phosphorous matrix.	Autocatalytic codeposition of PTFE in a nickel-phosphorous matrix. Chemical reduction process.	195°F (90°C)	Most metals. Steels, stainless steel, copper, brass, bronze, cast iron, aluminum.	Shiny silver/gray surface.	32–35 R_c as applied. (42–46 R_c if heat treated.)	0.0003–0.0008 in. for release. 0.0005–0.003 in. for corrosion.	0.17–0.21 (0.07–0.10 when lubricated with white oil.)
	NYE-TEF Nickel-phosphorous matrix in which submicron particles of PTFE are dispersed.	Autocatalytic (electroless) codeposition.	190°–200°F (90°–95°C)	Steels, copper, brass, aluminum, associated alloys.	Matte pewter gray.	48 R_c (68 R_c if heat treated.)	0.0005–0.001 in. typical.	0.1–0.3 (vs. steel)
Metallic Platings	INDUSTRIAL HARD CHROME Chromium plus trace amounts of oxides and hydrogen.	Electrolytic deposition. High density.	130°–140°F (55°–60°C)	Most metals, many other surfaces.	Shiny, silver.	68–70 R_c	0.0002–0.040 in. Can be applied at varying thicknesses.	0.16–0.43 (lubricated vs. steel)
	ELECTROLIZING High chromium alloy. Nonmagnetic.	Uniform deposition using proprietary methods which result in ultrapure chromium on surface.	90°F (30°C)	Tool steels, 4100 series steel, stainless steel, aluminum.	Silver. Very shiny and smooth.	70–72 R_c	0.000025–0.001 in. Maintains surface appearance.	0.10 against itself.
	ARMALOY Thin, dense chromium alloy applied such that the surface is nodular.	Hard precision chromium electrocoating.	<140°F (<60°C)	Ferrous and nonferrous metals. Not aluminum.	Satiny-silver matte or micro "orange peel" finish.	72 R_c	0.00001–0.0006 in. Normally 0.0001–0.0002 in.	Static 0.17 Dynamic 0.16 (vs. steel)
Surface Hardening Treatments	ELECTROLESS NICKEL Nickel alloy, with varying phosphorous content to provide needed properties.	Deposited by chemical reduction without electric current.	180°F (80°C) Up to 400°F (205°C) in postbake.	Ferrous metals, aluminum and copper alloys.	Shiny, silver appearance.	48 R_c as applied. Heat treated up to 70 R_c.	0.0001–0.0005 in. for lubricity. 0.0005–0.001 in. for corrosion.	0.38 vs. steel 0.45 vs. itself
	MELONITE Salt bath nitriding. Can incorporate *QPQ*-oxidizing salt bath step.	Thermochemical diffusion. Nitrogen-bearing salts react with substrate, creating nitrides.	1,075°F (580°C)	Ferrous-based metals.	Black and very dark. Shiny and smooth.	550–1,000 HV 55–70 R_c, depending on parameters.	0.0004–0.0008 in compound. 0.002–0.025 in. diffusion layer.	0.35 dry (0.05 lubricated with SAE 30 oil on itself.)
	ION NITRIDING Plasma glow surface hardness treatment.	Ionized nitrogen electrons form nitride compound (optional) and diffusion zones in the substrate. Impurities removed via vacuum.	900°–950°F (480°–510°C)	Most ferrous metals. Not usually applied to stainless steel.	Dull silvery-gray surface can be buffed out, leaving original appearance	Up to 70 R_c. 60–64 R_c typical.	No change Diffusion zone 0.005–0.025 in.	Improves, but not as slick as chrome.
Thin-Film, High-Hardness Coatings	TITANIUM NITRIDING Thin-film, high-temperature coating.	Negative ionization process. Physical vapor deposition.	900°–950°F (480°–510°C)	Most steels, stainless steel, beryllium copper.	Gold-yellow.	85 R_c 2,300 Vickers 0.03 HV	0.0008–0.00020 in. typical.	0.4 vs. steel 0.13–0.2 against itself.
	DIAMOND BLACK Boron carbide alloy.	Low-temperature deposition. Applied by magnetic sputtering.	<250°F (<120°C)	Steels, stainless steel, carbide, and aluminum. Some plastics.	Dark gray. Smooth texture.	93 R_c	0.0008 in.	Dry 0.2 0.08 with *Microseal* tungsten disulfide lubricant.

Table 17.39 Guide to tool surface enhancements and coatings commonly used (courtesy of Eastman Chemical Co./431)

Temperature Range	Corrosion Resistance	Removal Process	Advertised Advantages	Notes
−350° to 1,000°F (−210° to 540°C)	Minor delay of corrosion only.	Fine abrasive powder polishing.	Faster cycle times, fewer rejects, aids release, reduces mold wear. Compatible with releasing agents. More consistent performance.	Can be removed only by heavy abrasion or shearing of the substrate.
−460° to 1,200°F (−275° to 650°C)	Minor delay of corrosion only. Impervious to most solvents and refined fuels. Attacked by some acids and alkalines.	Polish out with diamond paste. Can be removed with high caustic solution, but not recommended.	Will not flake, chip, or peel off. Becomes part of the substrate. One year total continuous operation. Compatible with most lubricants.	Once applied, cannot be removed without removing part of the substrate. Used on wear plates, gibs.
−65° to 500°F (−55° to 260°C) continuous use.	Salt spray resistance +300 hr (ASTM-B-117). Absence of corrosion claimed.	Can be chemically stripped without damaging parent material.	Reduces cycle time and reject rate, reduces or eliminates need for release sprays. Prolongs tool life and increases hardness.	If two treated surfaces are in contact, mismatch surface hardness by 10 R_c or more. Can mask off areas.
500°–550°F (200°–290°C) maximum use temperature.	Better chemical resistance than plain hard chrome. Partially dependent upon substrate.	Chemically stripped, unless applied to aluminum, when it is stripped electrolytically.	Permanently locks PTFE into chrome. Gains sliding characteristics of PTFE and hardness, conductivity, and wear of chrome. Good release properties.	The surface may take on a white appearance after use (caused by light etching), but coating still maintains lubricity.
−250° to 550°F (−155° to 290°C)	Salt spray resistance 500 hr (ASTM-B-117). Resists etching from PVC by-products.	Can be stripped off chemically without damage to the substrate. Can be recoated repeatedly if necessary.	Precise control of thickness for close tolerance parts, even in deep cavities. Resists wetting by most liquids. Low friction.	Can produce different surface enhancement characteristics by applying to polished or glass beaded surfaces.
Up to 500°F (260°C)	Salt spray resistance >2,000 hr (ASTM-B-117) using 6061 alloy with smooth finish, 0.002 in. coating.	Chemical stripping process removes the coating plus the substrate material to the penetration thickness.	Allows substitution of aluminum for steel in many situations. Uniform thickness in all areas. Improved flow and release characteristics.	Does not adhere well to welded surfaces. Usually doubles the RMS value of finished surfaces.
Up to 750°F (400°C)	Resists most organic and inorganic environments. Salt spray resistance 1,000 hr per ½ thousandth thickness.	Can be chemically stripped in a process similar to removing nickel plating.	Can withstand 1–2 years of continuous operation. PTFE is evenly distributed through the matrix. Good surface duplication.	Cannot maintain a #1 diamond surface polish.
500°F (260°C) maximum.	Salt spray resistance 500–1,000 hr (ASTM-B-117). Highly resistant to alkaline and process acid media.	Proprietary chemical stripping process. Varies according to substrate.	Extreme uniformity of thickness. Can coat complex parts with thinner coatings for release. Use thicker coatings for corrosion resistance.	Surface contaminates and imperfections must be removed before plating.
550°F (285°C) maximum use temperature.	Good chemical resistance, but due to porous nature of *NYE-TEF* an underlying coat of electroless nickel is usually applied.	Alkaline or nitric acid stripping baths are used. The method depends upon the substrate material.	PTFE particles are dispersed in the range of 23%–25% and remain throughout the life of the coating.	Chemical bonding ensures strong adhesion to the substrate.
Begins softening at 400°F (200°C).	Good corrosion resistance, but all Cr deposits have microcracks. A sublayer of electroless nickel is suggested to prevent corrosion.	Caustic solution with electrical current, or hydrochloric acid.	Can salvage worn or mismachined parts by restoring dimensions. Good adhesion. Hard, slippery surface.	Stresses develop in the Cr during deposition, causing microcracks which can spread to the substrate if not applied carefully.
Up to 1,600°F (870°C).	High-purity chromium is applied without microcracking, providing improved chemical resistance vs. regular hard chrome.	Chemically stripped in a reverse alkaline chemical bath.	Bonds absolutely to substrate. Increases wear resistance, reduces friction, aids release, can be used to repair mold surfaces.	Involves cleaning and removal of the base metal's surface.
−400° to 1,800°F (−240° to 980°C)	Resists attack by most organic and inorganic compounds, except sulfuric and hydrochloric acids. Resistance will be enhanced.	Alkaline chemical stripping process. Does not affect most base materials.	Better release and improved mold wear. Good thin maintenance coating for PVC molds. Good for use with glass- and mineral-reinforced resins.	Does not affect base metal. Release properties enhanced by nodular structure of surface, resulting in low contact area.
600°–800°F (315°–425°C) suggested. 1,630°F (890°C) melting point.	Low porosity provides benefits over electroplated nickel. Good performance in alkaline media. Resistant to molding by-products.	Chemical stripping method. BeCu and brass harder to strip, more susceptible to damage. No damage to steel substrates.	Adheres well. High hardness and corrosion resistance make it popular for use in molds. Uniformly deposited to close tolerances. Good lubricity.	Abrasion resistance after heat treatment approaches that of hard chromium. Surface finish <16 RMS suggested for low friction.
570°F (300°C) maximum continuous use for most tool steels.	Improves environmental corrosion resistance. Outperforms hard chrome and electroless nickel (ASTM B-117).	Can be stripped chemically in acidic bath or mechanically, but must remove substrate to depth of diffusion zone.	Excellent sliding and running properties. Extremely resistant to wear—better than hard chrome. Improves fatigue properties by 20%–100%.	Thickness of compound and diffusion zone depends on treatment time and the carbon and alloy content of the substrate.
>1,000°F (540°C) Depends on the substrate and application.	Improves corrosion resistance. Reduces oxidation and rusting of molds.	Grinding, machining, or burning.	Low cost. Uniform case depth, regardless of geometry. Used to improve fatigue resistance, wear, and lubricity.	Can vary case hardness and depth to suit needs. Masking available. No cyanides used in process. Can be plated over ion nitriding.
>1,000°F (540°C)	Not soluble in most acids and alkalines.	Coating can be removed by grinding, polishing or EDM, or chemically stripped. Repaired molds can be recoated.	Reduces friction. Uniform thickness of thin coating will not affect dimensional tolerances. Good for glass-reinforced resins.	Burned surfaces will not coat. "White" EDM layer must be removed before coating.
2,200°F (1,200°C) oxidation. Limit to 750°F (400°C) with *Microseal*.	Resistant to attack by most chemicals. Protects mold from oxidation and acidic action of molding by-products.	Strippable without damage to the base material or finish. No heat involved in stripping process.	Can apply to micro finishes. Hardness increases tool life and eases cleaning. Provides good lubricity and improves corrosion resistance.	Molds must be clean and free from other coatings.

Table 17.39 Guide to tool surface enhancements and coatings commonly used (courtesy of Eastman Chemical Co./431) *(continued)*

PLATING

The term *electroless* describes a chemical composition/deposition process that does not use electrodes to accomplish the plating. Plating by the use of electrodes is called electrodeposition.

Electroless nickel provides a good protective treatment for tool components, including the holes for the heating media, and so on. Its deposit of up to 0.001 in of nickel uniformly prevents corrosion; also, any steel surface that is exposed to water, PVC, or other corrosive materials or fumes benefits from this minimal plating protection. The surface hardness of electroless nickel is 48 Rockwell C, and the hardness can be increased by baking to 68 Rc. Plated components will withstand temperatures of 370°C (698°F).

Electroless nickel has the characteristic of depositing to the same depth on all surfaces, which eliminates many of the problems associated with other metallic platings. Grooves, slots, and blind holes will receive the same thickness of plating as the rest of the tool. This allows close tolerances to be maintained.

The popular chromium plating is a hard, brittle, tensile-stressed metal that has good corrosion resistance to most plastics. Hard chrome can be deposited in a rather broad range of hardnesses, depending on plating-bath parameters. Average hardness is in the range of 66 to 70 Rc. A deposit of over 0.001 in (0.0025 cm) thickness is essential before chromium will assume its true hardness characteristics when used over unhardened base metals. Over a hardened base, however, this thickness is not necessary because of the substantial strength/backing provided. Precise control of thickness tolerances can be achieved in a particular type of chrome plating generally referred to as thin, dense chrome. This kind of plating provides excellent resistance to abrasion, erosion, galling, cavitation, and corrosion wear.

Chrome plating that is hard protects polished surfaces against scuffing and provides a smooth release surface that will minimize adherence of the plastic in the tool. Some precautions are necessary. Hydrochloric acid created in the processing of PVC will attack chrome. Chrome that is stressed and cracked under adverse conditions will permit erosion from water and gas penetration into the steel. To deal with hydrogen embrittlement created when hydrogen is absorbed by steel during the plating process, chrome-plated components should be relieved of stress within a half hour of completion of the plating. To protect against galling, chrome should not be permitted to rub against chrome or nickel.

As the chrome becomes thicker, it develops a pattern of tiny cracks because the stresses become greater than the strength of the coating. These cracks form a pattern that interlaces and sometimes extends to the base metal. A corrosive liquid or gas could penetrate to the base metal. This action can be prevented in three ways: A nickel undercoat can be applied to provide a corrosion-resistant barrier; the chrome plating can be applied to a maximum thickness; or thin, dense chrome can be substituted.

In some instances, cracks in the base material that are not visible through normal inspection techniques may become apparent only after plating. This phenomenon is attributed to the fact that grinding of steel often causes a surface flow of material, which spreads over cracks and flaws.

However, this cover is dissolved during the preplating treatment, and the cracks become apparent as the coating thickness increases. The deposited layer of chromium, although extremely thick, will not bridge a large crack.

It is an undesirable to sample a tool under pressure before plating without plastic in the tool. The effects of moisture on the steel can cause chrome plating to strip. Carelessness can also result in scratching an unplated tool. Dimensional checks can readily be made outside the press using wax or other sampling materials. Chrome can be stripped from a tool after sampling in order to make essential changes. Periodic checks after a chrome-plated tool is in full production are desirable to find evidence of wear, which will show up first in the tool corners and areas of high melt flow. Swabbing a copper-sulfate solution in the tool areas can make a simple check. If the copper starts to foam, the chrome is gone and must be replaced.

To improve the surface hardness, thus making the surface more resistant to wear, steels are nitrided and carburized. Nitriding will penetrate the surface from 0.008 to 0.051 cm (0.003 to 0.020 in), depending on the steel, and can result in hardnesses of 65 to 70 Rc.

Another process for imparting a surface hardness is carburizing. In this process, carbon is introduced into the surface of the cavity or core steel, and the inserts are heated to above the steel's transformation temperature range while in contact with a carbonaceous material. This process frequently is followed by a quenching operation to impart the hardened case. Hardness as high as 64 Rc and a depth of 0.075 cm (0.030 in) are possible with this process.

COATING

Different coatings provide special protection for tools (Table 17.40). Metal/ceramic alloys and plastic coatings are used. Standard alloy matrices have been developed with quench-hardenable tool steel, high-chromium stainless steels, high-nickel alloys, and age-hardenable alloys. Special alloys can be formulated for even the most corrosive conditions.

Coating	Typical Thickness (inches)	Max Thickness (inches)	Hardness (Rc)	Coeff Friction	Approx Cost
Chromium	0.0001 - 0.0005	0.035	70	0.15 - 0.30	$
Nickel	up to 0.100 +	0.100 +	45	0.2 - 0.3	$
Electroless Nickel	0.0001 - 0.002	0.003	48 - 70	0.2 - 0.3	$
Nickel PTFE	0.0002 - 0.0005	0.003	48 - 70	0.07 - 0.16	$$
TiN	0.00008 - 0.00016	0.00016	82	0.4 - 0.5	$$$$
Diamond BLACK®	0.00008	0.00008	94	0.200	$$$$$

Table 17.40 Examples of coating materials for tools

An example of this is a composite of ultrahard titanium carbides distributed throughout a steel or alloy matrix that is used very effectively for coating. The carbides are very fine and smooth, and the coatings are applied by sintering, a process in which the component to be coated is preheated to sintering temperature, then immersed in the coating powder, withdrawn, and heated to a higher temperature to fuse the sintered coating to the component. Flexibility in selecting and controlling the composition of the matrix alloy makes it possible to tailor the qualities of the alloy according to the requirements of the application. When heat-treatable-matrix alloys are used, the composite can be annealed and heat-treated, permitting conventional machining before hardening to 55 to 70 Rc.

These types of coatings are effective in severe wear conditions that occur with abrasive plastic compounds. Ceramic/metal composites provide the hardness and abrasion resistance to withstand wear by the most damaging glass-fiber, mineral, or ferrite filled plastic compounds. With the right metal-matrix selection, resistance to corrosion and heat is also obtained.

There are treatments using chemical processes that utilize thermal expansion and contraction to lock PTFE particles into a hard, electrodeposited surface. An example of a metal used in such a treatment is chromium. Surfaces treated by the process have the sliding, low-friction, non-stick properties of PTFE, along with the hardness, thermal conductivity, and damage resistance of chrome. Tool components benefit from this surface treatment. The process builds a thickness of 0.002 in (0.005 cm) of electrodeposited chromium on component surfaces and is available for ferrous, copper, and aluminum-alloy parts.

There are impregnation processes that provide continuous lubrication to metal parts (cavity and component parts) by impregnating fluoropolymers into the surface pores of the metal. Benefits of these processes include reduced friction, wear, and corrosion; in addition, pin galling and metal-to-metal mold sliding friction on the core and ejector pins are also eliminated, and plastic melt flow is improved. Ejection of plastic products eliminates or minimizing the need for release agents for most hard-setting, molded thermoplastics (TPs) and thermosets (TSs).

Titanium carbonitride is commonly applied to provide wear resistance to tools and wear parts made from tool steels. The coating can often be applied in a layer thick enough to allow for a surface-finishing operation after coating. The coating inhibits galling and results in a favorable coefficient of friction. The process protects against wear from abrasive fillers and corrosion from unreacted plastics that release acids. The coating is said to be 99% dense—that is, it has virtually no porosity, and is inert to acids.

In processing corrosive reacting plastics such as chlorinated polyvinyl chloride (CPVC), the metal tool is subjected to corrosion and pitting. Certain steels, such as stainless steel, can provide a degree of protection without a coating. Hard coatings such as PVD of titanium nitride, titanium cyanide, and so on are extremely corrosion resistant and provide excellent abrasion resistance. More prevalent are chromium-based materials that can be applied at rather low temperatures to provide resistance to corrosion, abrasion, and/or erosion if needed. Examples include pure Cr, CrN (chromium nitride), and CrC (chromium carbide). Some coating processes, such as PVD and chemical vapor deposition (CVD), subject the tool steels to excessively high temperatures that reduce steel hardness. Systems have been developed with PVD and CVD (plasma CVD or PCVD) that operate at lower temperatures of 200°C to 400°C (400°F to 750°F). Popular is the so-called nitrided coating, which is actually a hardened nitride casing (nitrogen is absorbed into the surface of the steel).

HEAT TREATMENT

Surface treatments used on tools are generally plated, coated, or heat-treated to resist wear, corrosion, and avoid release problems. Plating and coating affect only the surface of a tool or its component, while heat treatment generally affects the physical properties of the entire tool. Treatments such as carburizing and nitriding are considered to be surface treatments, and although heat is applied in these processes, they are not considered to be heat treatments. Heat treatment is more often the province of the steel manufacturer and toolmaker than the fabricator. However, stress relieving is a heat treatment that the fabricator can perform.

Steels are hardened to increase tensile and compressive strengths as well as wear resistance, resulting in an overall increased durability. This prevents abrasive wear and/or damage or distortion during machining. Toughness, however, is typically reduced by heat treatment that results in higher hardness. Different steels have different resulting combinations of toughness and hardness. In general, steels do not have both great toughness and great compressive strength or wear resistance at same time. The compressive strength and wear resistance tend to go together, but when compressive strength is high, the toughness is usually less and vice versa. H-13 steel exhibits a good blend of all properties; thus it is in widespread use in the tool-building industry. An important tool property, particularly in molds, is compression strength that is required to take the plastic melt pressure loading.

During heat treatment, steel is heated above the critical temperature of 1330°F, where it changes from pearlite to austenite stages (metallurgically, carbon goes into solution with the iron). It is held for a time at the elevated temperature, then quenched rapidly. With slow cooling, the steel reverts back into pearlite stage whereas fast cooling (so-called quenching) results in martensite for most steels. These required temperatures are the hardening temperatures, and they vary for different types of steel. This action results in an optimum mix of toughness and hardness and yet still avoids excessive distortion from too much heat. The optimum temperature requirement is based on the steel supplier's recommendation.

Quenching may take place in air, water (plain or salt), oil, and so on. The cooling rate must be faster than the critical cooling rate for the given steel. These critical cooling rates vary for each steel. During quenching, quickly bypass the point at which the austenite reverts to pearlite; this results in the formation of a hard martensite.

Tempering is done after the initial hardening heat treatment and required quench. Tempering is performed to both reduce the hardness and increase the toughness. It will also relieve many internal stresses and should be done as soon as possible after initial heat treatment. Tempering includes reheating parts up to some specified temperature. Higher tempering temperatures result in more martensite being transformed and resulting lower hardnesses; thus a wide range of hardnesses can be accomplished by varying the tempering temperature. It is important to note that any form of reheating, such as grinding, welding, and coating processes, after the initial heat treatment and tempering influences tempering. Processors should avoid using a torch on tool surfaces to the point of heating steel above its respective tempering temperature for hardness in use. Some high-hardness steels may be tempered as low as 350°F; thus if one heats an insert up past 350°F, then the hardness and tool performance are reduced. Most tool steels are tempered between 400°F and 1100°F. Obtain tempering temperatures from the steel supplier.

CRYOGENIC PROCESSING

Cryogenic processing (also known as cryogenic tempering or treating) is a process that holds much promise for manufacturers to increase wear resistance of metals and their alloys by startling percentages in the range of 300% to 400%. In contrast to conventional surface treatments, cryogenic treatment is a one-time process that affects the entire structure of the material being processed, not just the surface. It is therefore known as a "through treatment" as opposed to a surface treatment. Originally developed by NASA, it was used to extend the life of tooling. Wear resistance of tool steel is improved by slowly cooling the tool steel to cryogenic temperatures of –315°F and soaking the steel at this low temperature (436).

The controlled cryogenic process starts with the loading of a well-insulated treatment chamber with the materials to be processed. A microprocessor is programmed according to the weight of the part or parts being treated. Liquid nitrogen (LN_2) is used as a cooling medium to lower the temperature within the chamber to –315°F. The temperature is lowered at a very slow rate, under precise control, because a rapid change in temperature can induce stresses caused by thermal shock. This descent can take anywhere from seven to fourteen hours.

Once the temperature reaches –315°F, the process enters the soak phase, which maintains this temperature for twenty-four hours. The long soak ensures that the entire cross-section of the material in the chamber is completely treated. The final step in the process is the ascent phase, in which heaters raise the internal temperature of the chamber to ambient.

Residual stress relief, thermal and dimensional stability, greater machinability, and increased toughness are among the main benefits of the technique, which has long been known to extend the service life of end mills, drill cutting blades, punches, and dies. Now, with the popularity of the process on the rise, engineers and others are becoming increasingly aware of its effectiveness on steel parts.

Despite impressive reports of its benefits, cryogenic processing is not the answer to every situation. Like any real science, it has specific limitations and applications. One limitation is that it can take significantly longer than other types of processing. High-volume production requires specific considerations, and the process may not work on some anodizing treatments. Degraded performance has been reported as a result of cryogenic processing on some anodized parts. However, the process can be used in conjunction with coatings and in many cases will actually improve the quality of the coating to be applied to a part.

Although cryogenics is not a substitute for conventional heat treatment, it can provide improvement in carbon, alloy, and stainless steels. Besides increasing resistance to wear, it can also diminish the propagation of cracks. Information on this process is available from Cryogenic Society of America Inc. (CSA).

MAINTENANCE/CLEANING

Tools are usually very expensive and delicate instruments. Exceptional care should be taken to protect tools when they are in a machine, during disassembly, when cleaning components, and during both operation and storage. Any protruding parts should be protected against damage during transfer from the fabricating machine to other areas. Schedules should be set and records kept to ensure that required maintenance and cleaning is accomplished on a regular schedule (Table 17.41).

When cleaning, disassembly should be done only when they have had sufficient time to heat-soak or at the end of a run so that it is at operating temperature. Experience has shown a temperature of 230°C (450°F) to be adequate for most nondegradable plastics. Cleanup of degradable plastics (PVC, etc.) should begin immediately after shutdown to prevent corrosion on the tool's melt flow surfaces.

During cleaning, all the fastening bolts should be loosened while the heat is still on. The heat should then be turned off and all electrical and sensors should be removed carefully. While the tool is still hot, it should be disassembled and thoroughly cleaned with "soft" brass, copper, or aluminum tools. All surfaces, especially melt flow openings and cavities, should be covered with a protective, easily removed coating when they are not operating as well as during maintenance, to protect against surface corrosion.

There are different methods to remove contaminated plastics so that tools are not damaged. These methods permit removal of plastics that are easy to remove as well as those that are difficult. Cleaning methods include aluminum oxide beds (fluidized beds), hot-air ovens, blowtorches, hot plates, hand scraping, burn-out ovens, hot sand, dry crystals, high-pressure water, ultrasonic chemical baths, heated oil, lasers, and Dry Ice spray (437–439).

A very old cleaning method uses molten hot salt in a container. Cleaning cycles with this method are as short as one minute, depending on the tool's shape and the plastic.

For minor cleaning, hand-working brass tools are used. The brass does not damage the metal, whereas the tool would be damaged if steel or other metals were used. Beryllium or aluminum tools are sometimes used, but they are harder than brass.

The solvent method may consist of wiping, immersion, spraying, or vapor degreasing. Wiping is the least effective process and may result in distributing contaminants over the surface rather than removing them. Immersion, especially if accompanied by mechanical or ultrasonic scrubbing, is a better process. It is even more effective if followed by either another immersion or a spray rinse.

Solvent-vapor degreasing is carried out in a tank with a solvent reservoir on the bottom. The solvent is heated and vapor condenses on the cool, plastic surfaces. The condensate dissolves surface

Cleaning Method	Manual	Oven	Solvent	Salt Bath	Ultrasonic	Fluidized Bed	Vacuum Pyrolysis
Required auxiliaries	Acetylene and oxygen		Cooling water, ethylene glycol	$NaNO_2$ or 10% Na(OH) + 45% $NaNO_2$ + 45% KNO_3 or similar	Water, wetting agent	Al_2O_3 sand	Water
Cleaning temperatures, C	Locally to 600	400–500	280–285	320–300	20–70	400–550	360–525
Cleaning time, h		8–16; 6–12	8–10; 4–6	4–6; 4–6	1–2; 1–2	3–4; 2–3	5–7; 4–5
Recommended number of cycles		1	2	1	1	1	1
Postcleanings		Labor-intensive	1 to 2 baths, ultrasonic	3 baths, ultrasonic	O	Ultrasonic	Ultrasonic
Waste products	Burnto plastics	Burnt plastics, ash	Contaminated ethylene glycol	Contaminated salt	O	Contaminated Al_2O_3 ash	Contaminated plastic
Environmental pollution	Heavy	Very heavy due to soot and combustion gases	Very low contaminated triethylene glycol, reprocessed through distillation	Very heavy due to soot and combustion gases		Very heavy due to soot and combustion gases, moderate if postburning is used	Hardly any
Cleaning effect	Dependent on personnel	Brushoff, postclean, carbonized residues	Very clean, including interior parts without dismounting, very gentle, carbonized polymers are not dissolved	Very good, wash off salt crusts and neutralize	Very good as postcleaning	Very good externally	Very good with following final cleaning stage
Suitable for	Only as auxiliary	All steel and stainless steel parts contaminated with plastics	Multilayer and sinter metal filters, vanadium sand, mold interior, also aluminum parts, only for removal of PET and PA	All steel and stainless steel parts contaminated with plastics	Final cleaning after all processes	All steel and stainless steel parts contaminated with plastics	Steel and vanadium parts, filters resistant to 400°C gaps of 0.2-mm diameter
Not suitable for		Multilayer filters	Other polymers require specific solvents	Sintered metal, sintered web, aluminum		Sintered metal, sintered glass, aluminum	Open aluminum parts, soldered parts

Table 17.41 Examples of cleaning methods

impurities and carries them away. It is an effective process because the surfaces do not come in contact with the contaminated solvent bath. Cleaning action is as fast as a minute.

Cleaning by abrasion removes surface contamination and tends to increase surface roughness. This method includes the use of hand or machine sanding; dry blasting with nonmetallic grit (sand, flint, silica, aluminum oxide, plastic, walnut shell, dry ice, etc.); wet-abrasive blasting with a slurry of aluminum oxide; and scouring with tap water and a scouring powder.

The dry-ice blasting method uses frozen carbon dioxide (CO_2). This process using rice-sized, dry ice (CO_2) pellets uses shop-compressed air is often compared to sandblasting without the sand. CO_2 ice crystals exert a strong thermomechanical force on impact. Parts can be cleaned while hot and in the machine. With heat, they clean faster.

The cryogenic deflashing method uses basket-sealed enclosed tumblers with about 1.6 MPa (225 psi) shot blasts of plastic. Liquid nitrogen (LN_2) or dry ice is used at –196°C (–320°F). With or without vibrating devices, this method eliminates tool damage and produces no solvents or other environmental hazardous by-products. Other advantage includes accurate deflashing, and no tool dimension changes occur such as warpage, and other problems associated with heating methods.

The vacuum pyrolysis cleaner utilizes heat and vacuum to remove the plastic. Most of the plastic is melted and trapped. The remaining plastic is vaporized and appropriately collected in a trap. Pyrolytic ovens are a variation to cleaning with burn-off ovens. Instead of sealing the chamber and starving oxygen, they pull a vacuum over parts to remove combustible air. A vacuum oven is a safe cleaning method since it removes all oxygen from the heating chamber.

The ultrasonic method permits thorough cleaning, particularly for tools with electrical and mechanical components. A transducer mounted on the side or bottom of a cleaning tank is excited by a frequency generator to produce high-frequency vibrations in the cleaning medium. These vibrations dislodge contaminants from crevices and blind holes that normal cleaning methods would not affect.

The PVD method is an example used to optimize cleaning that has little effect on the contour of product. PVD coatings lead to an insignificant roughening of the surface. Tools that have been finish machined can be improved. No expensive posttreatment is necessary. In its vacuum chamber (10^{-2} to 10^{-4} mbar [1 to 0.01 Pa]), metals are converted to a gaseous state by the introduction of thermal (electron beam or arc) or kinetic energy (atomization). They condense on the surface being cleaned.

MOLD

INTRODUCTION

Following the product design is designing and producing a tool (mold or die) around the product (Fig. 17.10). Mold operation are summarized in Figure 17.11. Figure 17.12 provides an introduction to layouts, configurations, and actions of molds. Figure 17.13 provides the sequence of mold

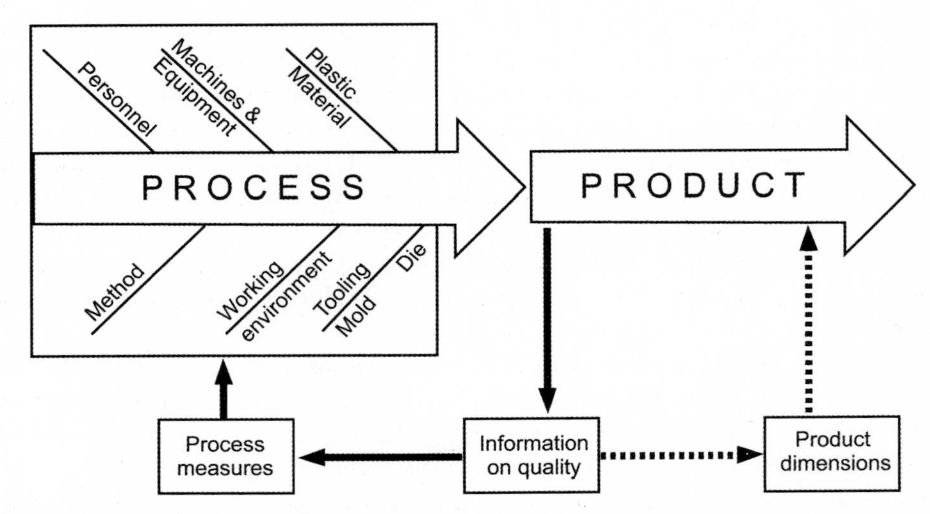

Figure 17.10 Flow of the molding from the process that includes the mold to the product.

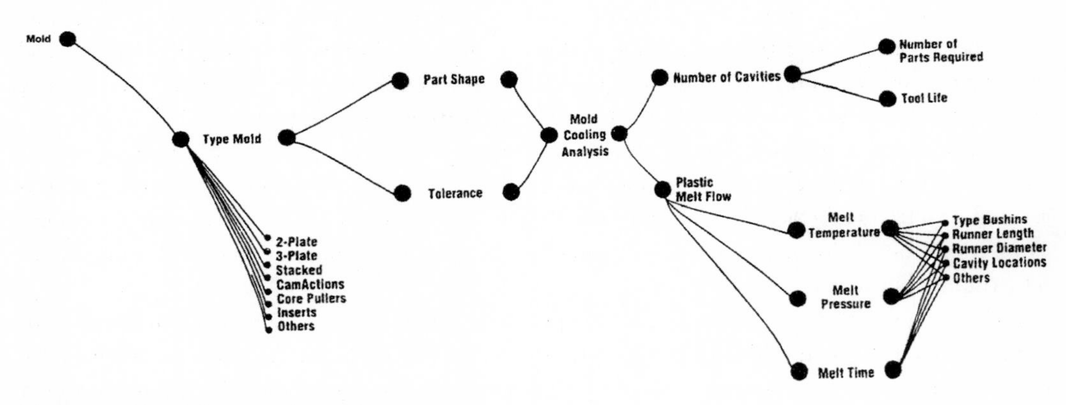

Figure 17.11 Mold operation and types.

operations. Figure 17.14 provides information on mold action during a fabricating molding cycle. Alignment of mold halves during their opening and closing actions requires precision mold parts to fabricate quality parts. Figure 17.15 provides a simplified approach to mold alignments. When it is possible, mold cavity walls are tapered to permit ease of separating molded parts from the cavity. Table 17.42 provides desirable tapers for cavity sidewalls.

There are approaches to simplifying mold design and its actions (Fig. 17.16). These approaches should be considered as parts are designed. Examples of different actions in molds are shown in Figure 17.17. There are different approaches used to mold threaded parts such as bottle caps, medical components, mechanical and electrical connectors, and so on. To date most of these molds

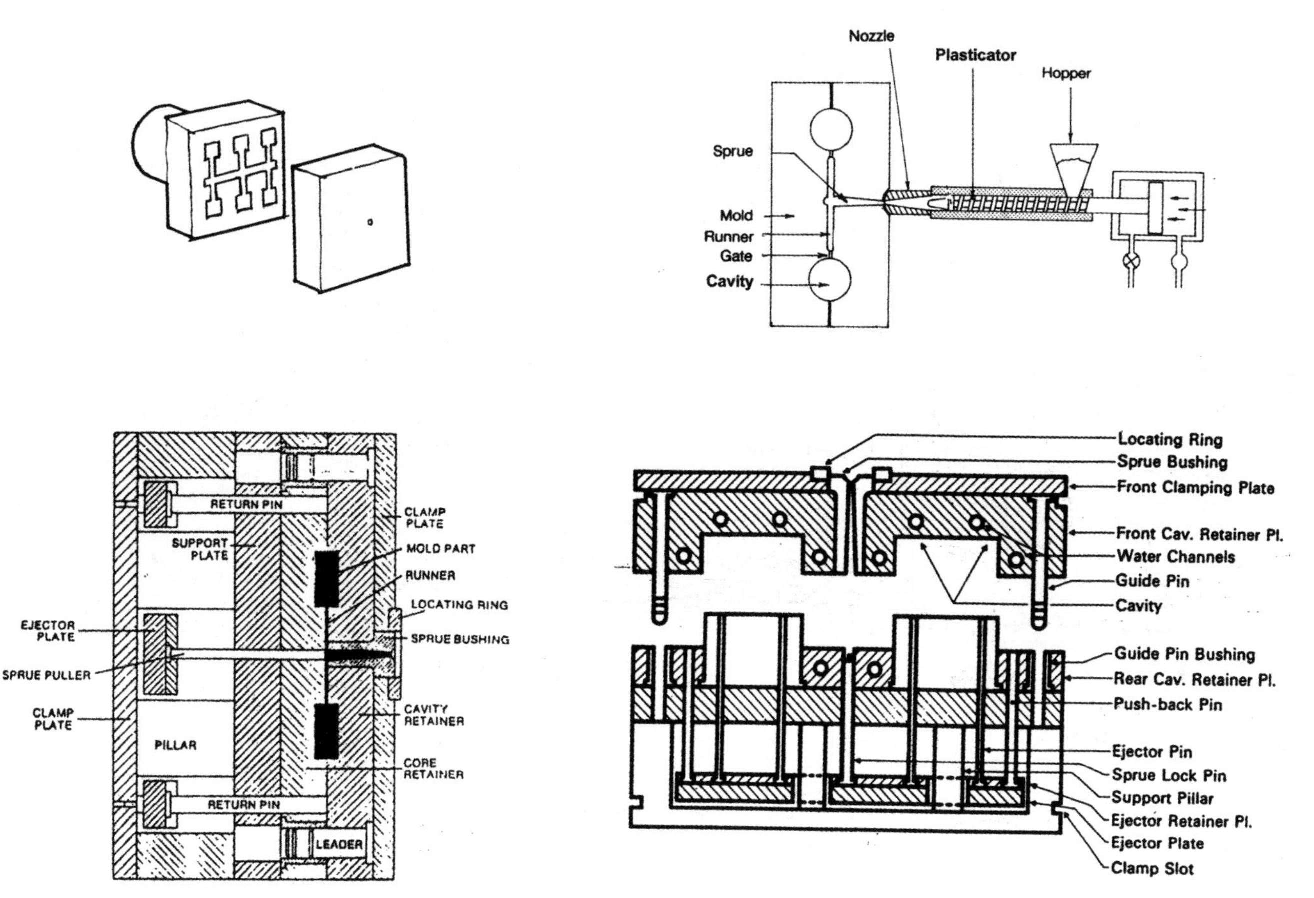

Figure 17.12 Examples of mold layouts, configurations, and actions.

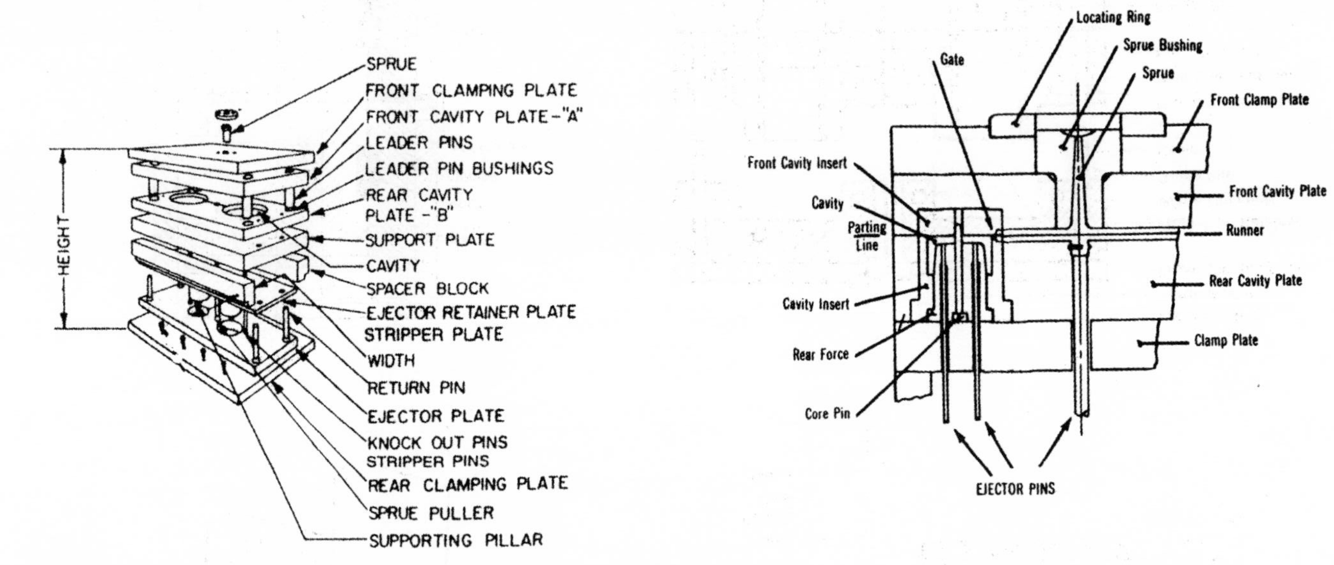

Figure 17.12 Examples of mold layouts, configurations, and actions *(continued)*.

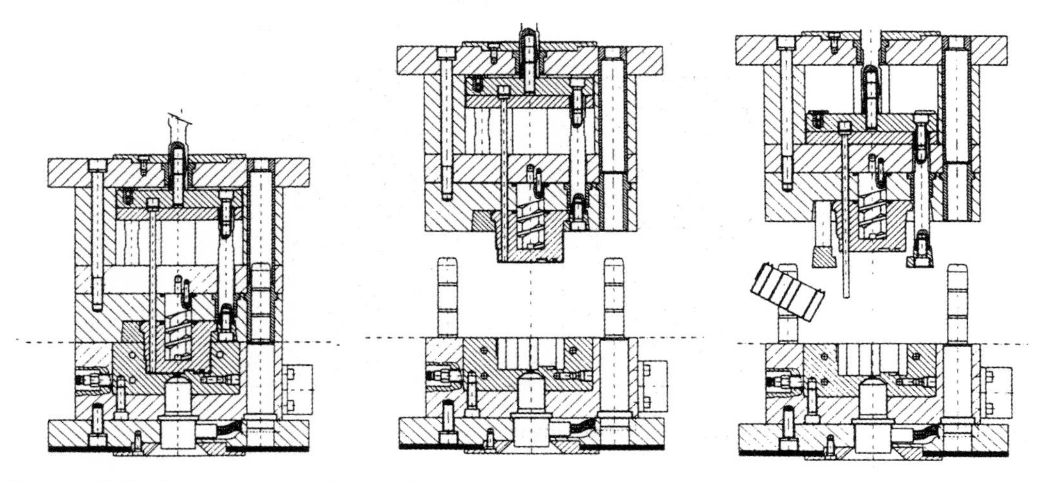

Figure 17.13 Sequence of mold operations.

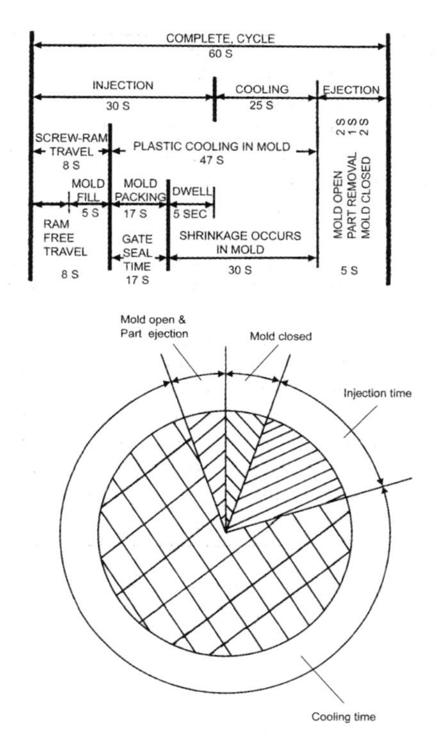

Figure 17.14 Mold action during a fabricating molding cycle.

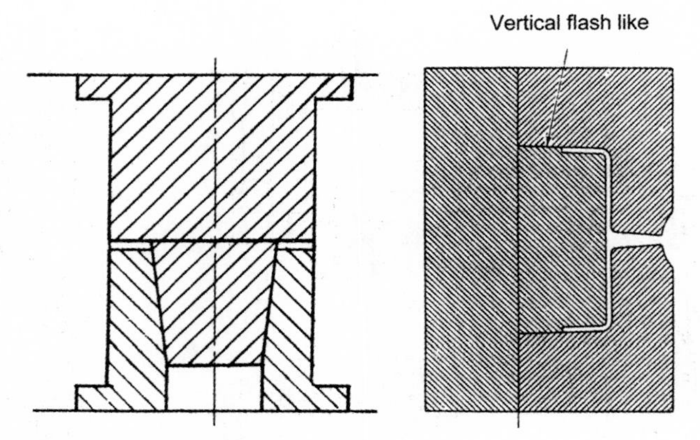

Figure 17.15 Examples of precision mold half alignment.

Standard draft angles.

Depth	$\frac{1}{4}$°	$\frac{1}{2}$°	1°	2°	3°	5°	7°	8°	10°	12°	15°
1/32	0.0001	0.0003	0.0005	0.0011	0.0016	0.0027	0.0038	0.0044	0.0055	0.0066	0.0084
1/16	0.0003	0.0006	0.0011	0.0022	0.0033	0.0055	0.0077	0.0088	0.0110	0.0133	0.0168
3/32	0.0004	0.0008	0.0016	0.0033	0.0049	0.0082	0.0115	0.0132	0.0165	0.0199	0.0251
1/8	0.0005	0.0010	0.0022	0.0044	0.0066	0.0109	0.0153	0.0176	0.0220	0.0266	0.0335
3/16	0.0008	0.0016	0.0033	0.0065	0.0098	0.0164	0.0230	0.0263	0.0331	0.0399	0.0502
1/4	0.0011	0.0022	0.0044	0.0087	0.0131	0.0219	0.0307	0.0351	0.0441	0.0531	0.0670
5/16	0.0014	0.0027	0.0055	0.0109	0.0164	0.0273	0.0384	0.0439	0.0551	0.0664	0.0837
3/8	0.0016	0.0033	0.0065	0.0131	0.0197	0.0328	0.0460	0.0527	0.0661	0.0797	0.1005
7/16	0.0019	0.0038	0.0076	0.0153	0.0229	0.0383	0.0537	0.0615	0.0771	0.0930	0.1172
1/2	0.0022	0.0044	0.0087	0.0175	0.0262	0.0438	0.0614	0.0703	0.0882	0.1063	0.1340
5/8	0.0027	0.0054	0.0109	0.0218	0.0328	0.0547	0.0767	0.0878	0.1102	0.1329	0.1675
3/4	0.0033	0.0065	0.0131	0.0262	0.0393	0.0656	0.0921	0.1054	0.1322	0.1595	0.2010
7/8	0.0038	0.0076	0.0153	0.0306	0.0459	0.0766	0.1074	0.1230	0.1543	0.1860	0.2345
1	0.0044	0.0087	0.0175	0.0349	0.0524	0.0875	0.1228	0.1405	0.1763	0.2126	0.2680
$1\frac{1}{4}$	0.0055	0.0109	0.0218	0.0437	0.0655	0.1094	0.1535	0.1756	0.2204	0.2657	0.3349
$1\frac{1}{2}$	0.0064	0.0131	0.0262	0.0524	0.0786	0.1312	0.1842	0.2108	0.2645	0.3188	0.4019
$1\frac{3}{4}$	0.0076	0.0153	0.0305	0.0611	0.0917	0.1531	0.2149	0.2460	0.3085	0.3720	0.4689
2	0.0087	0.0175	0.0349	0.0698	0.1048	0.1750	0.2456	0.2810	0.3527	0.4251	0.5359
Depth	1/4°	1/2°	1°	2°	3°	5°	7°	8°	10°	12°	15°

Table 17.42 Examples of tapers for cavity sidewalls

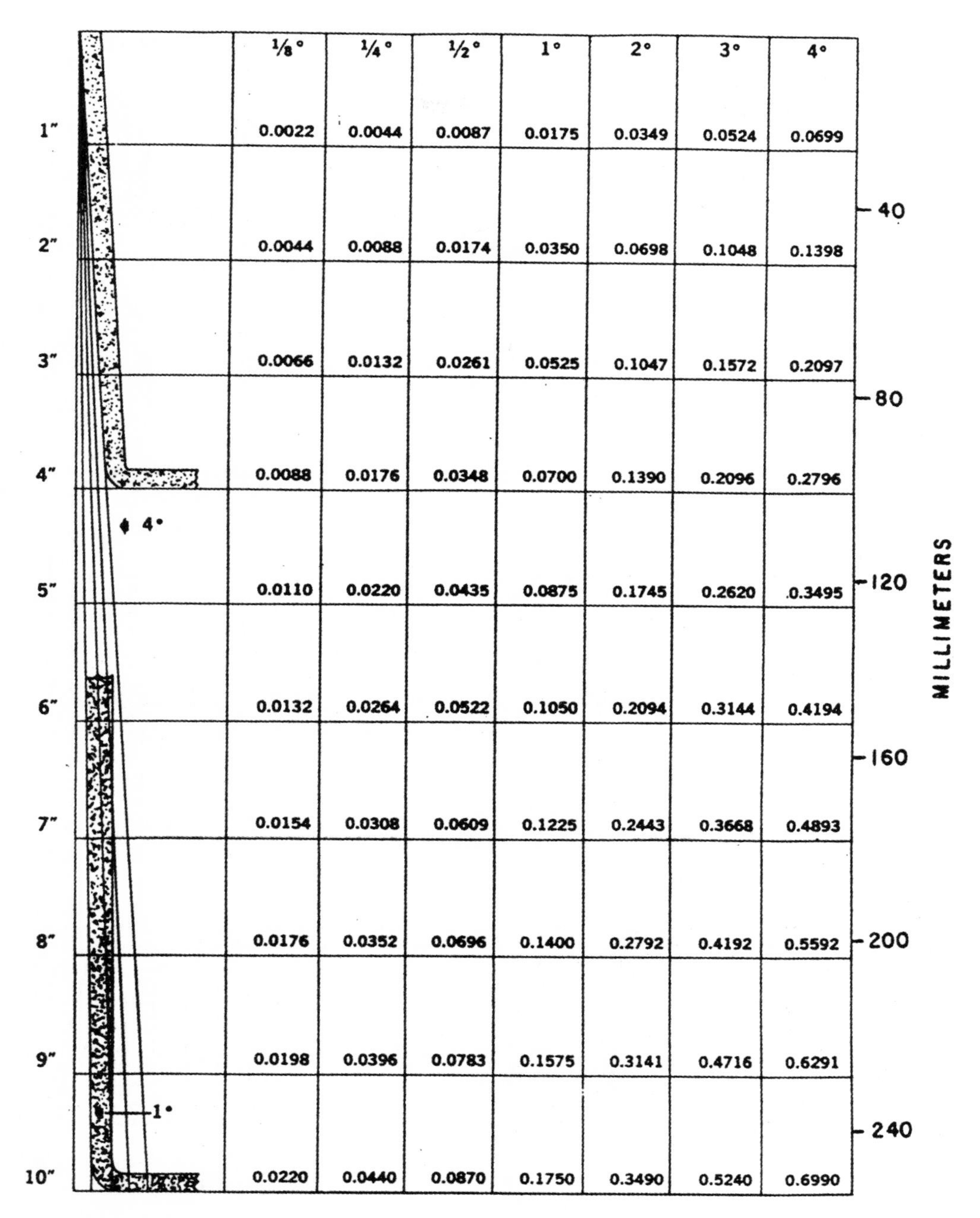

	1/8°	1/4°	1/2°	1°	2°	3°	4°
1"	0.0022	0.0044	0.0087	0.0175	0.0349	0.0524	0.0699
2"	0.0044	0.0088	0.0174	0.0350	0.0698	0.1048	0.1398
3"	0.0066	0.0132	0.0261	0.0525	0.1047	0.1572	0.2097
4"	0.0088	0.0176	0.0348	0.0700	0.1390	0.2096	0.2796
5"	0.0110	0.0220	0.0435	0.0875	0.1745	0.2620	0.3495
6"	0.0132	0.0264	0.0522	0.1050	0.2094	0.3144	0.4194
7"	0.0154	0.0308	0.0609	0.1225	0.2443	0.3668	0.4893
8"	0.0176	0.0352	0.0696	0.1400	0.2792	0.4192	0.5592
9"	0.0198	0.0396	0.0783	0.1575	0.3141	0.4716	0.6291
10"	0.0220	0.0440	0.0870	0.1750	0.3490	0.5240	0.6990

Table 17.42 Examples of tapers for cavity sidewalls *(continued)*

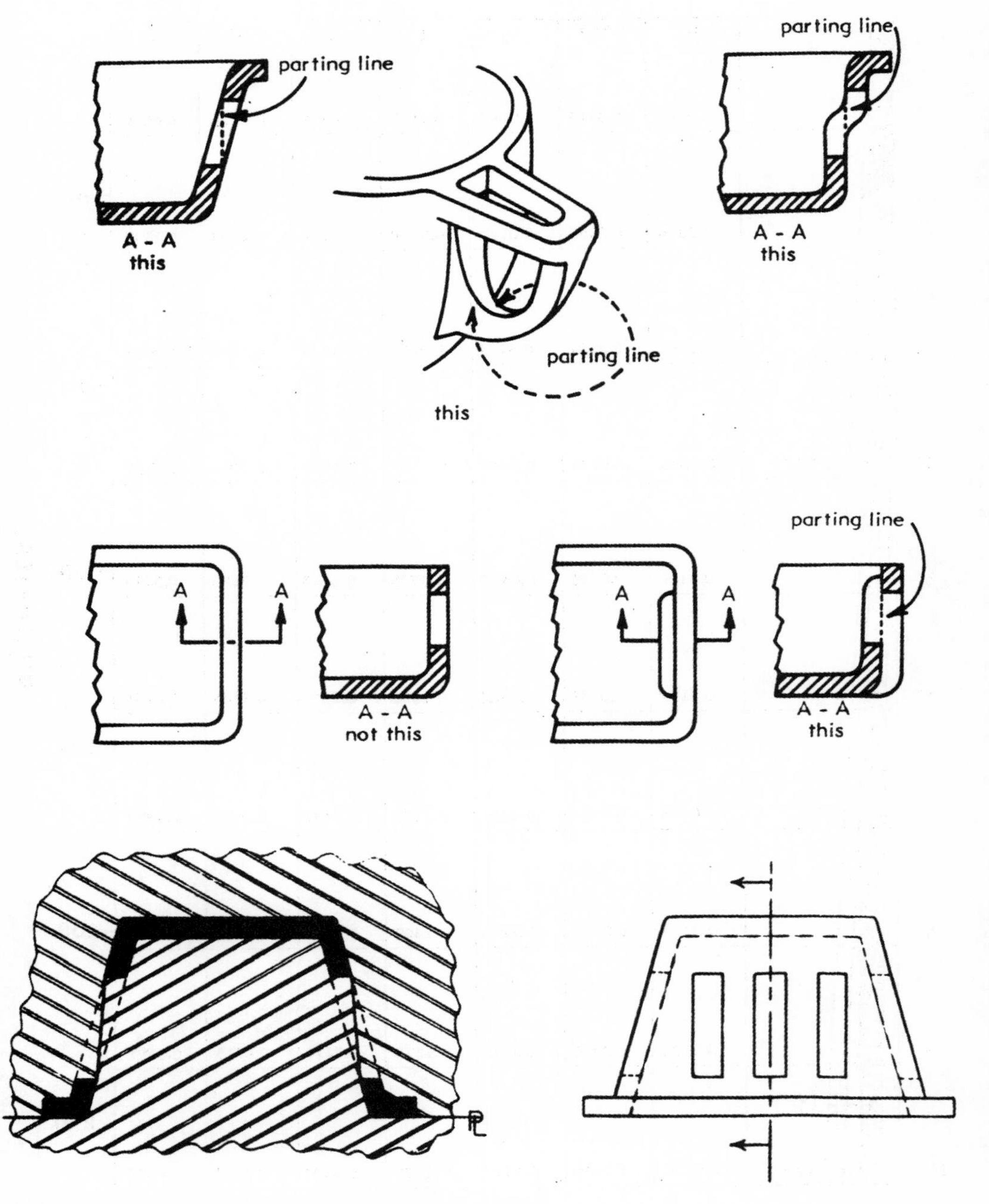

Figure 17.16 Examples to simplify mold design and action.

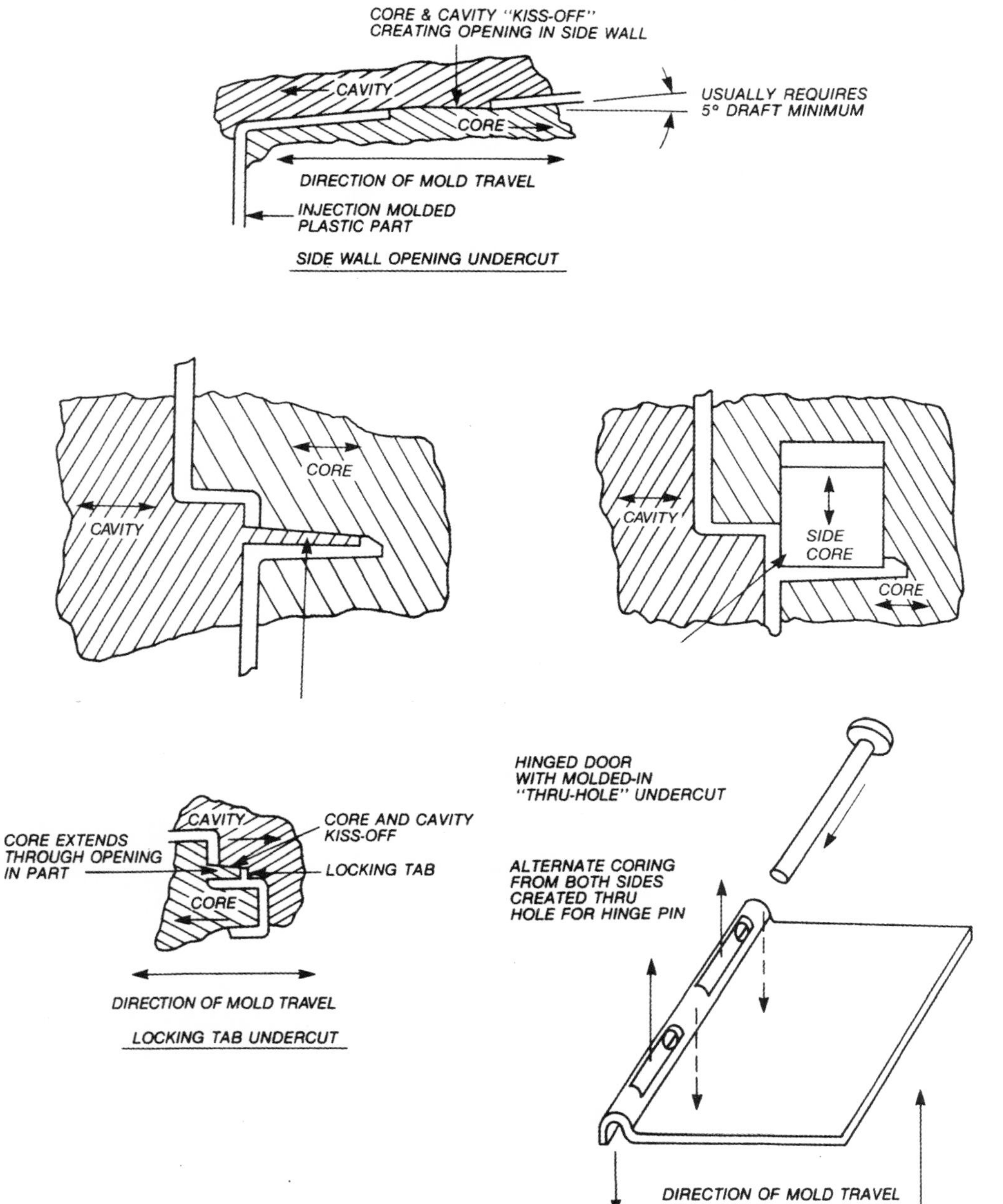

Figure 17.16 Examples to simplify mold design and action *(continued)*.

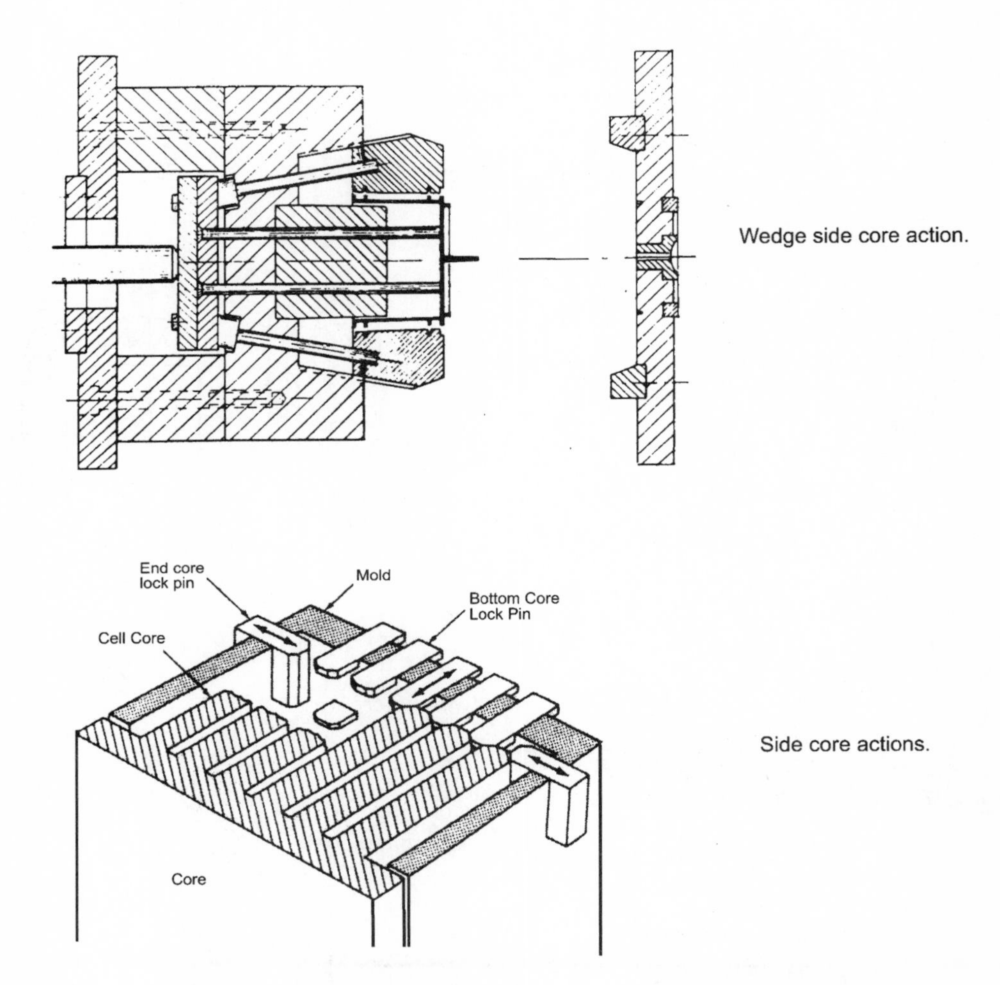

Figure 17.17 Examples of different actions in molds.

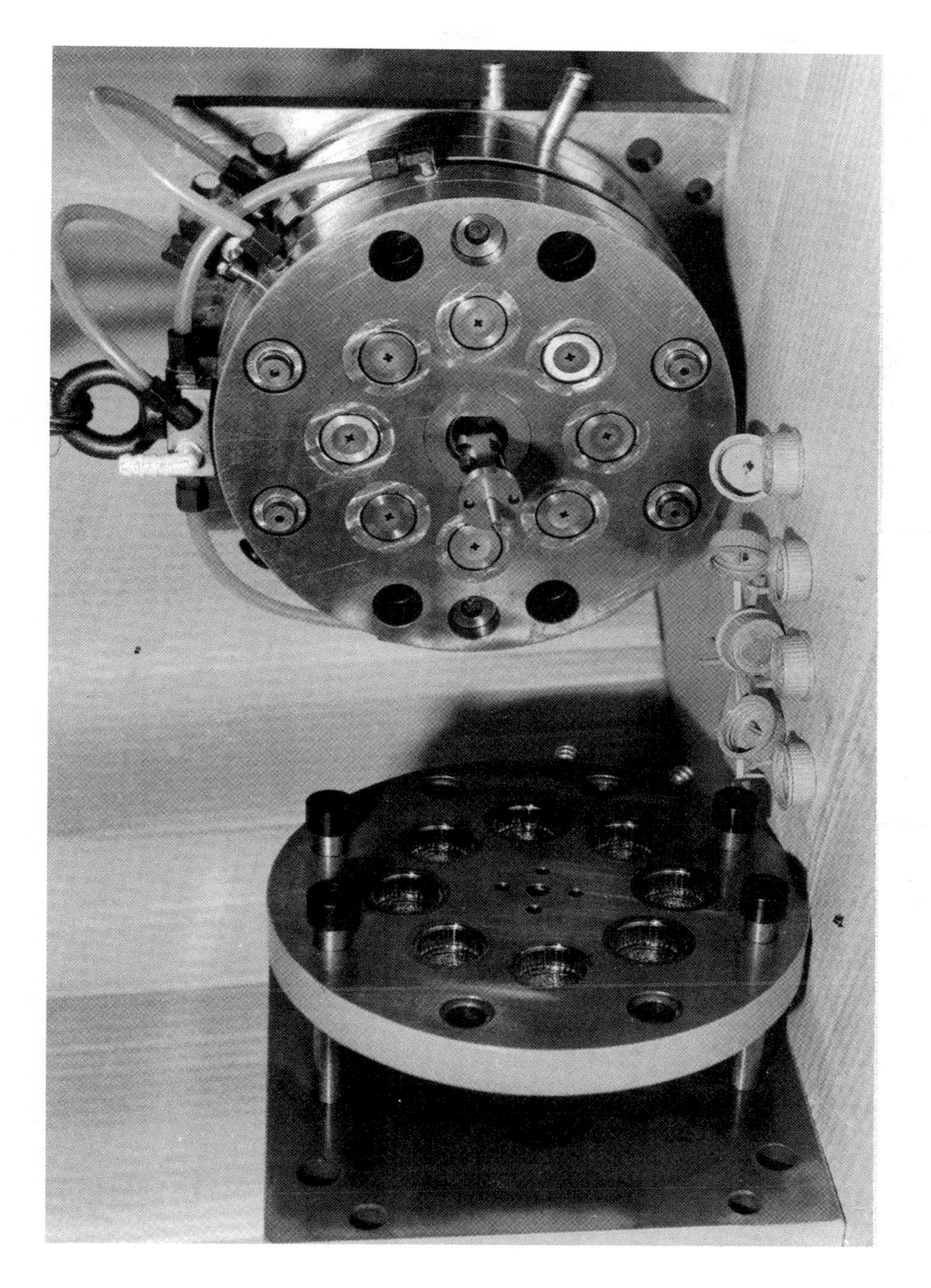

Satellite gear type hot runner mold

Figure 17.17 Examples of different actions in molds *(continued)*

Figure 17.17 Examples of different actions in molds *(continued)*

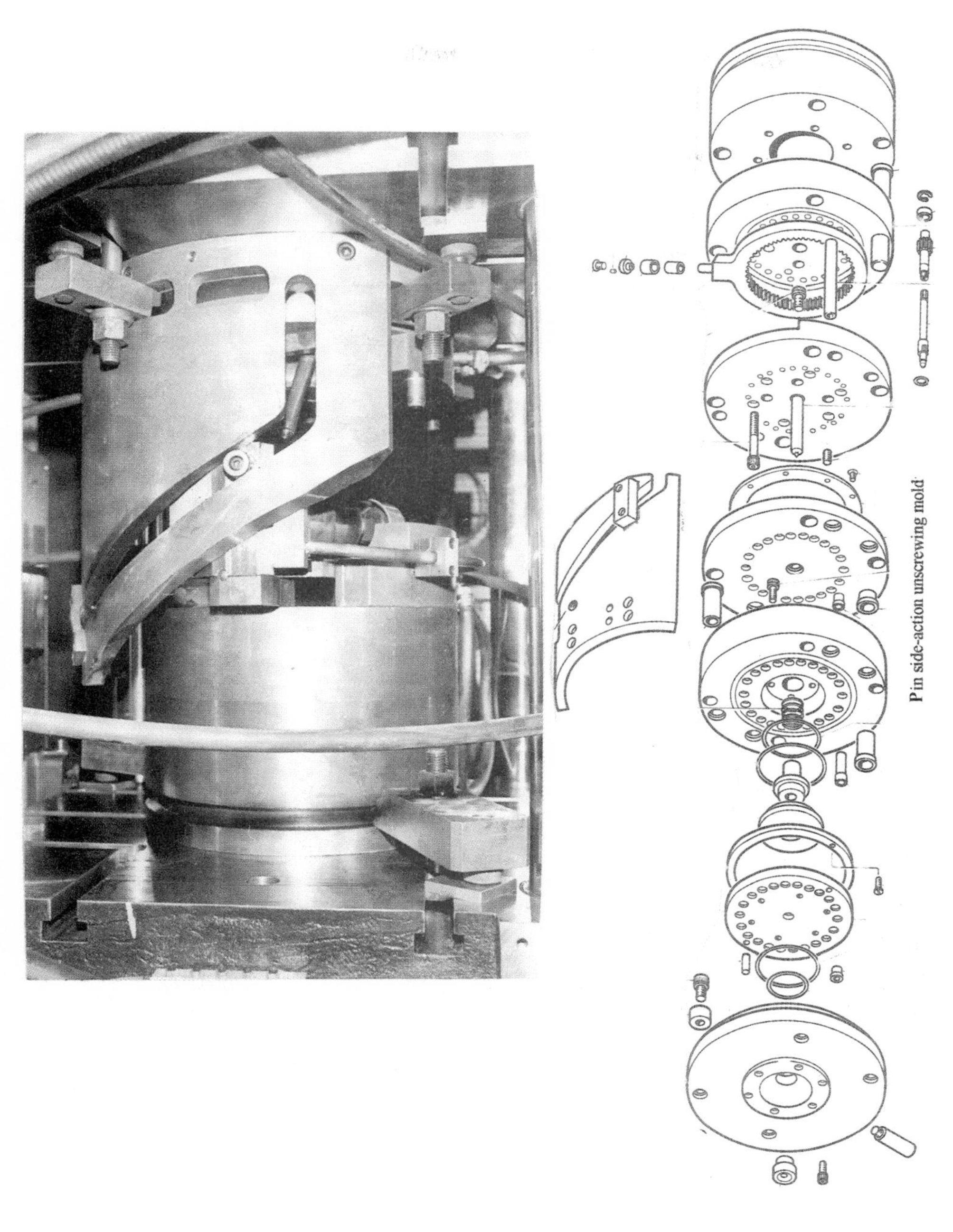

Figure 17.17 Examples of different actions in molds *(continued)*

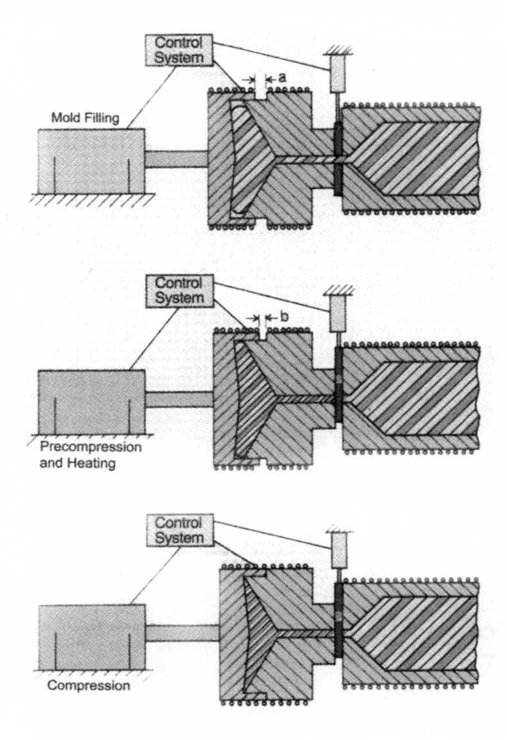

Injection-compression molding

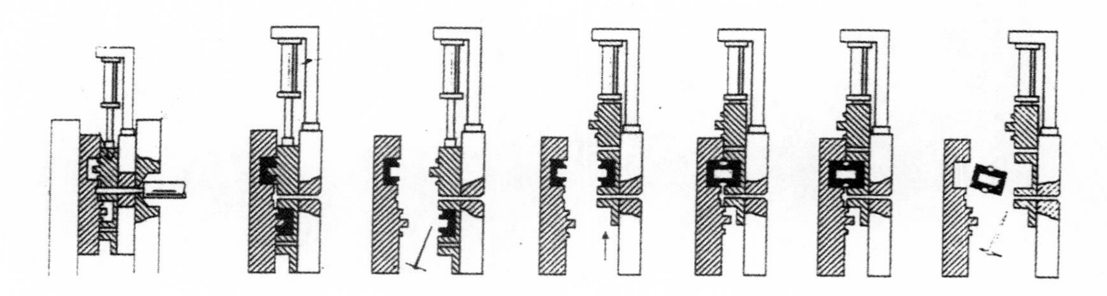

Two-part molding-patented

Figure 17.17 Examples of different actions in molds *(continued)*

Tandem (two IMMs) molding (cortesy of Husky)

Figure 17.17 Examples of different actions in molds *(continued)*

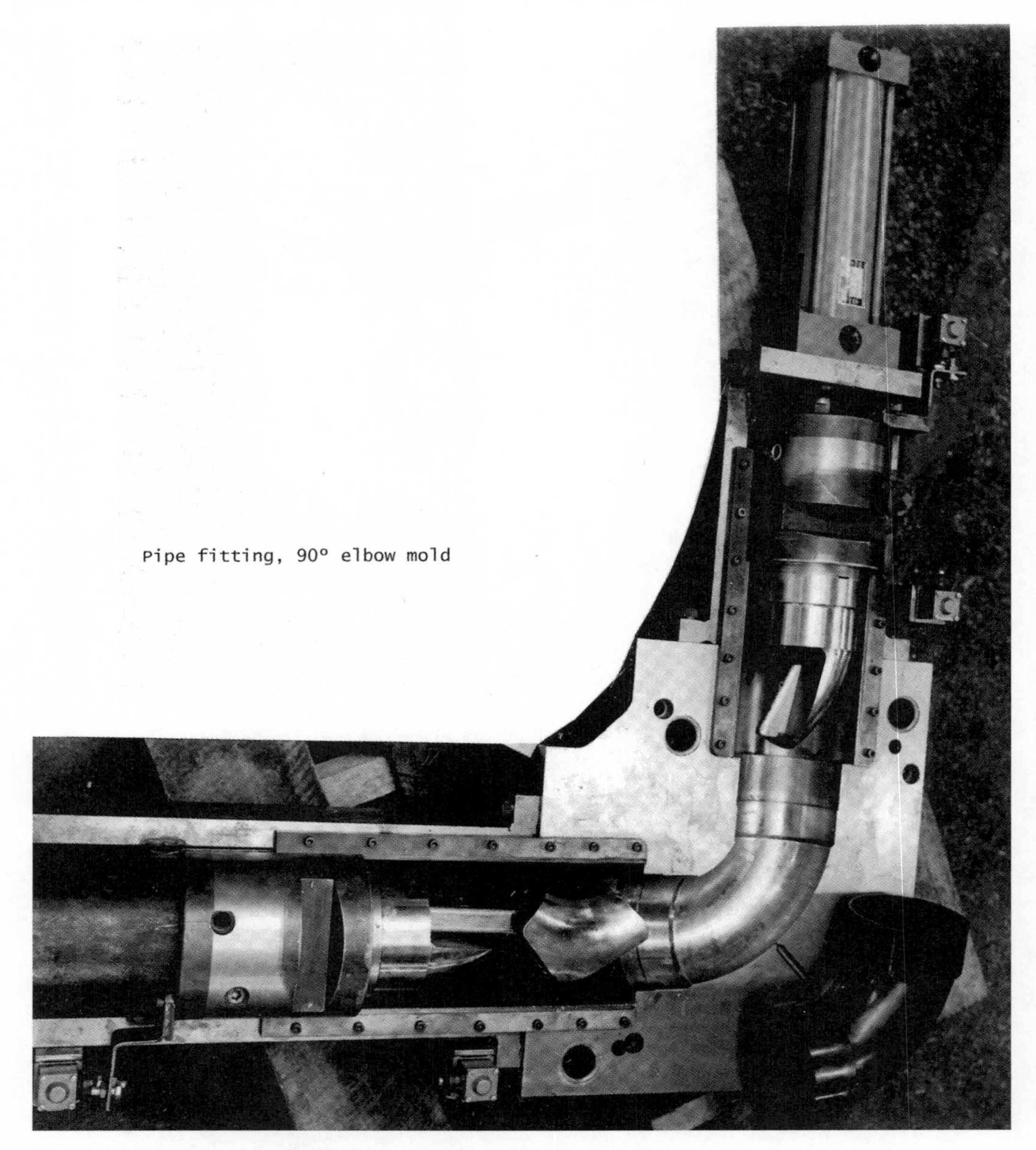

Figure 17.17 Examples of different actions in molds *(continued)*

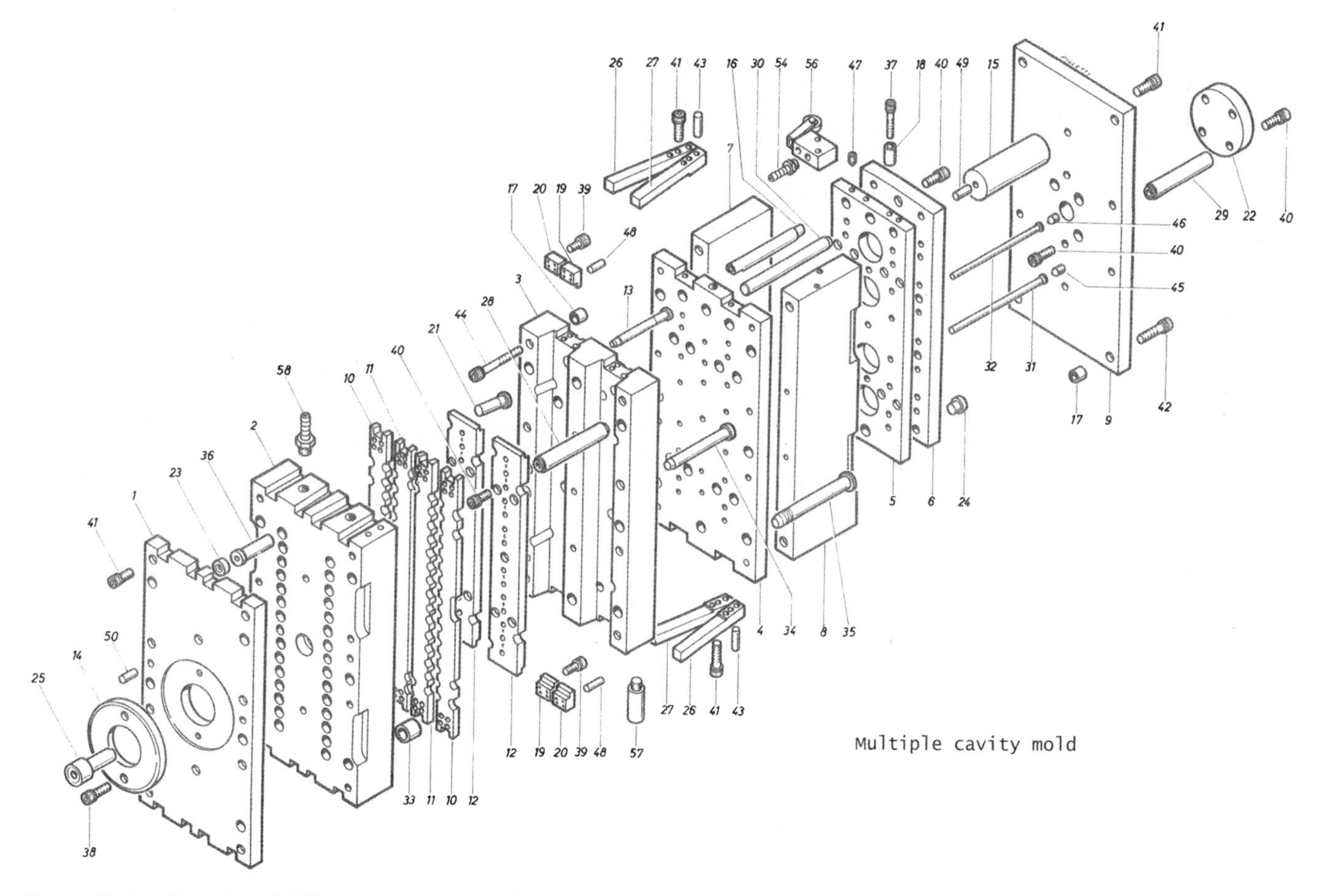

Figure 17.17 Examples of different actions in molds *(continued)*

use mechanical and/or hydraulic (toothed racks, spur-type gears, etc.) unscrewing drive systems (Fig. 17.18). Unscrewing cores are driven electrically with servomotors. This has many advantages, including space savings and the ability to create products with more precise dimensions and more compactness in a faster cycle with no oil contamination. These Programmable Electric Rotating Core (PERC) systems use small motors mounted on the mold (445).

Examples of molds—some including plastic products—are shown in Figure 17.19. Depending on the design approach, all kinds of variables can be used and problems can develop (chapter 27). Toolmakers are like medical doctors and lawyers; not all have the same specialties. In fact, individual toolmakers usually are experts in specific areas of tool design and manufacturing. Thus familiarity of the toolmaker's capabilities is helpful. All one has to do is find the best toolmaker for a project if those desired special capabilities do not exist in-house.

The mold designer matches the end-use requirements with the properties of the selected material using a practical or engineering technique (chapter 19). The aim is to achieve the basic three general requirements of design success. The design has to be (1) economical, (2) functional, and (3) attractive in appearance. In turn the functional aspect relates to the product's three environment conditions of (1) load, (2) temperature, and (3) time. The production method to be used will often set limitations on designs and vice versa. The way in which a product is manufactured has a profound influence on its design (446).

There are many different processing factors that could influence the repeatability of mold performance requirements (Figs. 17.20 and 17.21 and Tables 17.43 and 17.44). Some products may require only the compliance of one or two processing factors, but most others will require many. Computer programs have been developed to provide the capability of integrating all the applicable factors that can eliminate or at least reduce the traditional trial-and-error methods. These programs improve with time because providing improved controls on melts is an endless effort.

Even though it is possible to fabricate simple to complex shapes, consideration should be given for certain detractors and constraints. These types of problems are presented in chapter 27.

Molds are used in many plastic processes. Many of the molds have common assembly and operating parts with the aim to have the tool's cavity designed to form desired final shapes and sizes. Preengineered mold components are available to complete molds from various suppliers worldwide. Computer software guides that aid in designing molds are also available (chapter 25). Information contained here will explain how the mold components fit together to produce a useful mold.

The following information reviews IM molds, which represent most of the molds produced worldwide. There are different designs such as those for the EPS process (chapter 8), rotational molding (chapter 13), compression molds (chapter 14), RPs (chapter 15), and so on. The information presented here can be related to molds used by other processes. To understand and appreciate the use of molds it is best to understand the processing methods of interest to the reader that are reviewed in chapters 3, 4, 6 to 8, and 11 to 16. In the different chapters that review the different methods of processing plastics, there will be information on their specific mold requirements.

The function of a mold (Table 17.45) and the basic melt flow into a mold (Fig. 17.12) are important to understand. Melt moves from the injection unit (plasticator; chapter 3), though the

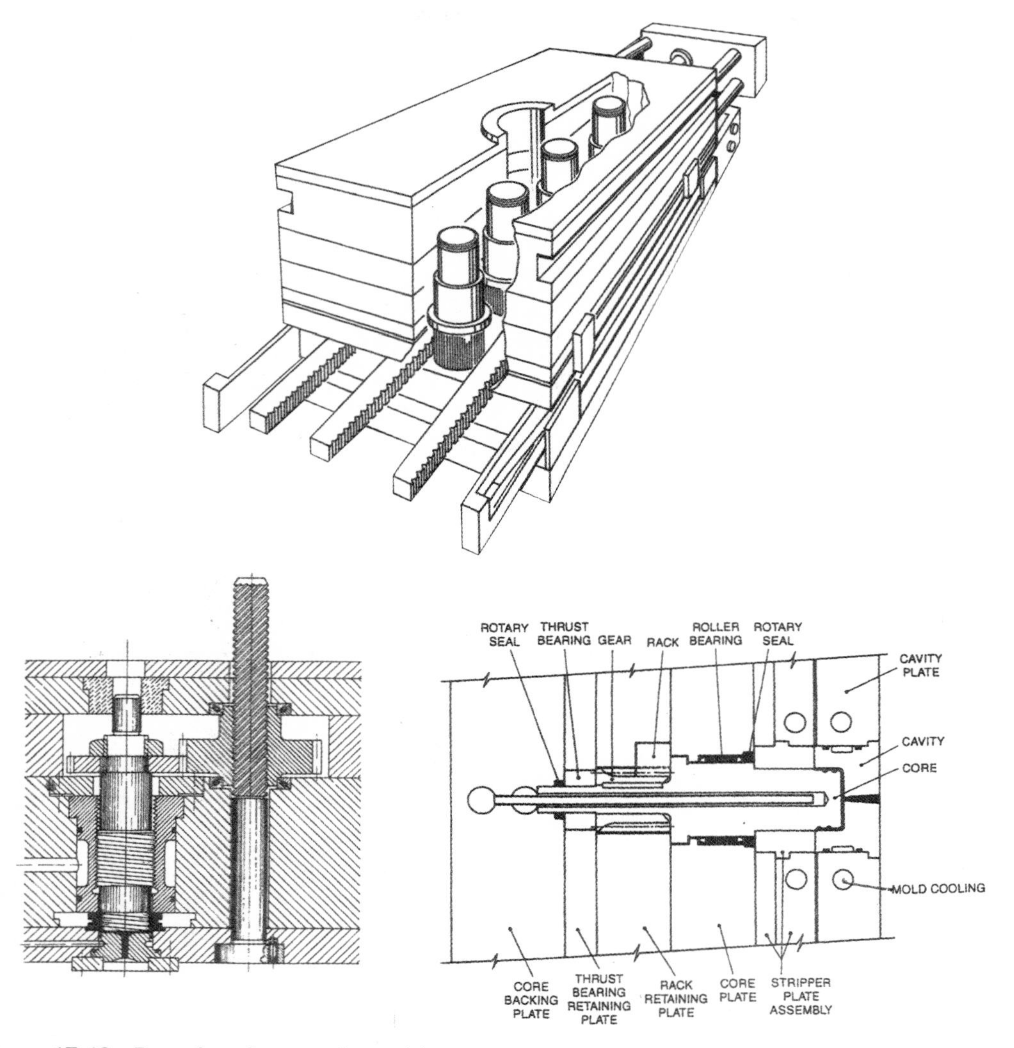

Figure 17.18 Examples of unscrewing molds.

mold passageways (sprue, runner, and gate), and into the cavities. The mold cavity containing the plasticized, melted plastic produces the desired shape and solidifies the molded product. TP melt is cooled for solidification. The TS plastic hot melt enters the mold cavity that provides the higher heat to solidify the plastic (Fig. 1.20; chapter 1).

The mold determines the size, shape, dimensions, finish, and often the performance properties of the molded product. Melted plastic from the IM machine (IMM) travels from the machine's nozzle

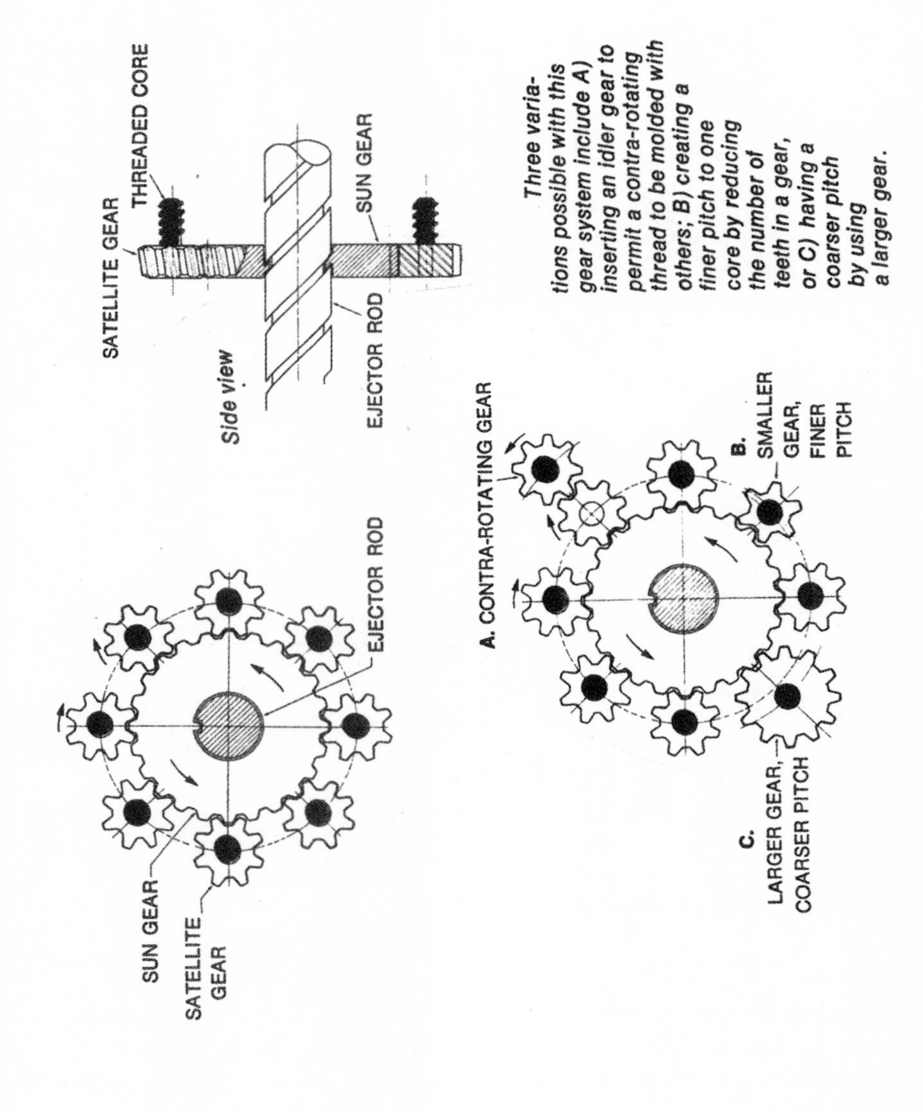

Figure 17.18 Examples of unscrewing molds *(continued)*.

Mold opens with caps remaining on cores.

Core plates rotate 180° on the third tiebar.

Molded parts are accurately alligned with the unscrewing chucks.

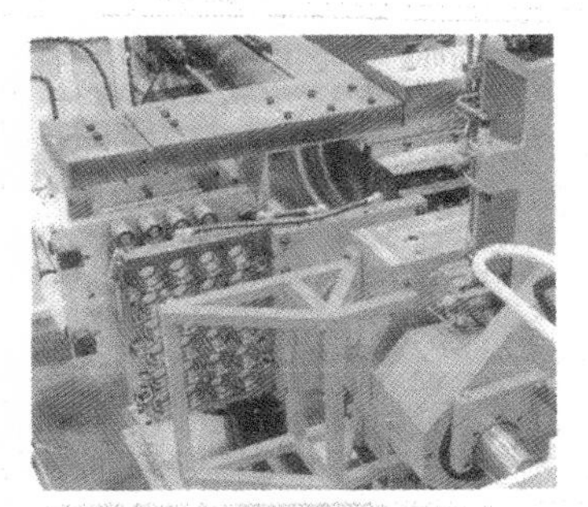

Mold closes to begin another cycle.

Figure 17.18 Examples of unscrewing molds *(continued)*.

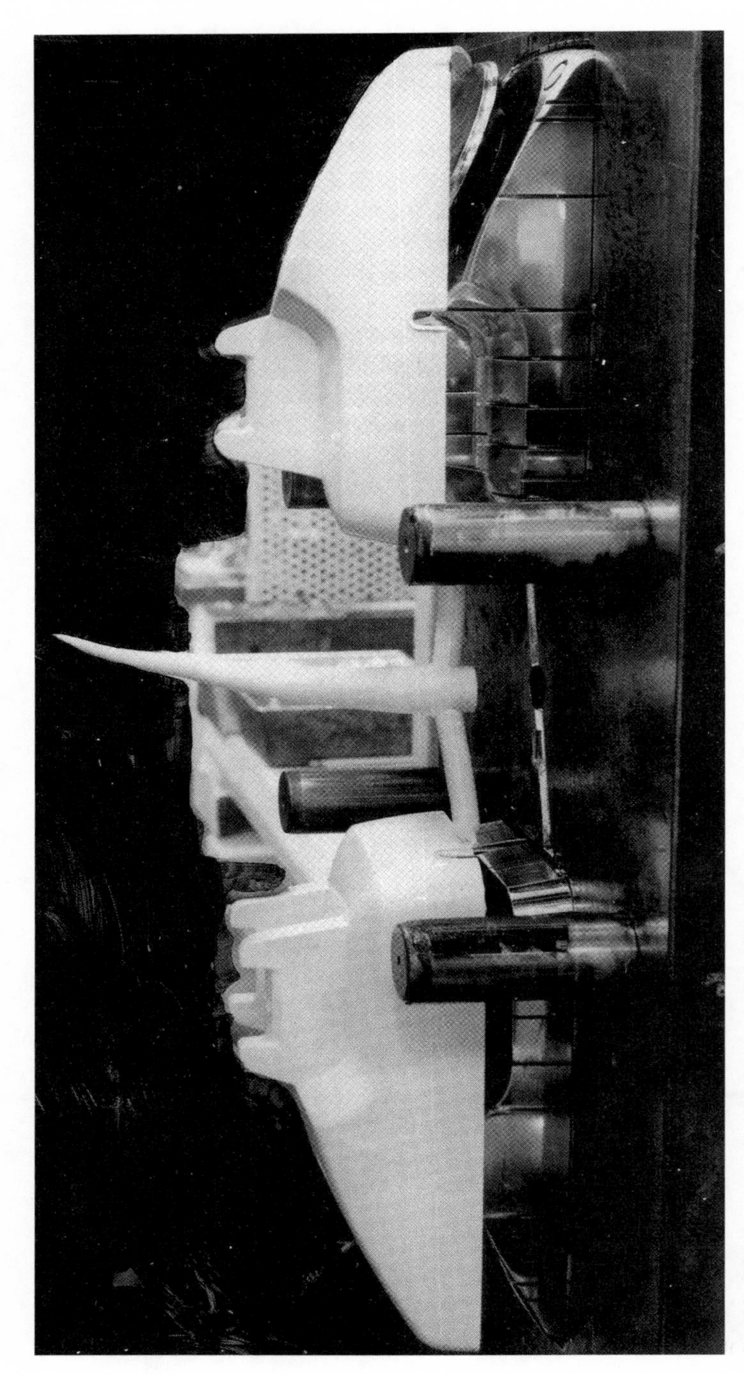

Figure 17.19 Examples of mold parts and molds.

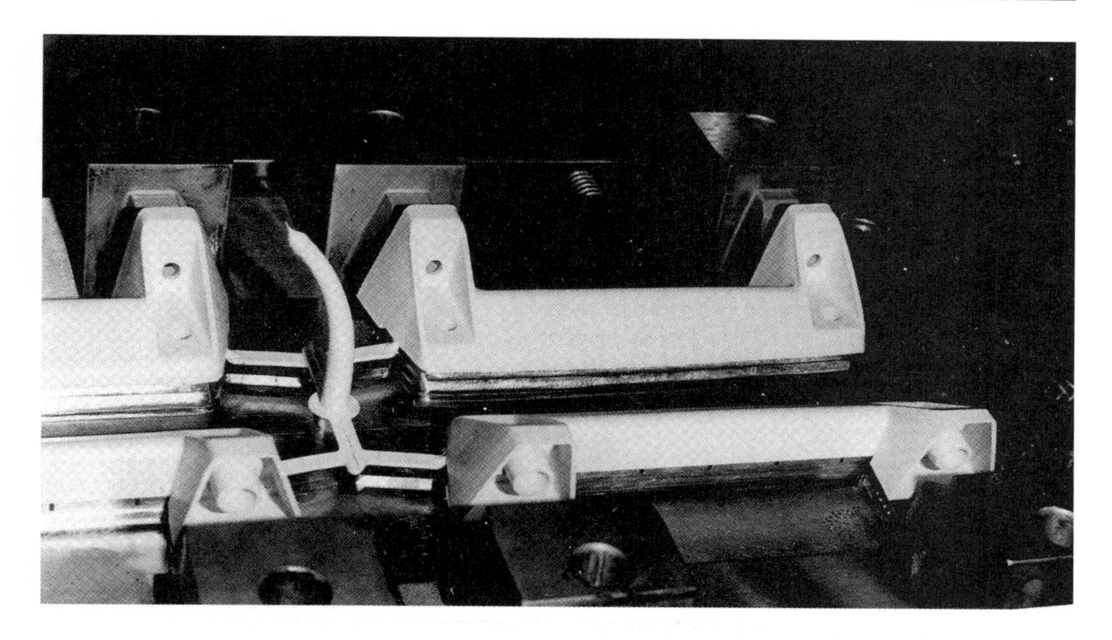

Figure 17.19 Examples of mold parts and molds *(continued)*.

Figure 17.19 Examples of mold parts and molds *(continued)*.

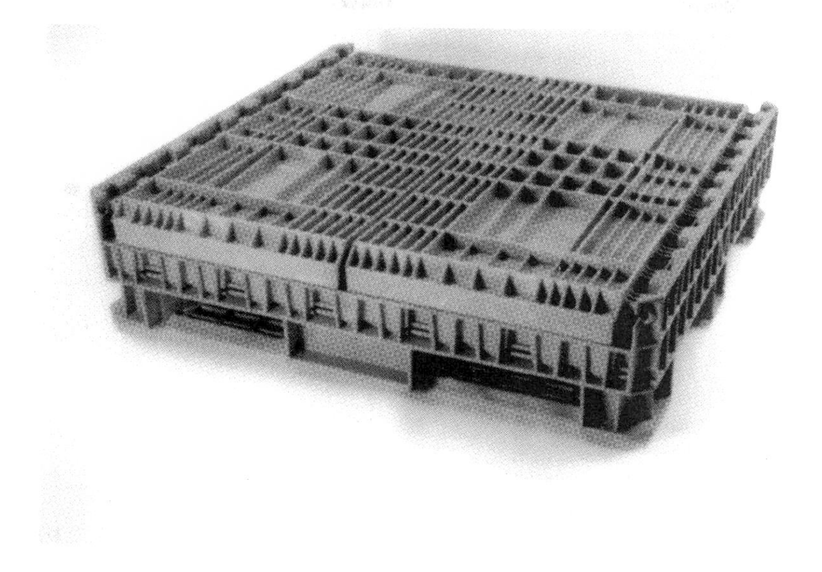

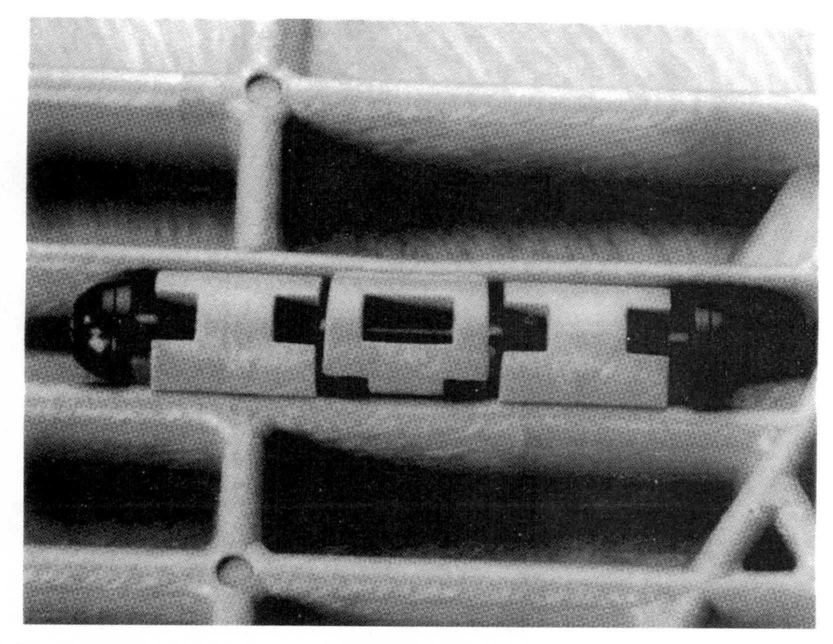

Figure 17.19 Examples of mold parts and molds *(continued)*.

Figure 17.19 Examples of mold parts and molds *(continued)*.

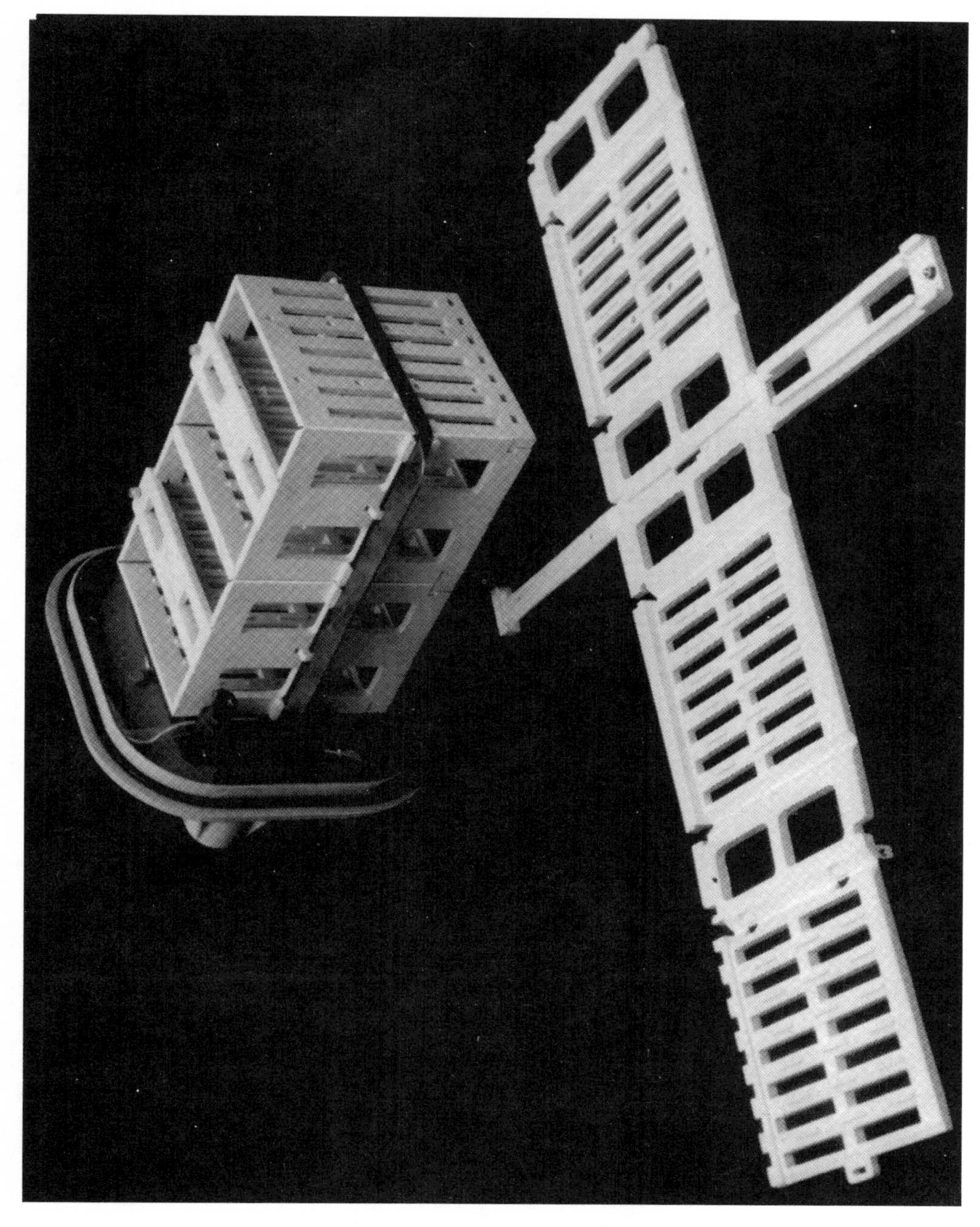

Figure 17.19 Examples of mold parts and molds *(continued)*.

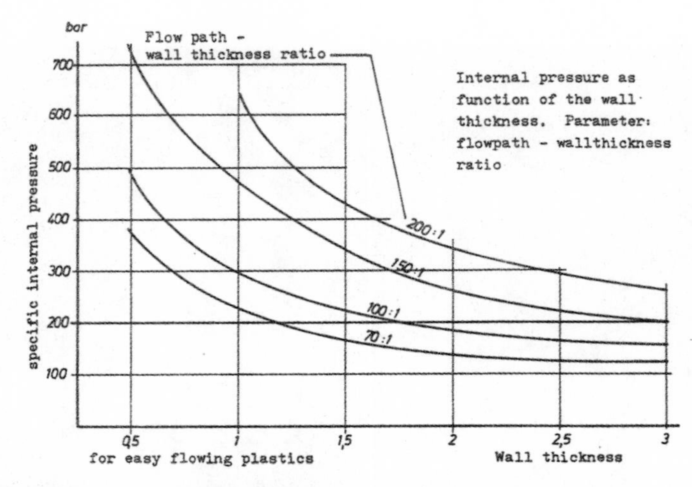

Figure 17.20 Examples of mold force based on determining clamp force required for melt flow.

Bolt size (in.)	Engagement in platen (in.)	Slot in clamp (in.)	Holding power in clamp (lb)	Torque wrench (in./lb)
$\frac{1}{2}$	0.75÷1.0	$2\frac{13}{16}$	32	210
$\frac{5}{8}$	$\frac{15}{16}-1\frac{1}{8}$	$3\frac{3}{8}$	45	340
$\frac{5}{8}$	$\frac{15}{16}-1\frac{1}{8}$	5	35	340
$\frac{3}{4}$	$1\frac{1}{8}-1\frac{5}{16}$	$3\frac{3}{8}$	50	450
$\frac{3}{4}$	$1\frac{1}{8}-1\frac{5}{16}$	5	40	450
1	1.5–1.75	$5\frac{5}{16}$	80	900

Table 17.43 Examples of pressures applied to molds

MATERIAL	BARREL TEMP. (°F)	MOLD TEMP. (°F)	INJECTION PRESSURE (PSI)
Polysulfone	710/740	325	13,500
Polyetherimide	740/770	308	17,500
Polyether sulfone	680/700	305	14,000
Polyphenylene sulfide	670/670	200	11,000

Table 17.44 Examples of plastic mold temperatures and pressure requirements

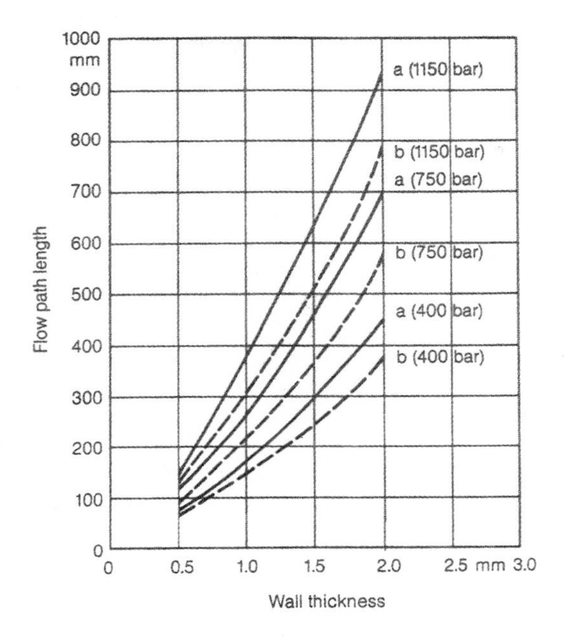

Material factors for clamp force determination.

Polymer	Material Factor
LD polythylene	1
Polystyrene	1
Polypropylene	1.0-1.2
HD polyethylene	1.0-1.3
Polyamide 11	1.2-1.4
Polyamide 12	1.2-1.4
Polyamide 6	1.2-1.4
Polyamide 6/10	1.2-1.4
Polyamide 6/6	1.2-1.4
ABS	1.3-1.5
PMMA	1.5-1.7
Polycarbonate	1.7-2
PVC-U	2

Figure 17.21 Examples of melt flow's path length as a function of part wall thickness and injection pressures.

Component	Operation
Mold base	Hold cavity (cavities) in fixed, correct alignment relative to machine nozzle.
Guide pins	Maintain proper alignment of two halves of mold during molding
Sprue bushing (sprue)	Provide means of melt entry into mold interior.
Runners	Convey molten plastic from sprue to cavities.
Gates	Control flow into cavities.
Cavity (female) and force (male)	Control size, shape, and surface texture of molded article.
Water channels	Control temperature of mold surfaces, to chill plastic to rigid state.
Side (actuated by cams, gears, or hydraulic cylinders)	Form side holes, slots, undercuts, threaded sections.
Vents	Allow escape of trapped air and gas.
Ejector mechanism (pins, blades, stripper plate)	Eject rigid molded article from cavity or force.
Ejector return pins	Return ejector pins to retracted position as mold closes for next cycle.

Table 17.45 Basic mold component operations

through a central feed channel, called the sprue, to feed the cavity or cavities. This sprue, which is located in a sprue bushing, is tapered to facilitate mold release. In single-cavity molds, the sprue usually feeds the plastic directly into the mold cavity, whereas in multicavity molds it feeds the plastic melt to a runner system (cold or hot), which leads into each mold cavity through a gate or gates.

The mold with its cavity (or cavities) and core (or cores) uses two basic parts to contain the melt. They are the usual stationary mold half (with its female cavity) on the side where the plastic is injected and a moving half (with its male cavity) on the closing or ejector side of the machine. The separation between the two mold halves is called the parting line.

Molds range from simple and low-cost designs to highly sophisticated and expensive pieces of equipment. They can be composed of many parts requiring the high-quality metals, precision machining, and protective surfaces previously reviewed. Production molds are usually made from steel for pressure molding that requires heating and/or cooling channels, strength to resist the forming forces, and/or wear resistance to withstand the wear from plastic melts. To capitalize on its advantages, the mold may incorporate many cavities, adding further to its complexity; however, this provides the means to reduce the cost of fabricated molded parts.

The basic mold is the male plug with a female cavity. The usual procedure is to clamp the male plug to the moving platen and the female cavity to the fixed platen of the molding machine. The male part of the cavity is designed as a plug that fits into the female part of the cavity. When these two parts close, the result is a cavity that will meet the geometric configuration of the product to be molded.

Different designs of molds permit processing plastics by different IM techniques that permit fabricating product shapes ranging from simple to very complex designs (chapter 4). Each design has advantages and disadvantages based on factors such as control of the melting process of the processing machine; required production rate, life, and quantity; type of plastic to be processed; product tolerance requirements; mold delivery time; and/or mold cost.

A mold is usually required to operate with the fastest cycle time, continually around the clock. In order to be productive it must meet certain requirements that involve its own capability and that of the fabricating machine. The maximum size or weight of a single-cavity or multiple-cavity mold is based on the machine's plasticating capacity. With large products, such as an auto bumper, the large exterior dimensions of the mold must fit within the machine's tie bars or platens (as in tiebarless IMM). The machine's daylight opening will permit the mold to eject the part when the operation is completed.

The overall height of the mold (known as the stack height) should correspond to the horizontal open space in between the usual horizontal IMM platens. In the moving mold half, spacers (known as side rails) are used to create space for the ejector system, which consists of the ejector plate and the ejector retainer plate, which holds the ejector pins. The open space should permit the ejector pins to complete their ejection stroke. Note that the mold height, or stack height, in the usual horizontal operating machine is the vertical dimension of the mold when the mold is removed and placed upright on a workbench; its mold height is then vertical.

It is important to design a mold that will safely absorb the pressure forces of clamping, injected plastic melt, and product ejection. The melt flow conditions of the plastic path must be adequately proportioned in order to obtain uniformity of melt to the cavity, which will result in balanced filling. Effective heat absorption from the TP melt to the mold that will result in controlled rate of solidification prior to removal from the mold is important. Practical access for maintenance is required. More details on these subjects as well as others will follow.

A mold is an efficient heat exchanger. If not properly designed, handled, and maintained, it will not be an efficient operating device. Hot melt, under pressure, moves rapidly through the mold. Air is released from the mold cavity or cavities to prevent the melt from burning, prevent voids in the product, and/or prevent other defects including the molded product's service operating performances. In order to solidify the TP hot melt, water or some other media circulates in the mold to remove heat from TPs; heat is used for this process with TS plastics.

The melt flow is largely governed by the shape and dimensions of the product and the location and size of the gate. A good flow will ensure uniform mold filling and prevent the formation of layers. Jetting of the plastic into the mold cavity may give rise to surface defects, flow lines, variations in structure, and air entrapment. This flow effect may occur if a fairly large cavity is filled through a narrow gate, especially if a plastic of low melt viscosity is used.

The hot TP melt entering the cavity solidifies immediately upon contact with the relatively cold cavity wall. The solid outer layer thus formed will remain in situ and form a tube through which the melt flows to fill the rest of the cavity. This accounts for the fact that a rough cavity wall adds only marginally to flow resistance during mold filling. Practice has shown that only very rough cavity walls (sandblasted surfaces) add considerably to flow resistance.

BASIC OPERATION

The injection mold performs two vital functions. It defines the shape of the molded part, and it acts as a heat exchanger to cool TP or heat TS plastic from melt temperature to ejection temperature. Since over 90 wt% of all plastics molded are TPs, the usual reviews that take place worldwide regarding IM only concern TPs; this review will follow the same approach but significant differences will be presented. Overall the mold must be designed to withstand injection and clamping forces, it must operate automatically at high speed, and it must be built to very high standards of precision and finish (chapter 4; 1, 6, 8, 10, 77, 205, 221, 274, 330, 420, 441, 442).

The mold also has other less obvious influences on the finished part. The dimensions and properties of the molding are greatly affected by shear rates, shear stresses, flow patterns, and cooling rates. Both mold and IMM affect some of these; others are almost exclusively a function of the mold. These factors combine to make the injection mold a costly and what can be considered a delicate tool. Modem technologies in the form of concurrent engineering, computer-aided design (CAD) of flow and cooling, high-speed and computer-controlled machining can help to keep the costs down. So too can the use of standard mold components that enjoy economies of scale, specialization, and performance. Even so, mold costs frequently dismay purchasers of moldings, and there can be pressure to cut costs by reducing mold quality and performance. This is a false economy and will almost always result in a more expensive fabricating process. There is no substitute for a quality mold.

MOLD COMPONENTS

Figure 17.22 illustrates the principal components used in the construction of a mold.

The sequence of operations (Fig. 17.13) for a typical mold is the following:

1. Plastic material is injected into the closed mold.
2. The mold remains closed while the molding cools for TPs or is heated for TSs.
3. The mold temperature is controlled by a coolant fluid (generally water with or without ethylene glycol), which is pumped through cooling channels for TPs; electric heaters are used for TSs. Even if the mold is heated for TPs relative to ambient temperature it is still cool in relation to the plastic's melt temperature.
4. After sufficient solidification occurs, the mold opens, leaving the molding attached by shrinkage to the core side of the mold.
5. During mold opening, any existing side action is retracted by an action, such as a cam, to release an undercut on the molding.

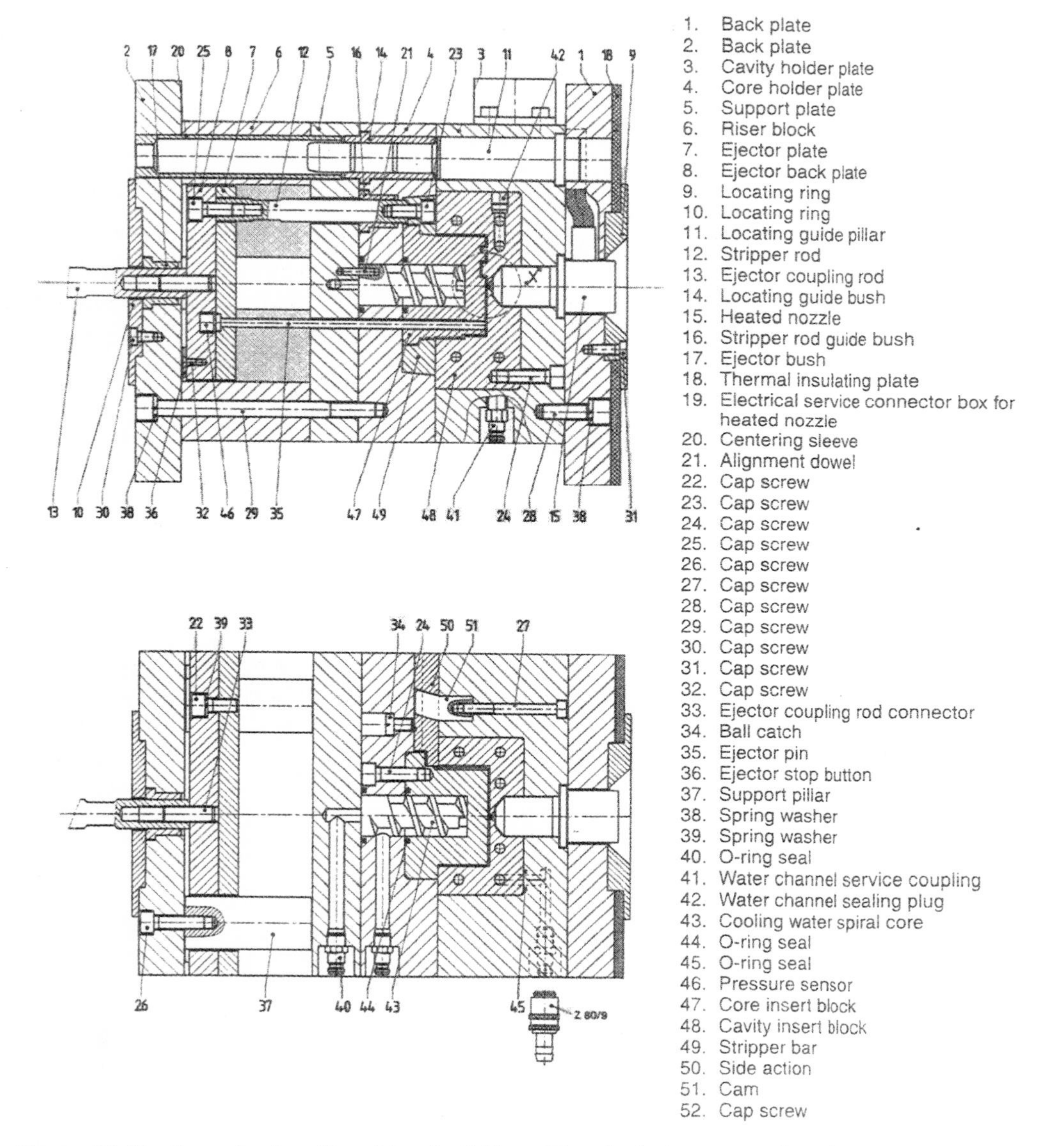

Figure 17.22 Example of an IM mold and a listing of its principal component parts.

6. The mold's ejector plate is moved forward, causing ejector pins and stripper bars to push the molding off the core.
7. The ejector plate returns and the mold closes, ready for the next melt shot in the molding cycle.

MOLD TYPE

Different molded parts are fabricated, with each having their distinguishing standard features and types. The principal types are two-plate, three-plate, and stack molds. A further distinction concerns the feed system that can be either the cold or hot type. These classifications overlap. Three-plate molds will usual have a cold runner feed system, and a stack mold will have a hot runner system. Two-plate molds can have either feed system.

TWO-PLATE MOLD

The two-plate mold has more than two plates in its construction. The description means that the mold opens or splits into two principal parts. These are known as the fixed or injection halves that are attached to the machine's fixed platen, and the moving or ejection half that is attached to the moving platen. This is the simplest type of injection mold and can be adapted to almost any type of molding.

Figure 17.23 shows schematics of two-plate mold, where in the first view the letters identify: A-Knockout plate, B-Ejector retainer plate, C-Ejector travel, D-Support plate, E-Sleeve-type ejector, F-Rear cavity plate, G-Cavity, H-Front cavity plate, I-Locating ring, J-Sprue bushing, K-Clamping plate, L-Parting line, M-Core or force, N-Ejector pin, O-Sprue puller, P-Clamping plate, O-Retainer plate, R-Knockout pins, S-Core or force pin. The non-self-degating and self-degating schematics are courtesy of Husky Injection Molding Systems Inc. The last view shows the mold halves separated with knockout (KO) pins expanded outwardly.

The cavities and cores that define the shape of the molding are so arranged that when the mold opens, the molding remains on the ejection half of the mold. In the simplest case, this is determined by shrinkage that causes the molding to grip on the core. Sometimes it may be necessary to adopt positive measures such as undercut features or cavity air blast to ensure that the molding remains in the ejection half of the mold.

THREE-PLATE MOLD

The three-plate mold splits into three principal linked parts when the machine clamp opens (Fig. 17.24). As well as the fixed and moving parts equating to the two-plate mold there is an intermediate floating cavity plate. The feed system is housed between the fixed injection half and the floating cavity plate. When the mold opens it is extracted from the first daylight formed by these plates parting. The cavity and core are housed between the other side of the floating cavity plate and the moving ejection part of the mold. Moldings are extracted from the second daylight when these plates separate.

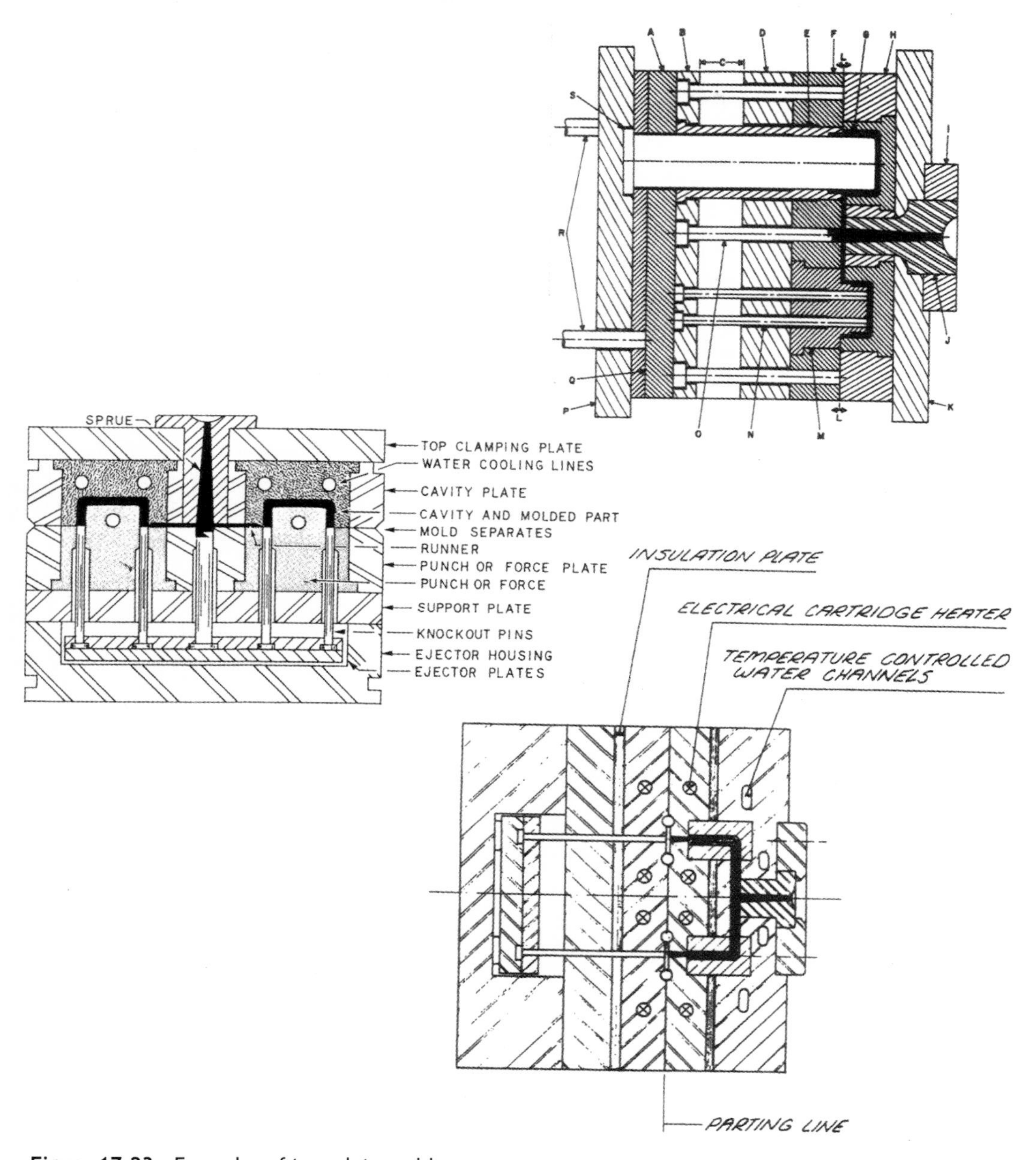

Figure 17.23 Examples of two-plate molds.

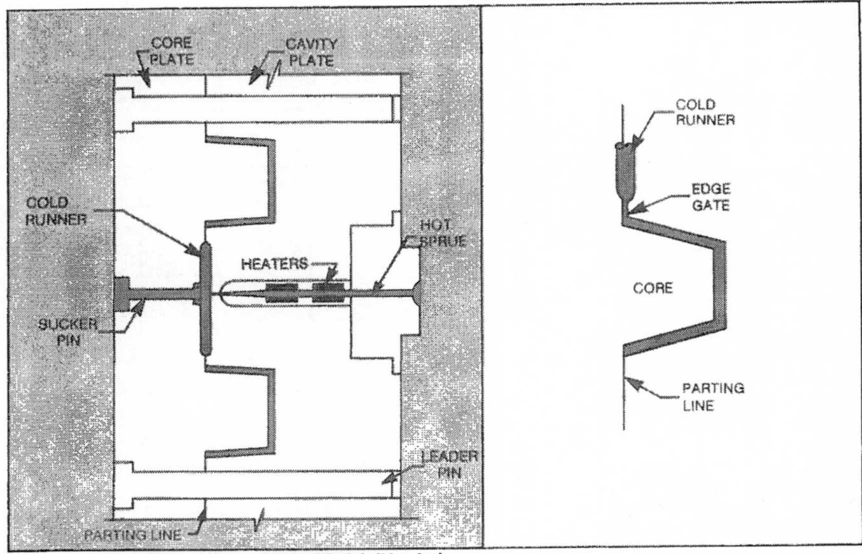

Non Self-Degating Two Plate Mold (Husky)

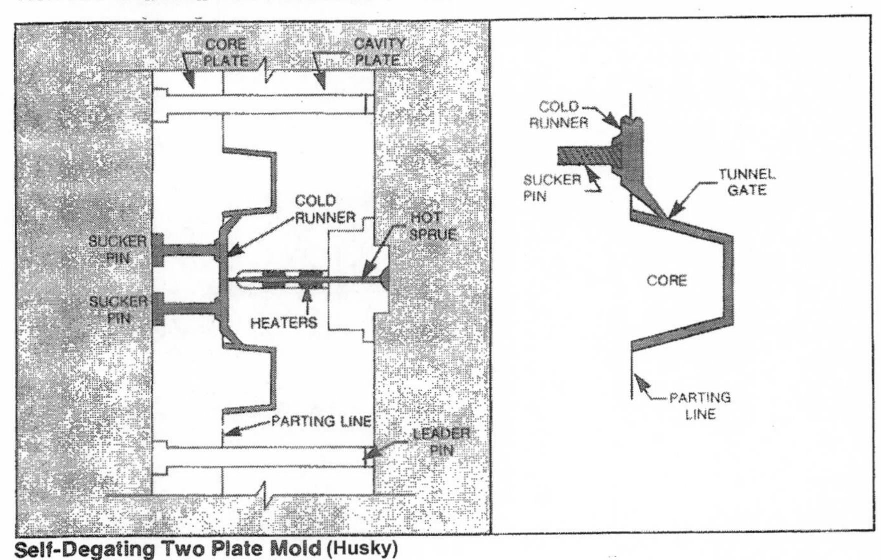

Self-Degating Two Plate Mold (Husky)

Figure 17.23 Examples of two-plate molds *(continued)*.

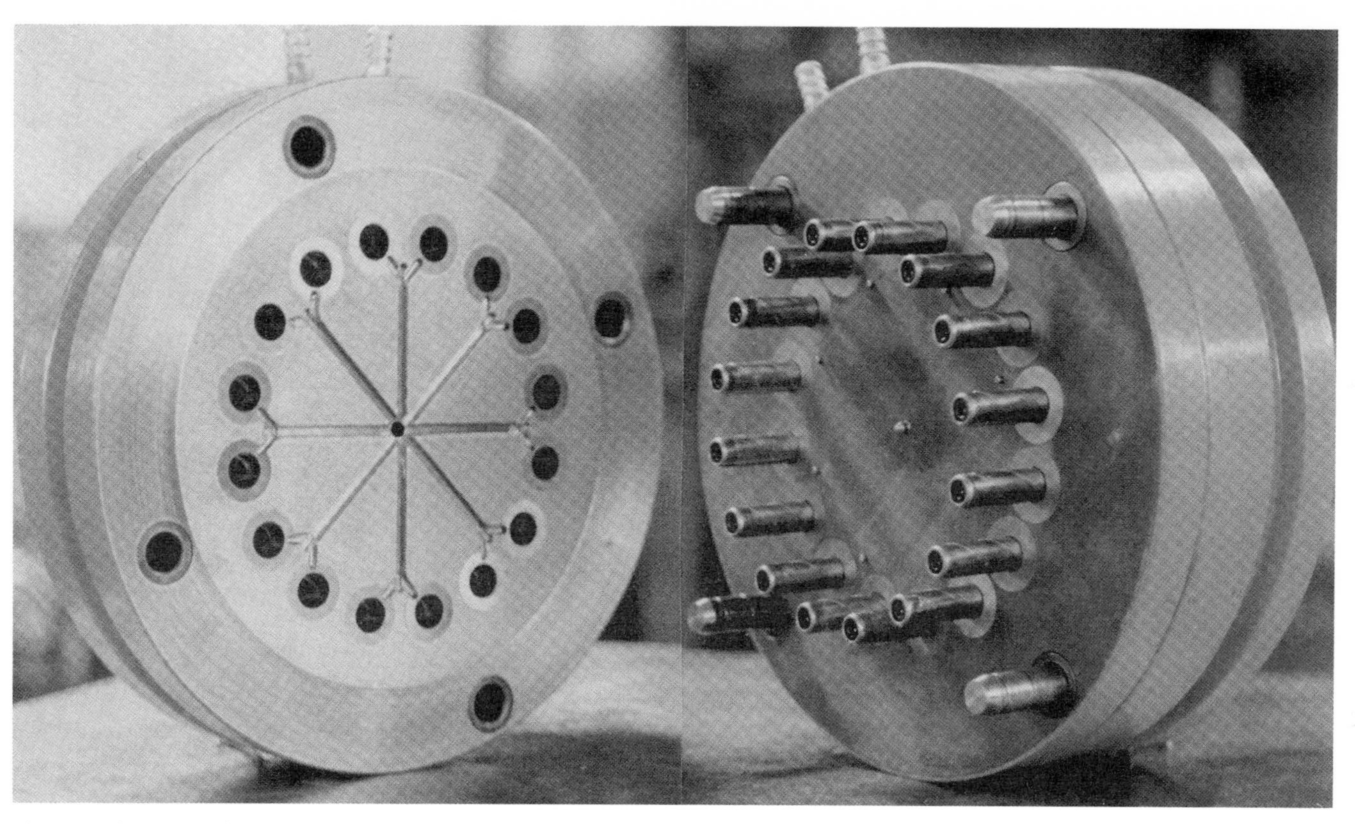

Figure 17.23 Examples of two-plate molds *(continued)*.

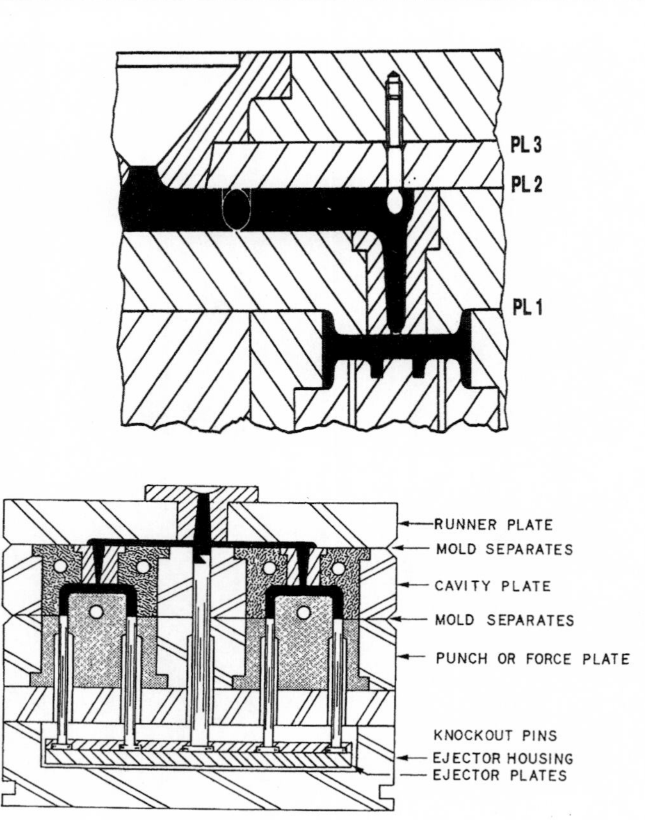

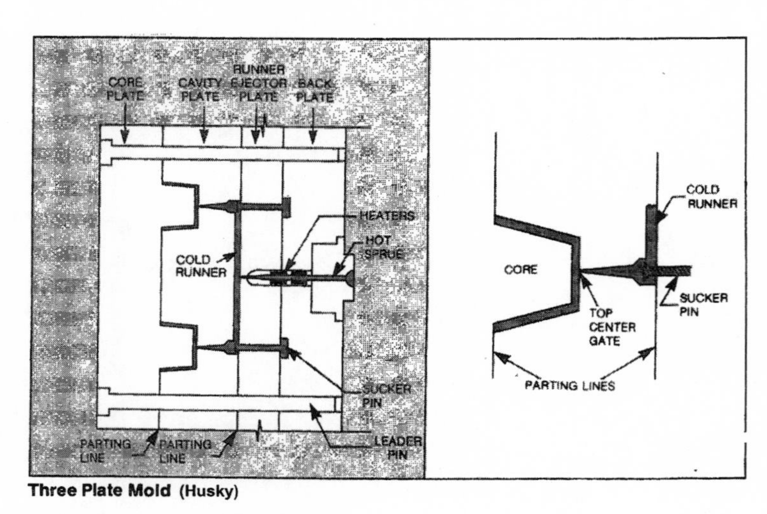

Three Plate Mold (Husky)

Figure 17.24 Examples of three-plate molds.

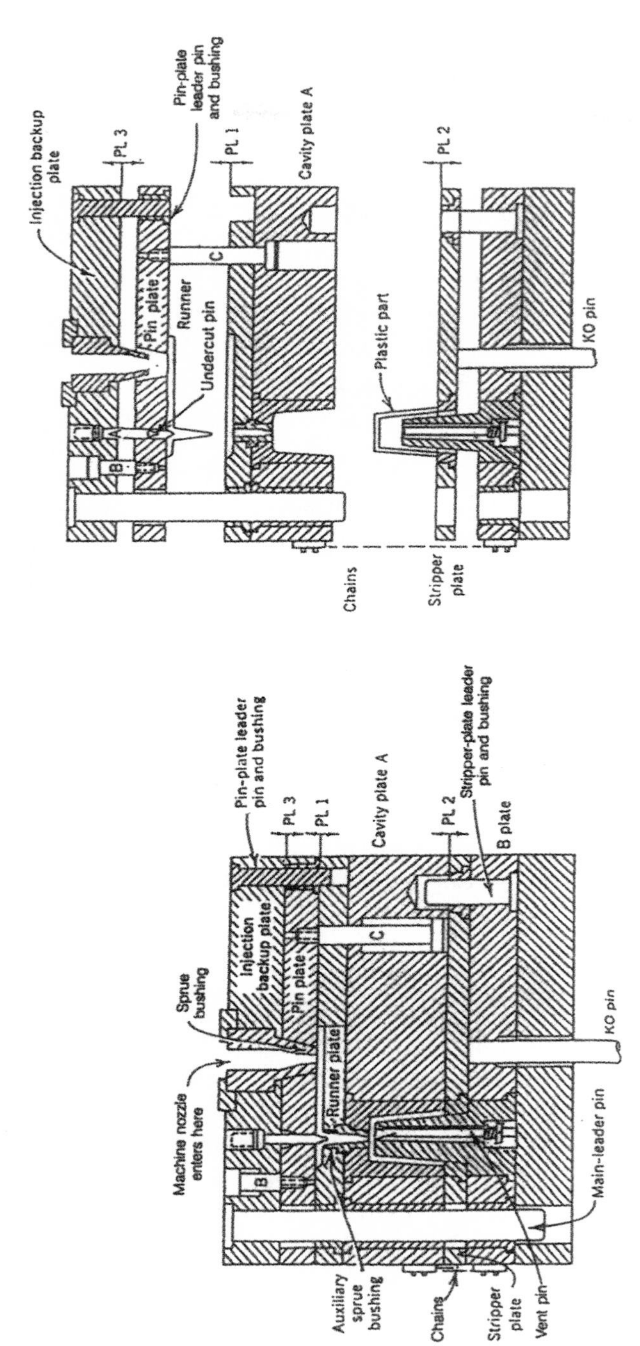

Figure 17.24 Examples of three-plate molds *(continued)*.

The mold needs separate ejection systems for the feed system and the moldings. Motive power and opening time for the feed-system ejector and the movement of the floating cavity plate is derived from the clamp-opening stroke position by a variety of linkage devices. The molding ejection system can be powered normally by the injection machine's ejection system or hydraulic, electrical, or a combination of the two.

The three-plate mold is normally used when it is necessary to inject multiple cavities in central rather than edge positions and/or to increase the production rate. This is done for flow reasons, to avoid gas traps, ovality caused by differential shrinkage, or core deflection caused by unbalanced flow. This type of mold also has the advantage of automatically removing (degating) the feed system from the molding. The disadvantages are that the volume of the feed system is greater than that of a two-plate mold for the same component and that the mold construction is more complicated and costly.

STACK MOLD

The stack mold also features two or more daylights in the open position. Two daylights are the normal form (Fig. 17.25), but up to four (sometimes more) are also used. The purpose of the stack mold is to increase the number of cavities in the mold without increasing the projected area, so that the clamp force required from the IMM remains the same. Providing cavities and cores between each of the daylights does this.

The projected area at each daylight must be the same; opposed components of opening thrust from each daylight cancel out, leaving the total mold opening force no greater than that developed in a single daylight. The cavities are fed by a hot runner system deployed in the floating cavity plate. Separate ejection systems are required for each daylight. The mold engineering and hot runner control systems are complicated and become much more so when there are more than two daylights present. Stack molds are normally used for high-volume production of relatively small and shallow components such as closures for packaging applications.

MICROMOLD

Microscale mold fabrication continues to expand its capability from recent developments in electronic signal sensing, part measurement, and process control. These improvements allow moldmakers to produce molds with extremely small cavities (Fig. 17.26) and hold tolerances of ±10 nm while cutting mold steel (chapter 4; 381, 447).

To make tiny features, moldmakers can use an unconventional technique, such as reactive ion etching as developed at the Georgia Institute of Technology in Atlanta. The moldmakers use reactive ions to knock metal atoms out of a mold surface. Moldmakers can use lasers to create extremely small features, such as small holes, that can not be made with a conventional electronic discharge machine (EDM).

New technologies in manufacturing micromolds continue to develop. For example, there is the LIGA. It is a lithography/electroplating technique developed in Germany. Companies are

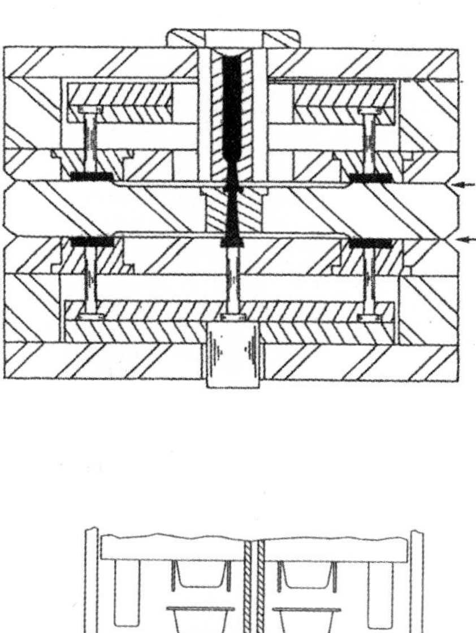

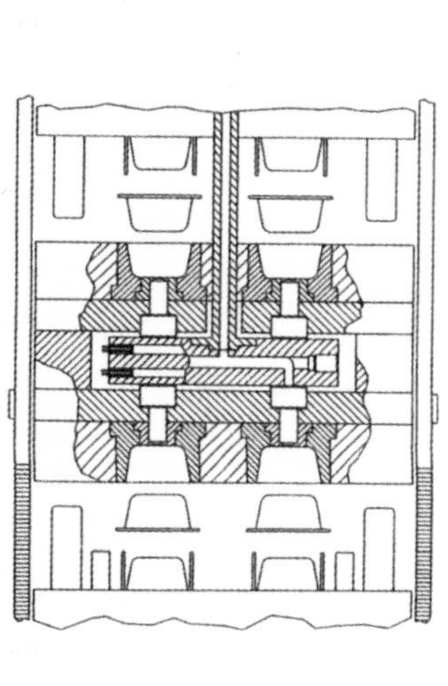

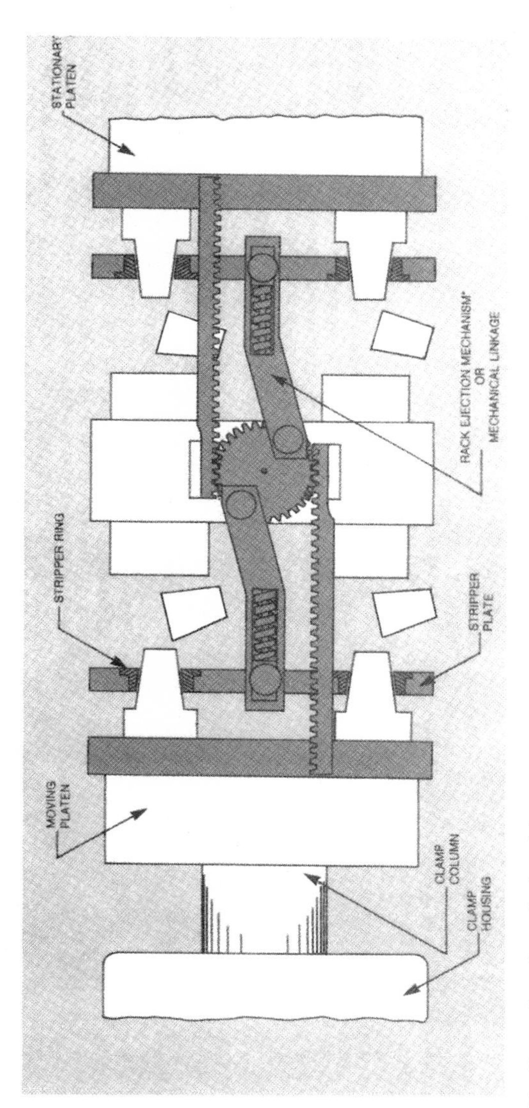

Figure 17.25 Examples of stacked molds.

Figure 17.25 Examples of stacked molds *(continued)*.

Figure 17.26 Examples of micromolded products compared to a US coin.

producing LIGA structures that could be converted to molding cavities. This technology allows molds to be minuscule.

High-volume molding operations using molds with sixty-four cavities limits control of the individual cavities, so the parts are not the same. Use is made of two- or four-cavity molds that produce more identical parts.

Microscale parts can be damaged by a rough mold surface, so moldmakers must pay close attention to surface finish. An EDM surface that looks smooth from a distance will probably appear much rougher up close. To smooth these surfaces, moldmakers often polish them with brass or copper tools. Another option is plating the mold surface with nickel or tin; this is used particularly for molding TPs containing an abrasive additive material.

Once microparts have been formed in the mold, they must be removed and placed in containers. To manage this process efficiently, some micromolders have developed micropart-handling systems. They include automated venturi systems that suck microparts out of the molds and place them in tubes.

Regardless of the microparts on the drawing board, moldmakers should enter into the design process as early as possible. Often, an engineer will design something that cannot be tooled. Thus the moldmaker can help customers design parts that can be fabricated.

INJECTION MOLD FEED SYSTEM

The feed system is the melt flow passage in the mold, between the nozzle of the IMM and the mold cavity (Fig. 17.27). This feature has a considerable effect on both the quality and the economy of the molding process. The feed system must conduct the plastics melt to the cavity via a sprue, runner, and gate at the correct temperature, pressure, and time period, must not impose an excessive pressure drop or shear input and should not result in nonuniform conditions at the cavities of multi-impression molds.

The feed system is an unwanted by-product of the molding process, so a further requirement is to keep the mass of the feed system at a minimum to reduce the amount of plastic material used. This last consideration is a major point of difference between cold and hot runner systems. The cold runner feed system is maintained at the same temperature as the rest of the mold. In other words, it is cold with respect to the melt temperature. The cold runner solidifies along with the molding and is ejected with it as a waste product in every cycle. The hot runner system is maintained at melt temperature as a separate thermal system within the cool mold.

Plastic material within the hot runner system remains as a melt throughout the cycle, and is eventually used on the next or subsequent cycles. Consequently, there is little or no feed-system waste with a hot runner system. Effectively, a hot runner system moves the melt between the machine plasticizing system and the mold to a point at or near the cavities.

SPRUE

The sprue (Fig. 17.28) is often the thickest part in an IM shot and in extreme cases may influence the cycle time. Since the sprue is a waste part, this should never be allowed to happen. The reason that it sometimes occurs is that there is little control over the dimensions of the sprue. The length is fixed by the thickness of the fixed mold half while the diameter is largely a function of the machine nozzle bore and the necessary release taper.

Sometimes the sprue can serve a useful function as a gripping point for an automated shot-handling system, but its use in most cases is to eliminate the cold sprue by means of a heated sprue bush (Fig. 17.28) that serves as a limited hot runner device. In this case, the material remains as a melt while the cold runner system is injected through the hot sprue. This approach saves material and eliminates waste.

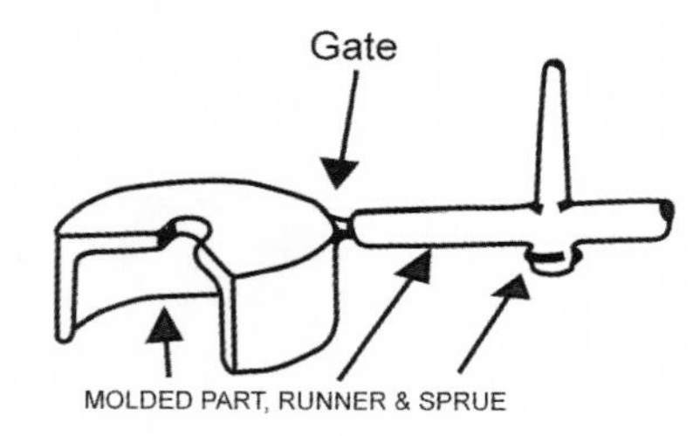

Figure 17.27 View of plastic flow from sprue to runner to gate to cavity.

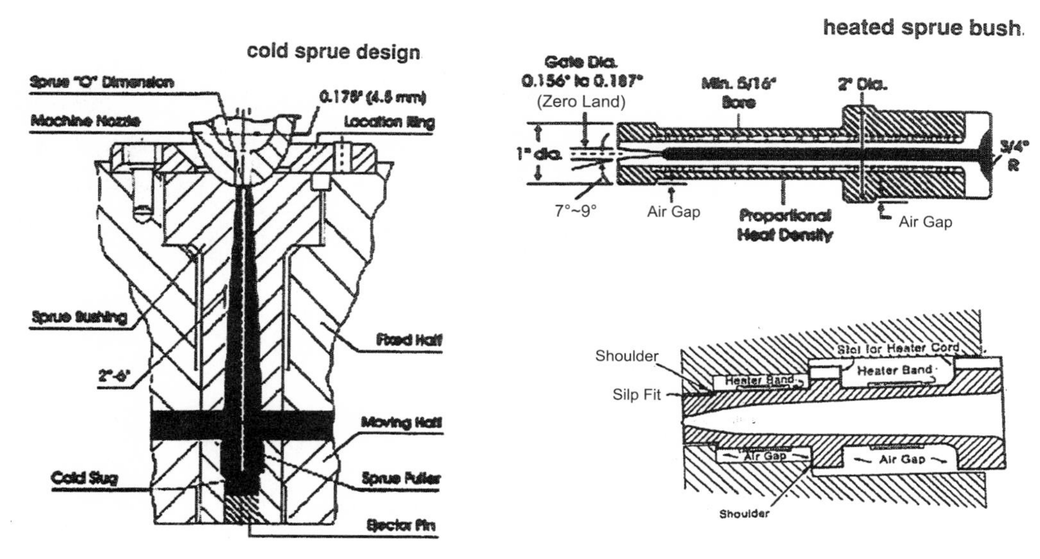

Figure 17.28 Examples of cold and heated sprue designs.

RUNNER

The following information explains the differences in TP and TS plastic runner terminology and serves as an introduction to runners. There are the terms cold runner and hot runner. With a TP, one refers to a cold runner when the TP runner solidifies in the mold, such as a two-plate mold (Fig. 17.29). If it remains hot, it is a hot runner. The terminology reverses when TSs are used. When the TS runner solidifies, it is also a hot runner. If it remains in a melt phase, the TS has a cold runner (Fig. 17.30). So the behavior of a TP cold runner is like that of a hot runner for a TS, whereas a TP hot runner identifies a cold runner for TS.

When processing TPs, both hot runners and cold runners provide benefits. To reduce cycle time, energy cost, and some labor and material costs, the usual choice is the hot runner system. However, in some applications, mold or part complexity, part size, heat sensitivity and the time-to-temperature plastic relationship can be a problem. A family of parts for handling and orientation purposes may need to be attached to a runner until they reach a downstream operation, and/or predetermined fixed gate locations may not allow direct gating with a hot runner nozzle of any type.

There is a hot-to-cold runner system, where a hot runner system with fewer nozzles feed small, balanced cold runners that in turn feed the cavity. The use of small runners considerably reduces the weight of the full cold runner and in certain molds can provide a more balanced flow of plastic. It provides many of the cost and part-quality advantages associated with molds using full hot runners.

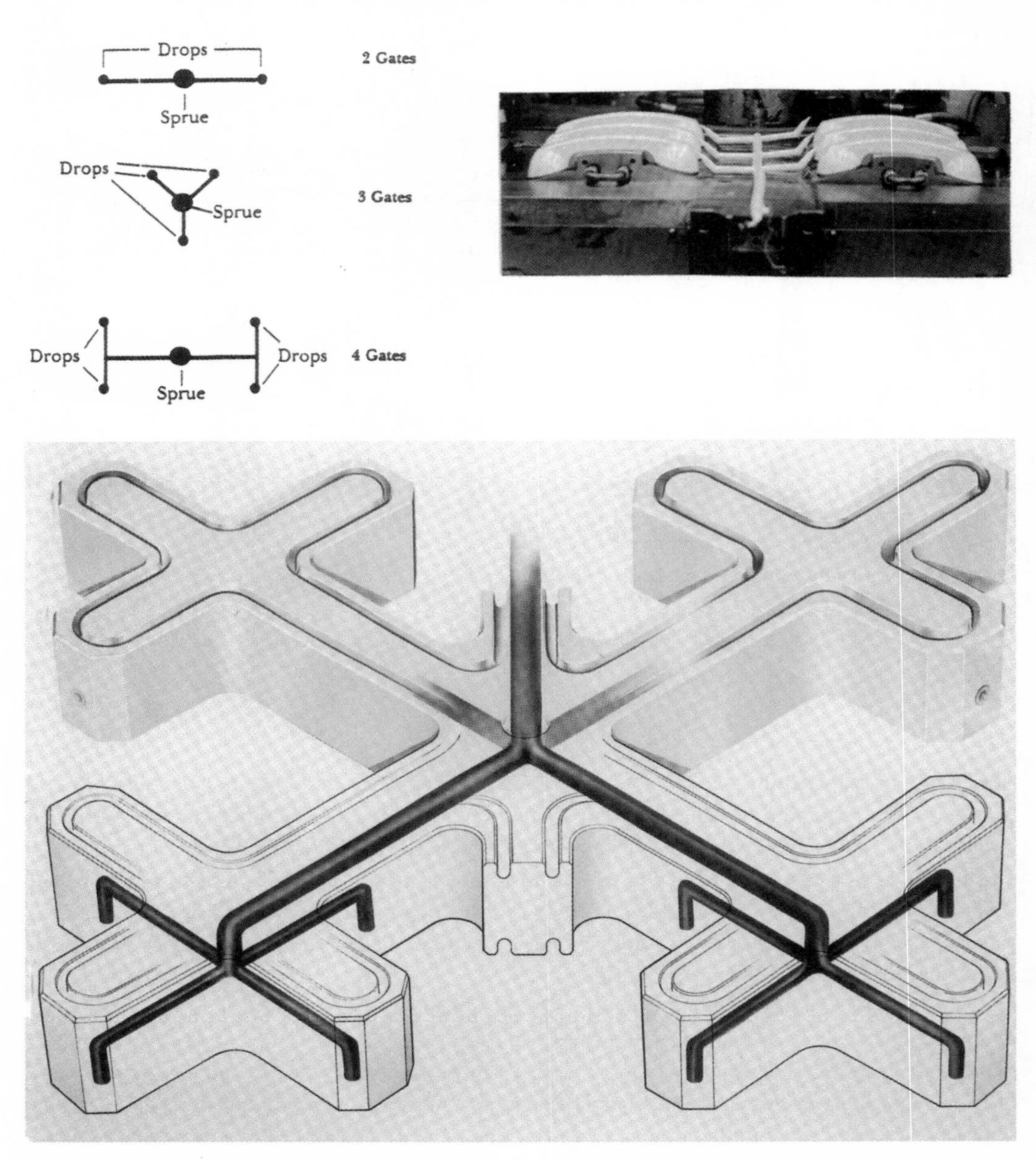

Figure 17.29 Examples of TP balanced cold runners that include primary and secondary runners.

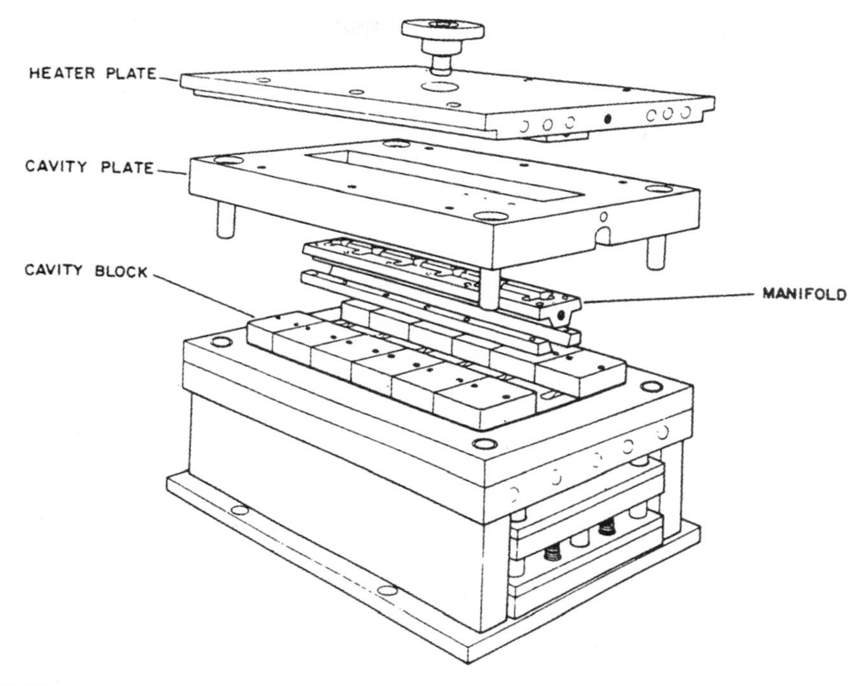

Figure 17.30 Example of a cold runner mold for processing TS plastics.

GATE

The gate is the region of the feed system between the runner and the cavity (Fig. 17.31). It is the entry point by which the melted plastic enters the cavity. Its position and dimensions have a considerable influence on the properties of the finished molding. The position of the gate or gates directly influences flow paths in the cavity, and so has a major bearing on issues such as filling pressure, weld-line quality, and gas traps.

Gate positions must be judged individually for each molding case. An experienced person can usually assess the best gate position for relatively simple part geometry, but for complex parts, computer-aided flow analysis is preferred for gate positioning. Gate positions should be chosen with the following guidelines:

1. Gates should be placed near a thick section to ensure that it can be packed out.
2. Position the gate where the gate scar or witness mark will not be cosmetically objectionable.
3. Position the gate where it can be easily removed with cutting tools.

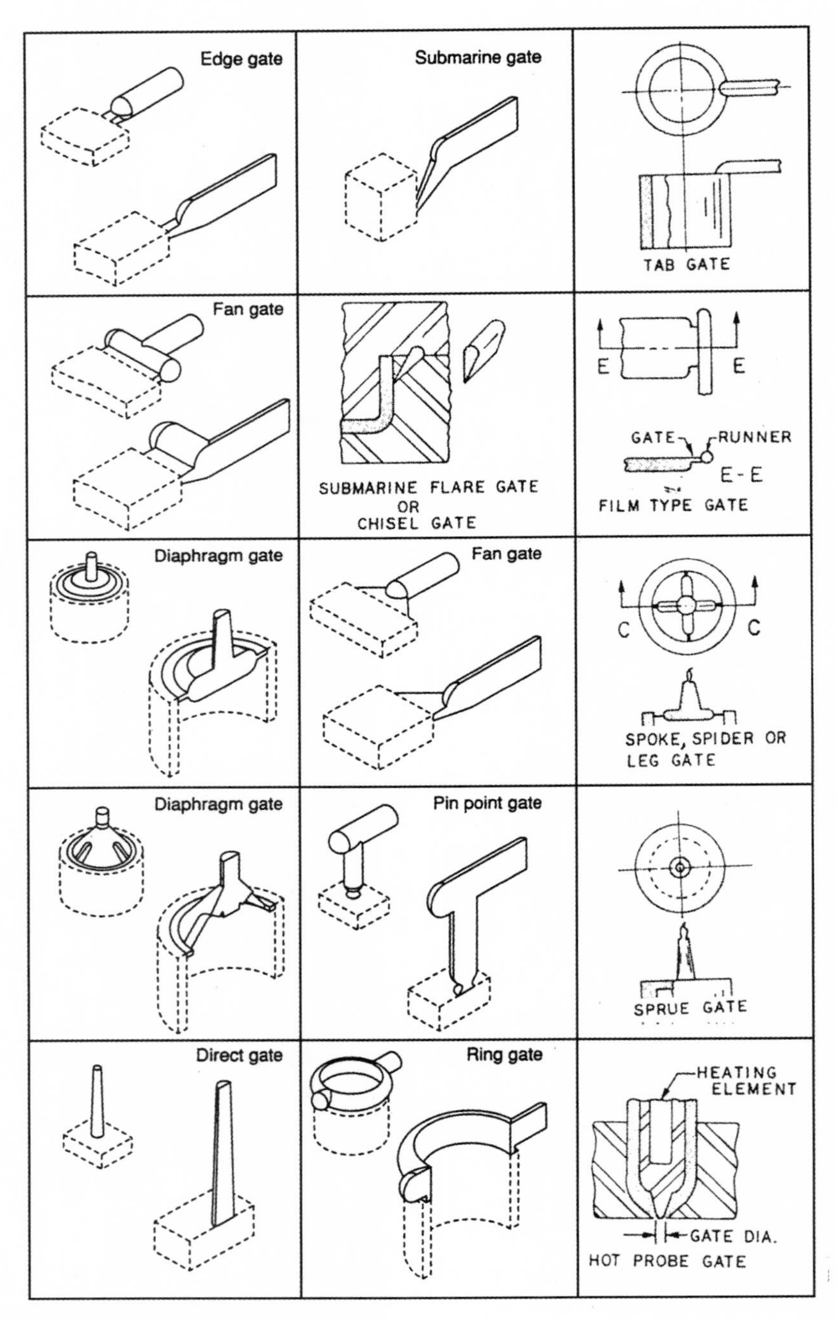

Figure 17.31 Examples of various gate types.

4. Position the gate for flow symmetry in symmetrical parts.
5. The gate should minimize gas traps and weld lines.
6. Place the gate so that unbalanced flow does not take place around cores.
7. Do not place the gate at a point where high stresses or high packing could cause problems.
8. Position the gate so that flow impinges on a mold surface rather than jets into a void.
9. Position the gate to minimize shrinkage differentials.

The gate is usually small in relation both to the molding and the upstream melt-feed system. There are two principal reasons for this approach. The gate acts as a thermal valve that seals the filled mold cavity from the feed system. When the gate freezes, no more flow can take place. A very large gate would be slow to freeze. It would possibly allow compressed melt to flow back out of the cavity and into the feed system, so one aim of gate design is to find a size that will remain open during the injection-packing phase and freeze off immediately thereafter (chapter 4). The other reason for a small gate is so that the feed system can be easily removed from the molding, leaving little trace, if any, of its presence.

Flow conditions in the gate are extremely severe (Fig. 17.32). The melt is accelerated to a high velocity and is subjected to a high shear rate. Figure 17.33 relates gate temperature/pressure/temperature relationships for amorphous and crystalline plastics, where

T2 = Gate temperature for crystalline plastics,
T2 = Gate temperature for amorphous plastics,
T1 = Hot runner (processing temp),
TC = Crystalline melting temperature,
TG = Amorphous glass transition temperature,
T3 (C) = Mold temperature (crystalline),
T3 (A) = Mold temperature (amorphous),
ΔTN (C) = Gate temperature elevation (crystalline),

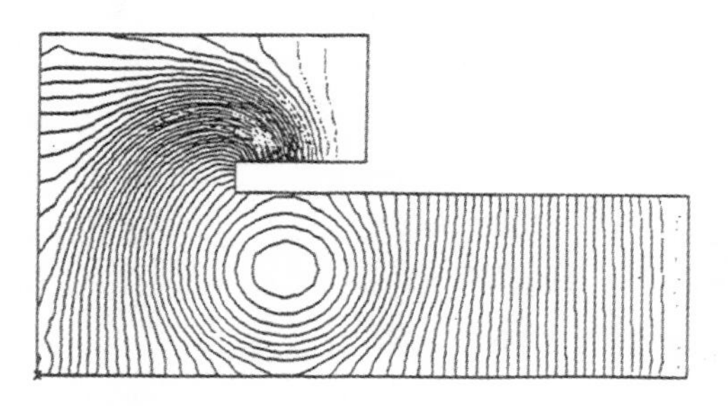
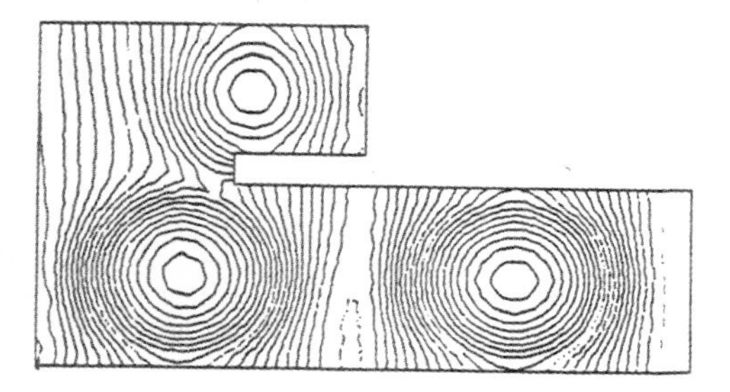

Figure 17.32 Melt flow pattern in cavity can relate to gate-flow pattern based on single gate (left) or multiple gates.

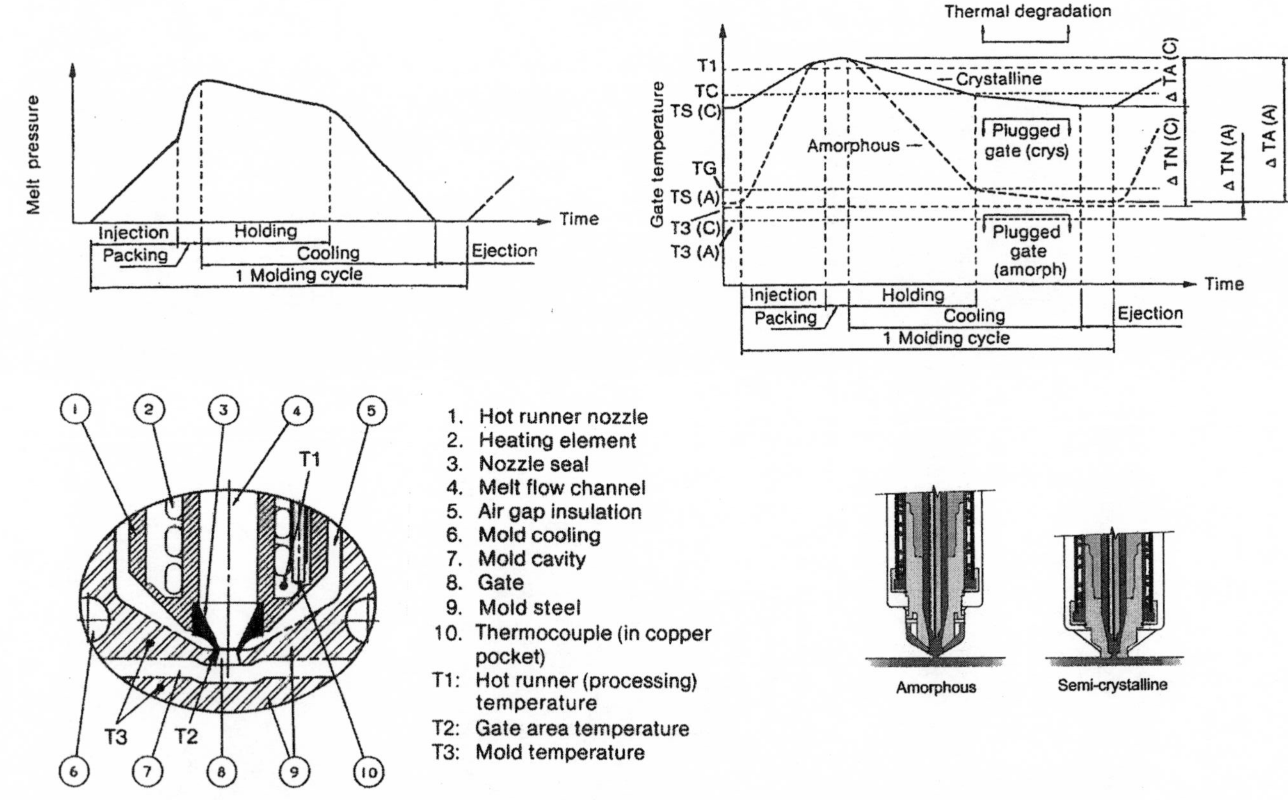

Figure 17.33 Gate temperature/pressure/temperature relationships for amorphous and crystalline plastics are shown.

ΔTN (A) = Gate temperature elevation (amorphous),
ΔTA (C) = Gate temperature addition (crystalline),
ΔTA (A) = Gate temperature addition (amorphous),
SST (A) = Steady-state gate temperature (amorphous).

The melt's high velocity with its high shear rate requires that gates be kept as short as possible (Fig. 17.34). The gate length is often referred to as the land length. If the gate land is very short, there will be a weak section in the mold between the runner and cavity, and there will be insufficient clearance to use cutters for gate removal. During the passage of the melt through the gate, frictional heating is likely to occur and indeed is sometimes exploited in the design of feed systems for heat-sensitive materials. Here the approach is to keep barrel temperatures low and generate additional heat at the last moment in the feed system.

Many different gate types (Fig. 17.30) have been developed to handle the various molding situations. Some of these are associated with particular geometries. The most common gate type used is the edge gate. Diaphragm and ring gates are usually used for parts with a cylindrical form. The pinpoint gate is normally used in three-plate molding. The edge gate is usually square or rectangular in cross section and is cut in just one parting face of the mold. Circular section gates are also used but must be cut equally in both parting faces. Semicircular section gates are not recommended.

Gate dimensions depend on product geometry and are frequently adjusted by trial and error during the testing of a new mold. The only practical way to do this is to start with a small gate and gradually enlarge it. As a rough rule, the gate size should be about half the maximum wall thickness of the molding, and should not be less than 0.030 in or 0.75 mm. This dimension is the diameter of a full-round gate or the inscribed circle of a square gate.

Running the sprue directly into single-cavity molds eliminates use of runners (Figs. 17.13 and 17.17). This is known as sprue gating or direct sprue gating. The gate must be removed in a subsequent machining operation that leaves a witness mark. The sprue gate is perhaps the least sophisticated way of dealing with this type of molding. Fan gates are preferred for thin-walled parts

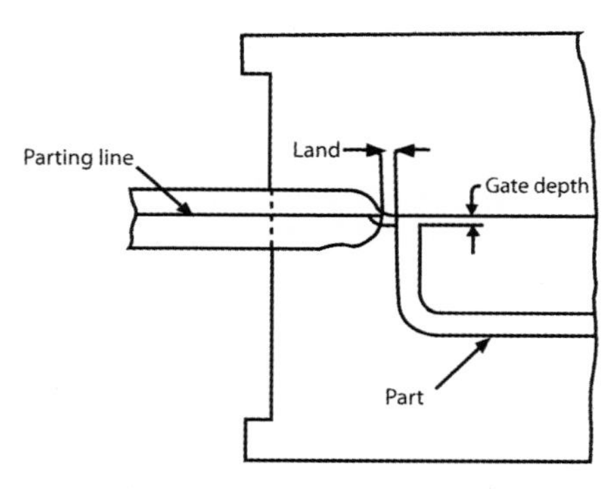

Figure 17.34 Schematic of gate land location.

of a relatively large area. The flash gate, a variant of the fan gate, is used for thin-walled parts that would be difficult to fill from individual gates at any point. The flash gate is very wide and shallow so that a large flow area is combined with a short freeze time. Ring and diaphragm gates are similar in concept to the flash gate. They are normally used to obtain cylindrical parts free of weld lines and core shift caused by unbalanced flow.

The submarine gate, also known as the tunnel gate, is cut into the one half of the mold rather than on a parting face. This design has an undercut that is freed on ejection by the runner and gate flexing. The advantage of the submarine gate is that the feed system is automatically separated from the mold by the act of ejection that shears the gate off. For plastics that are not very rigid, the inclination of the submarine gate is not critical; angles of 30° to 40° will be suitable. The so-called "cashew" gate is a variant in which the gate is machined in a curved form and is suitable only for flexible plastics. Another gate type known as the tab gate is used to prevent jetting into an open cavity or when some defect is expected close to the gate. The tab is a small extension on the molding that is fed at right angles to its axis by an edge gate; the tab is later removed from the finished part.

Because of high melt pressure, the area near a gate is highly stressed both by the frictional heat generated at the gate and the high velocities of the flowing material. If the product designers do not suggest gate locations, they should caution the tool designer to keep the gate area away from load-bearing surfaces and to make the gate size such that it will improve the quality of the product. It so happens that the product wall in the gate area develops the minimum tolerance due to the high melt pressure in that area.

If it were possible, molding without gates would yield significantly better products. The important action of the gate is that it opens to let the plastic melt squeeze through and into the cavity. It closes once the cavity is properly filled. It must not only permit enough material to enter and fill the cavity, but also must remain open long enough to allow packing in extra plastic to accommodate shrinkage. For example, nylon 6/6 has a volume contraction of about 8%. Therefore, in order to control product weight and tolerances, extra plastic must pass through the gate before it solidifies and closes.

As reviewed, many gating methods that meet different requirements are available, and they all have their advantages and disadvantages. The positions and dimensions of gates are critical. At times, the gates must be modified after initial trials with the gate design used to fabricate quality parts. The gate must be located in such a way that rapid and uniform mold filling is ensured. Thus the location of the gate and the type used must be given careful consideration

For example, feeding melt into the center of one side of a long narrow molding can result in distortion; the molding being distorted is concave to the feed. In a multicavity mold, sometimes the cavities closest to the sprue fill first and the farther cavities later. This condition can result in sink marks or shorts in the outer cavities. It is corrected by increasing the size of some gates so that the simultaneous filling of all cavities will result. The gate usually is located at the thickest part of the molding, preferably at a location where the function and appearance of the molding are not impaired.

Gate size has a tremendous effect on the success or failure of attempts to produce high-quality parts economically. The cooler the melt, the more viscous it becomes; the more viscous it becomes, the more difficult it is to move the melt through very small gate orifices. High injection pressure is then needed to move the melt through the gate. The higher pressure, the smaller the total area of the mold must be, otherwise high pressure can result in flash. Another approach to reducing melt viscosity is to raise the melt temperature from the plasticator or to use heated gates (Figs. 17.35 and 17.36). However, an increase in cycle time occurs due to the longer cooling time.

Gate size is usually the critical factor that dictates the final mold-filling speed. During the mold filling, plastics show a certain degree of molecular orientation in the flow direction of the melt that

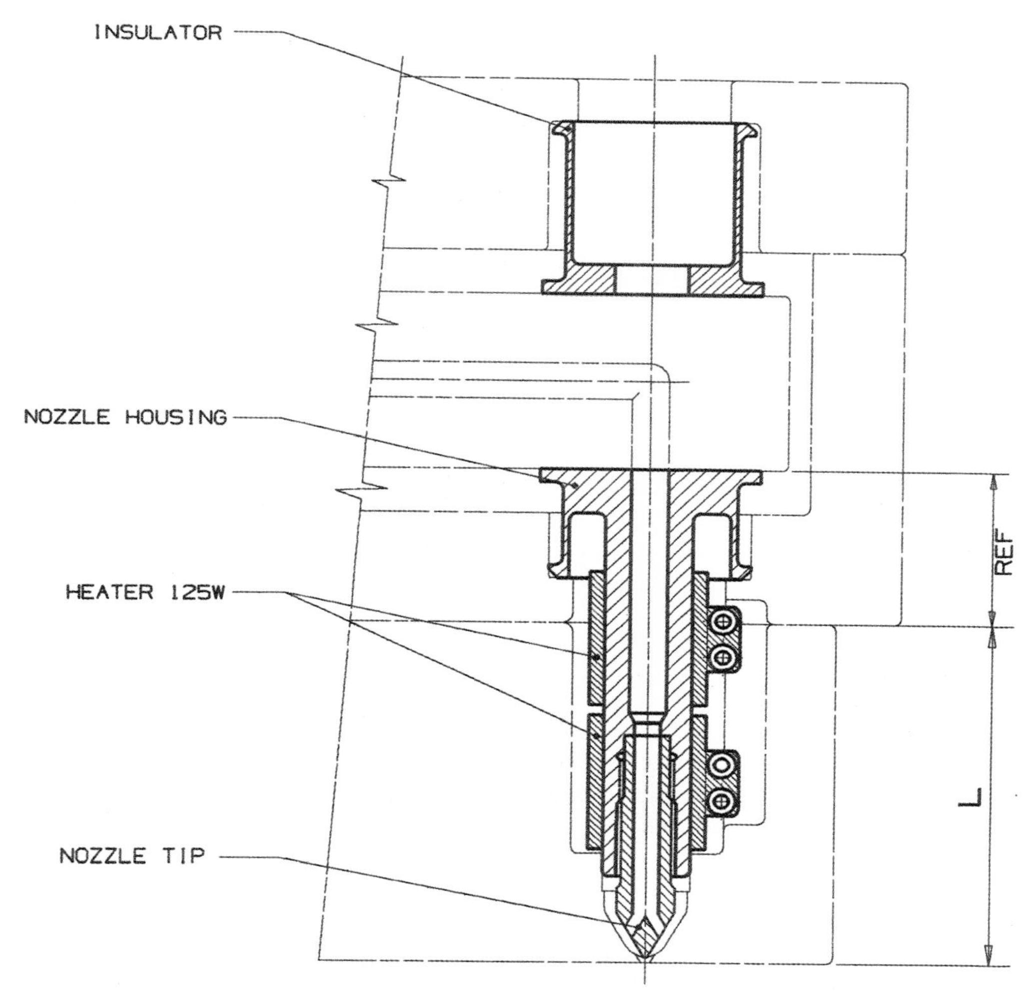

Figure 17.35 Schematic of heated single-edge gate.

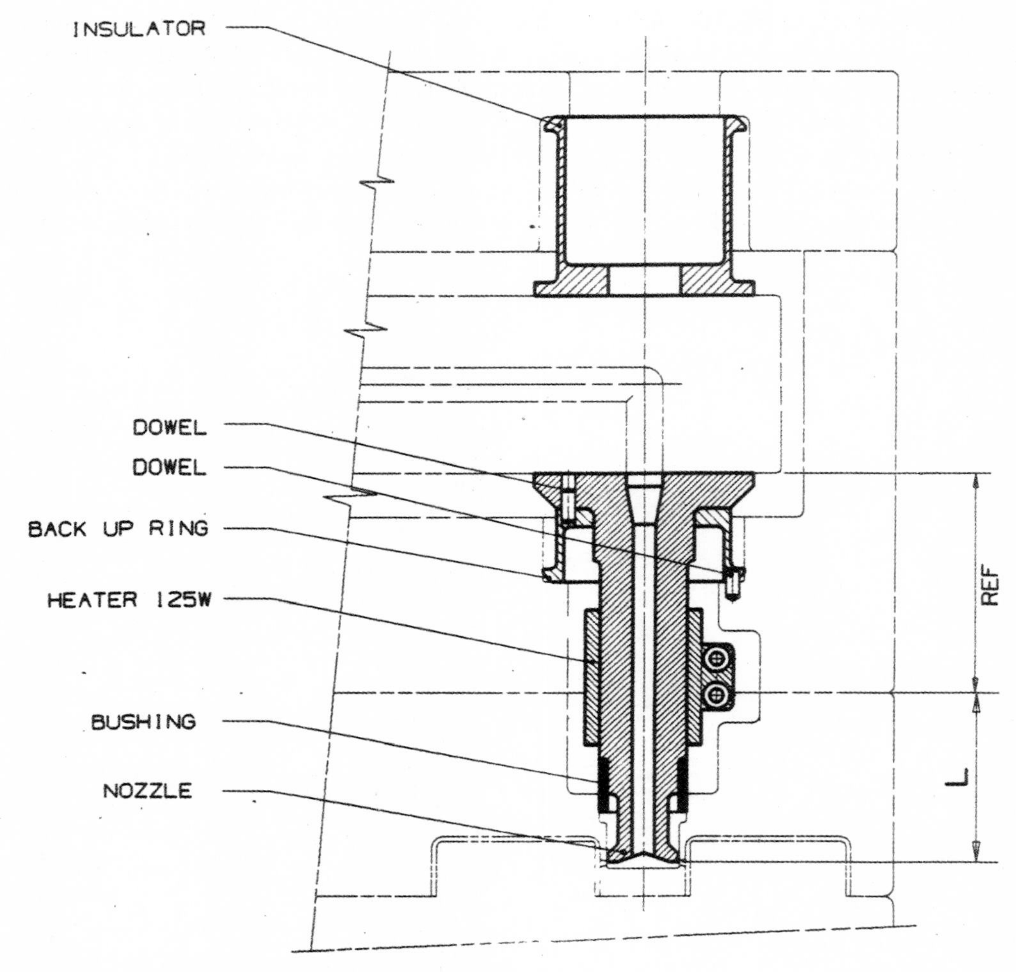

Figure 17.36 Schematic of heated double-edge gate.

affects the properties of the molding. Important factors in this respect are the location and type of gate (Fig. 17.37). Enlarging the gate or locating the gate in such a way that the flow is directed against a cavity wall can prevent jetting.

WELD LINE

For gate type and location, the points where two plastic flow faces meet must also be taken into consideration, as shown in the center diagram in Figure 17.37. If flow comes to a standstill in these places, which may be the case for flow around a core, premature cooling of the interfaces may cause

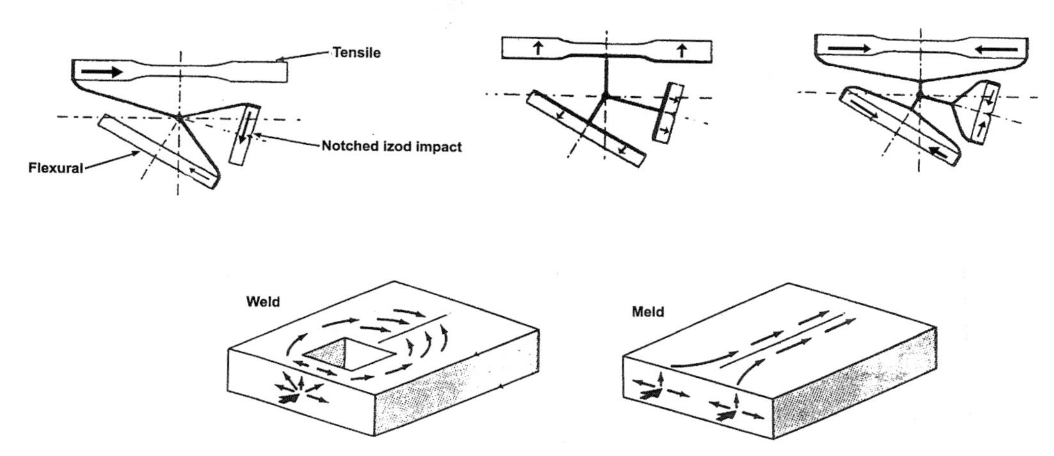

Figure 17.37 These molded test specimens highlight melt flow direction from a gate or gates.

weak weld lines. Although in practice sufficient strength (possibly up to 90% of strength) may be obtained in such cases by good mold venting, high injection speed, and proper plastic and mold temperatures. The main core weld line can only be eliminated entirely if a ring gate can be used. Partial improvement is provided by a design in which the weld line has been shifted to a removable tab on the molding.

When possible the location of the gate must be such that weld lines are avoided. Weld lines reduce the strength and spoil the appearance of the molding, particularly in glass-fiber RPs. If both flow fronts are cool, particularly with RPs, strength could be at or near zero. Under the most ideal processing conditions, the RPs could reach about a maximum strength of 85% compared to a non-welded part. The gate must be so located that the air present in the mold cavity can escape during injection. If this requirement is not fulfilled, either short or burnt spots on the molding and reduced strength will result.

Weld lines may also be formed at places where the plastic flow slows down, such as a place where wall thickness increases suddenly. In grid-shaped articles, weld lines are mostly inevitable. With correct gate location, the plastic flows may be arranged so as to meet on an intersection, in which case the plastic continues to flow, so that better strength is obtained than if the weld line were situated on a bar between two intersections.

DESIGN ASPECT

As reviewed, gates should always be made small at the start; they can easily be made larger but cannot so easily be reduced in size. Since the pressure drop in a system is proportional to the length of the channel, the land length of the gate should be as short as possible (land length is the area of those faces of a closed mold that come into contact with one another). The land-length strength of the metal may be a limiting factor, as may its machining method. On average, 0.10 to 0.15 cm (0.040 to 0.060 in) is a suitable length. The gate cross-sectional area for thin-walled products generally has a width and height of 50% to 100% of the runner cross-section. An example of a gate for thicker walls is shown in Figure 17.33. Based on the plastic shear rate and volumetric flow rate, equations are available for determining gate sizes of different shapes.

Gate dimensions can be estimated. The depth of the gate is 40% to 90% of the nominal wall thickness. The width of the gate is typically three times the depth but it can be made wider to assure that the maximum permissible shear rate of the melt is not exceeded. The gate length is taken equal to the gate depth. A longer gate length is more likely to cause jetting. Guidelines for shear rates are given in Table 17.46.

For round gates, the following equation can be used:

$$d_2 = d_1 \, (W_2/W_1)^{1/4}.$$

For rectangular gates (assume gate width is constant), the equation is

$$t_2 = t_1 \, (W_2/W_1)^{1/3},$$

where

d_1 = gate diameter of the first cavity (in/cm),
d_2 = gate diameter of the second cavity (in/cm),
t_1 = depth of gate in first cavity (in/cm),
t_2 = depth of gate in second cavity (in/cm),
W_1 = weight of first cavity component (oz/g), and
W_2 = weight of second cavity component (oz/g).

HOT RUNNER GATE

There are different hot runner gates that meet different requirements depending on plastic selection and the product design. The hot-tip design is the most common thermal gate style. It places a heated probe at the gate, supplying sufficient heat to keep the cold slug close to melt temperature and to remelt it prior to injection. On large parts the thermal gate delivers the plastic to the vicinity of the part and usually leaves a cold sprue. Thermal edge gating allows gating on the side of a part, similar to a tunnel or submarine cold runner gate. This type of gate shears itself off, leaving only a small mark.

Hot valve gating uses a valve stem to produce mechanical shutoff at the gate, as opposed to thermal gating, where the gate closes due to the adjustment of the gate tip heater and the cooling water flowing close to the outside of the tip. Valve gates may be activated by different techniques, such as hydraulics (with pneumatics) and high-temperature steel springs. With valve gating, the gate

Material	Description	Max allowable shear rate 1/sec
ABS	Acrylonitrile Butadiene Styrene	40,000
EVA	Ethylene Vinyl Acetate	30,000
PS	Polystyrene	40,000
PE	Polyethylene	40,000
PA6	Nylon 6	60,000
PBT	Polybutylene Terephthalate	50,000
PC	Polycarbonate	40,000
PES	Polyethersulphone	50,000
PMMA	Polymethyl Methacrylate	40,000
POM	Acetal	20,000
PPS	Polyphenylene Sulphide	50,000
PP	Polypropylene	100,000
PVC	Polyvinyl Chloride	20,000
SAN	Styrene Acrylonitrile	40,000
TPO	Thermoplastic Olefin	40,000

Table 17.46 Guidelines for melt shear rates (courtesy of Synventive Molding Solutions)

size is normally larger and allows easier fill, creates less molded-in stress, allows for quick color changes, and is less likely to plug.

In terms of gate freeze, TPs are divided into amorphous and semicrystalline types. They have differences in molecular structure that vary the solidification of plastics considerably (chapter 1). The amorphous materials, with their wider processing temperature range, take a longer time to cool down and to seal the gate. For these materials, thermal gating works well for small gates but for others takes a long time and gives poor gate seal. For that reason, one usually goes to valve gates for large gate diameters that are dictated by the shot size and the permissible shear rate through the gate.

The crystalline plastics have a very narrow processing range. Gate freeze is very quick and works even for large gates so thermal gating is easier to operate.

Amorphous materials do not crystallize during solidification under any processing conditions. The change in a crystalline plastic from a melt to a solid phase is rather sudden and easily discernible. In an amorphous plastic, the phase change is not so readily apparent, as the material remains in a softened state over a wide temperature range.

The temperature window available for processing crystalline TPs is much narrower than for amorphous materials. This can be calculated from Table 17.47, where the various IM parameters of amorphous and crystalline plastics are compared, including mold, average melting, and processing temperatures without and with glass fibers.

The range below the processing temperature (PT) at which the plastic remains a liquid is determined by subtracting the average melting temperature from the hot runner processing temperature. For example, let PT = hot processing temperature or the average melting temperature. Then, for

Thermoplastic short forms	Average melting temperature	Mold temperature	Hot-runner process temp.	% Glass fiber weight	Mold temperature	Hot-runner process temp.
PE	140	25	250	30	40	230
PP	170	35	255	30	40	245
PS	100	45	275	30	65	245
SAN	115	80	255	30	90	260
ABS	110	75	250	30	90	260
PMMA	100	70	245			
POM	181	100	200	30	105	210
CA	227	75	235			
CAB	140	55	215			
CAP	190	65	225			
PETP	225	140	280	30	140	285
PBTP	225	35	255	30	90	243
PC	150	90	300	40	120	310
PA 6	220	90	250	40	110	280
PA 6/6	255	90	285	30	110	300
PA 6/10	215	90	250			
PA 11	175	60	230			
PA 12	175	60	230			
PPO	120	80	300	30	105	325
PVC	100	35	195			
PUR	160	35	205	30	50	215
PSU	200	150	315	30	160	385
PPS	290	110	330	40	120	315
PES	230	150	350	40	150	360
FEP	275	150	315	30	230	255
PAI	300	230	365			
PEI	215	100	370	30	150	420
PEEK	334	160	370	30	180	380
LCP	330	175	400	30	180	400

Table 17.47 TP melt temperatures (°C)

amorphous acrylonitrile butadiene styrene (ABS), PT = 250°C – 110°C = 140°C; SAN = 140°C, and PSU = 115°C. The crystalline PA 6 = 30°C, POM = 10°C, and PPS = 40°C.

This temperature difference can be used to determine the style of gate, as it affects the rate of heat transfer required to optimize filling conditions under the shortest possible cycle time. Opening or closing the gate is, one way or another, thermally controlled. This includes mechanical shutoff gating, or valve gating, which is successful only because heat is transferred out of the pin, lowering the gate temperature.

The thermal control of gate solidification can become difficult and is definitely time dependent. Figure 17.33 shows that the greatest upward pressure on the temperature occurs in the gate area identified as T2 (the gate's nozzle section). This section is electrically heated and controlled, with its temperature set at the processing temperature. The mold cavity's walls are set at a lower temperature (T3) and must not be affected by the heated nozzle, but thermally controlled by means of sufficient mold cooling. All this action highlights the importance of precision temperature control. In the diagram, consider the gate area to be in a state of thermal equilibrium, with no flow through the gate. In this example, T2 is the steady-state temperature. Maintaining the steady-state temperature at a specific level provides a constant flow of heat from the nozzle to the mold-cooling channel. It is the function of mold-cooling action to control the rate of heat transfer from not only the plastic, but also the hot runner nozzle.

In the steady-state condition, the nozzle is the only heat source to the gate area that elevates the steady-state temperature above the mold temperature T3. The thermal gradient between two locations in this gate area can is expressed by the equation

$$\Delta T/L = Q/KA,$$

where

Q = the rate of heat flow,
L = the length of the heat-flow path,
K = the thermal conductivity, and
A = the cross-sectional area.

With the gate in a steady-state condition, Q, L, and the gate diameter are constant. Therefore, the thermal gradient between the gate the steady-state temperature and gate nozzle T1 is a function of maximizing thermal separation, where the mold-to-nozzle contact area A must be minimized. Also required is the thermal conductivity of the nozzle seal and the nozzle tip. For a large thermal gradient, the thermal conductivity K of the seal or tip must be low. The gate material should have a high K (thermal conductivity) to give adequate heat flow from the material in the gate. This results in short cycle times.

In this example rheological influences destroy thermal equilibrium as plastic begins to flow. By forcing melt through the gate, its velocity increases, which causes a corresponding rise in both shear rate and kinetic energy. With a smaller gate, the increase in shear is greater. Some of this kinetic energy is converted into heat, which raises the local gate-area temperature T2. The T2 also increases because of contact with the hot melt flowing from the nozzle runner channel. Thus the temperature rise is a function of flow rate, velocity, and the diameter of the gate. A rise in gate temperature T2 by an amount TA occurs when these two transient rheological conditions are created. Total increase in the gate temperature occurring during injection must not place T2 above the point at which thermal degradation could occur. The temperature must not drop below the point at which the gate becomes plugged (when melt solidification occurs) and the normal injection pressures cannot easily remove the plug with the next shot.

%op>Temperature-versus-time curves show that different gating techniques are required to process amorphous and crystalline plastics (ΔT >> ΔTN[A]). A transfer of heat takes place between the hot runner nozzle of the gate and the cooling mold. The result is establishment of an elevated steady-state gate temperature (SST = T3 + ATN). The hot runner nozzle end must supply more heat to the gate area for crystalline types than for amorphous types, giving crystalline materials much higher steady-state gate temperatures (SST[C] >> SST[A]). The curves also show that ΔTA(C) << ΔTA(A). Thus additional heat is added to the gate via rheological influences during injection, raising T2 by ΔTA.

It is essential that gate cooling for amorphous plastics be highly efficient in order to dissipate the high heat generated during injection. With poor cooling, cycle times may be unacceptably lengthy. In fact it is possible that no melt solidification will occur in the gate, resulting in stringing or drooling. However, if cooling in the gate area is too powerful for a crystalline plastic, it is possible that the gate will freeze off prematurely, resulting in short shots and inadequate packing.

The size of the gate is also an important consideration. Small gates generate more heat, solidify more quickly, and are easier to degate. This is an advantage in processing amorphous plastic because of the low ΔTN and high ΔTA required during injection. However, the required high ΔTN and low ΔTA necessitate a larger gate diameter for crystalline plastics.

CAVITY

The mold may consist of a single cavity or a number of similar or dissimilar cavities, each connected to flow channels (runners) that direct the flow of the melted plastic to the individual cavities (Fig. 17.38). Melt is forced into the mold cavity. During the initial mold-filling phase of the molding cycle, very high injection pressure may be needed in order to maintain the desired mold-filling speed. There are melt flow analyses that provide detailed information concerning the influence of mold-filling conditions on the distribution of melt flow patterns that basically relates to a fountain flow pattern (Fig. 17.39; chapter 4).

The object in filling the cavity is to achieve complete filling without short shots while avoiding sink marks, warpage, sticking in the mold, flash, and poor mechanical properties. This is accomplished by delivering the correct amount of melt to the cavity while avoiding overpressurization, high thermal stresses, and high residual orientation. Some of the factors that favor complete filling, however, also promote overpressurization and residual stresses, so care must be taken in selecting operating conditions for a given mold and plastic. Within differently shaped cavities, the direction of melt flow can vary and influence shrinkage as well as strength properties. Properly positioning the cavity in the mold is very important because it allows cooling lines to operate efficiently (a process that will be reviewed in this book). Cavities should be positioned as central and symmetrical around the mold centerline and within the mold base (448).

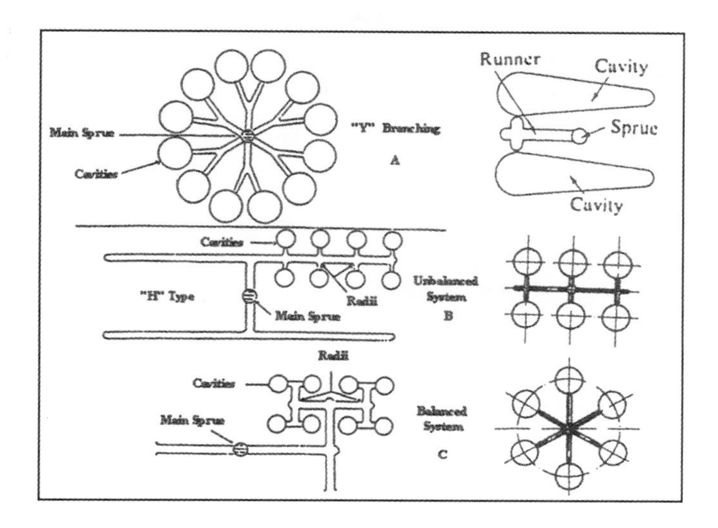

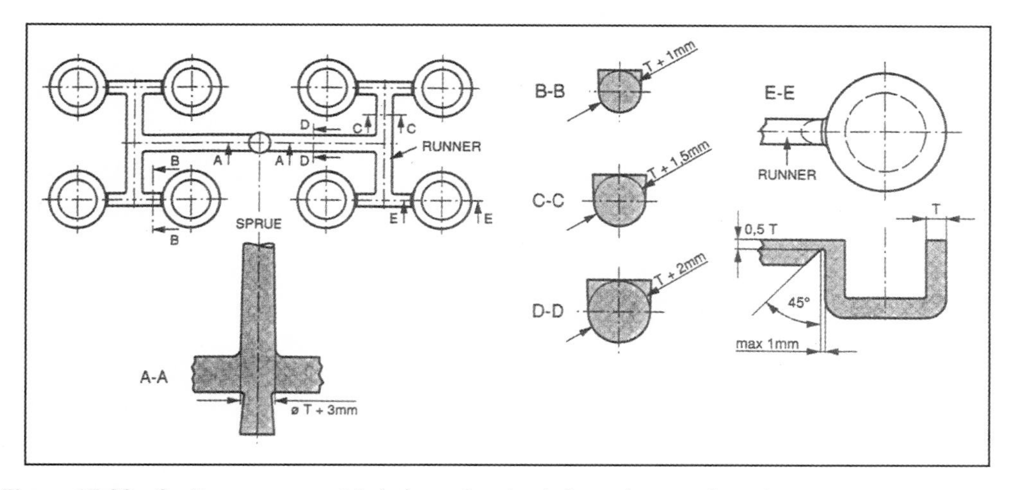

Figure 17.38 Cavity arrangement in balanced and unbalanced runner layouts.

CAVITY SURFACE

A significant advantage of the molding process is that different surfaces and combinations of surfaces can be molded into the product: high-gloss polishes, dull finishes, matte finishes, textures, and engravings. Different techniques are used to develop these surfaces. The surface of the mold cavity reproduces its surface condition on a molded product. No secondary surface-finishing operations are required unless special finishes are required such as plating, hot stamping, and so on. Another popular technique is in-mold labeling.

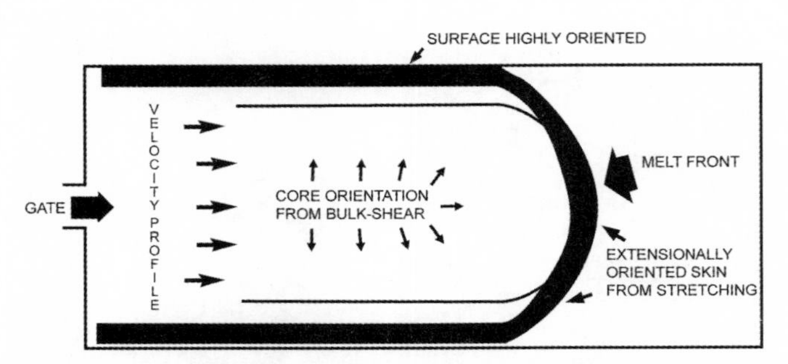

Figure 17.39 Example of a melt flow fountain (or balloon) pattern across the thickness in a mold cavity.

The important operation of polishing the cavity can account for up to 30% of the mold's construction cost. The specification of the surface finish must be carefully considered. The SPI sets surface finish standards, with A-1 being the smoothest and D-3 being the roughest.

Molded products can have features that must be cut into the surface of the mold cavity perpendicular to the mold parting line. To properly release the product from the tool, products almost always include a taper; the usual amount to start with is a 1/20 taper. The taper is also called draft in the direction of the mold.

The amount of mold taper required will depend on factors such as the type of plastic being processed, processing conditions, surface finish, and so on. For example, a highly polished surface will require less taper than an unpolished mold. Any cavity surface texture will increase the draft at least 1° per side for every 0.003 cm (0.001 in) depth of texture. Special mold cavity surface action can be used, such as on the cavity side when the part that is to be ejected is moved slightly parallel to the wall direction. With flexible or elastomeric material, it is possible to mold with zero draft. The part unites with its rubbery condition and does not need the required draft for ejection.

The cutting of a pattern or etching (usually electrochemical or by laser machining) on a mold cavity's surface can be reproduced on the molded product. Tool texturing can be used as a method to minimize the effect of flow lines, sink marks, and other flaws or functional needs on products. However, they are an integral part of the fabricated product design. Since the texturing used can influence the type of cavity material (type steel, beryllium-copper, etc.), it is important in the initial stages of mold design to specify what is to be used so that no problems develop when the surface is to be treated.

Sometimes steel grit or sand is blown onto the wall cavity to produce a rough surface. This surface treatment may be required to permit air to leave the mold during molding and/or provide a desired surface finish on the part. In honing, use is made of a fine-grained whetstone (or its equivalent) to obtain precise accuracy on the surface finish. Electric discharge-controlled sparking erodes or removes metal to make machine cavities.

The etching technique is also called photoetching, chemical blanking, or chemical machining. It is a controlled etching process that depends on the etching action of an acid or alkali that depends on the type of cavity material that uniformly attacks all exposed areas of the product. Factors that influence results on the different tool steels are grades, annealing, hardening methods, cleanliness, and hardness. Also important is grain-flow direction of the tool steel.

Nitriding must be done after etching. Flame hardening prior to etching should be avoided because the pattern will be etched differently in the flame-hardened zone. Welded steel can usually be etched if the same steel is used in the weld. Poor etching occurs on surfaces marred by residual traces of spark machining, grind material, or polish. Steels with a clean microstructure and low sulfur content give the most accurate and consistent patterns.

A mask or protective coating is used on those surfaces that are not to be etched. Electric fields (voltage) are applied to speed up and control the process to cause the metal to go into solution. A wide variety of molded products can be produced with a pattern or textured surface reproduced from a mold photoetching. The pattern ranges from leather or wood grain to line patterns with varying directions and depths. It is etched to the required depth by the application of an appropriate acid under closely controlled conditions. It can be performed either on complete tools or on any specific area.

The production or reproduction of a product by electrodeposition on a mandrel or mold that is subsequently separated from the pattern identifies the electroforming technique. It can produce small to very large cavities. This mold-cavity form can be supported or strengthened using a backup material, such as RP or a foamed compound, or by spraying low-melting metal alloys. The pattern material can be wax or a flexible material. The molded part usually employs low or moderate pressure.

Because the surfacing process can include many hand-applied techniques, access to the surfaces that are to be decorated is crucial and areas with restricted access should also be discussed in the design stage. Adjustment of the cavity dimensions may be necessary to compensate for the metal removal that occurs during the etching procedure. Knife-edge inserts and cams are of particular concern.

The decorating options available are numerous and elaborate. For example, there is microtexturing, which involves a mechanical abrasion process, commonly referred to as sandblasting, where usually glass or aluminum oxide are impacted against the surface of the tool, leaving a lightly scarred steel surface. The appearances that are available are limited to matte textures and stripes. However, their real value is the ability to reduce gloss levels. Graphic designs such as company logos and written or pictorial information may be etched onto tooling surfaces when mechanical means may not be suitable.

COLD RUNNER

Cold runner feed systems include three principal components: sprue, runner, and gate (Fig. 17.40). The sprue is a tapered bore in line with the axis of the injection unit that conducts the melt to the parting line of the mold. The runner is a channel cut in a parting face of the mold to conduct melt

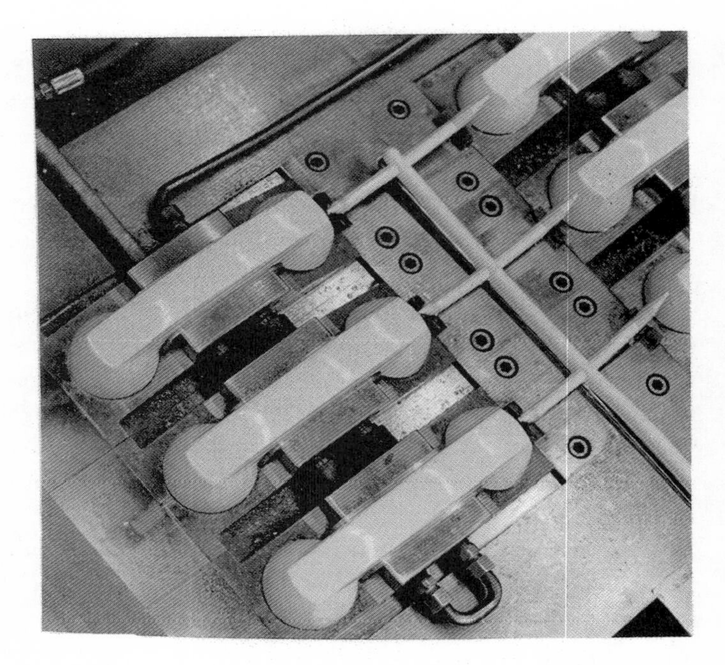

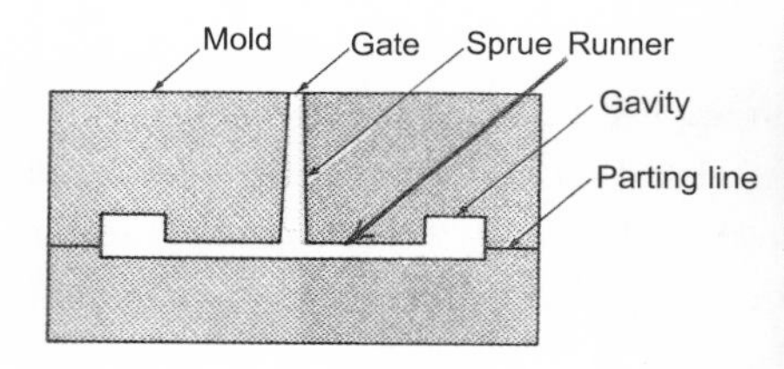

Figure 17.40 Examples of cold runner feed systems.

from the sprue to a point very close to the cavity. The gate is a relatively small and short channel that connects the runner to the cavity. The gate is the entry point of the melt into the molding cavity.

Runners are produced in a variety of cross-sectional configurations (Fig. 17.41), but not all of them perform equally well. The best shape for the runner itself is a full-round section, cut in both halves of the mold (Table 17.48). This is the most efficient form for melt flow to avoid premature cooling. There are some instances when it is desirable to cut the runner only in one half of the mold, either to reduce the machining cost or where it is mechanically necessary over moving splits. In this case, the preferred runner sections are trapezoidal or modified trapezoidal. The half-round runner

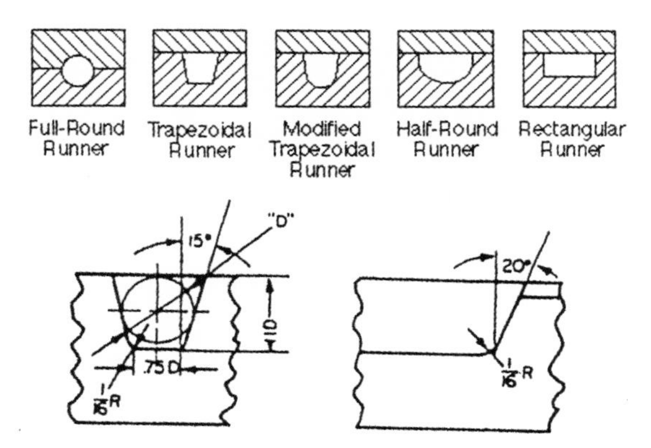

Figure 17.41 Common runner configurations.

Plastic	Diameter (in.)	Metric (mm)
ABS, SAN	0.187–0.375	4.7–9.5
Acetal	0.125–0.375	3.1–9.5
Acrylic	0.312–0.375	7.5–9.5
Cellulosics	0.187–0.375	4.7–9.5
Ionomer	0.093–0.375	2.3–9.5
Nylon	0.062–0.375	1.5–9.5
Polycarbonate	0.187–0.375	4.7–9.5
Polyester	0.187–0.375	4.7–9.5
Polyethylene	0.062–0.375	1.5–9.5
Polypropylene	0.187–0.375	4.7–9.5
PPO	0.250–0.375	6.3–9.5
Polysulfone	0.250–0.375	6.3–9.5
Polystyrene	0.125–0.375	3.1–9.5
PVC	0.125–0.375	3.1–9.5

Table 17.48 Guide to size of round runners

provides only a restricted flow channel combined with a large surface area for cooling and consequently is not recommended.

The concept of the hydraulic diameter (D_H) provides a quantitative means of ranking the flow resistance of the various runner configurations. Hydraulic diameter is calculated from an expression chosen to give a full-round runner a value of ID, where D is the runner diameter. The expression is

$$D_H = 4A/P,$$

where

A = cross-sectional area, and
P = perimeter.

The resulting values for equivalent hydraulic diameter (Fig. 17.42) demonstrate the superiority of the full-round design for runners cut in both halves of the mold and the modified trapezoidal design for runners cut in one half of the mold.

Runner layouts should be designed to deliver the plastics melt at the same time and at the same temperature, pressure, and velocity to each cavity of a multicavity mold. Such a layout is a balanced runner (Figs. 17.38 and 17.43). A balanced runner will usually consume more material than an unbalanced runner, but this disadvantage is outweighed by the improvement in the uniformity and quality of the moldings.

Balance in a multicavity mold with dissimilar cavities that is known as a family mold can be achieved by careful variation of runner diameter in order to produce equal pressure drops in each flow path (Fig. 17.44). Such balancing can only be achieved efficiently by the use of computer flow simulations, and this method really should now be the norm for injection mold design.

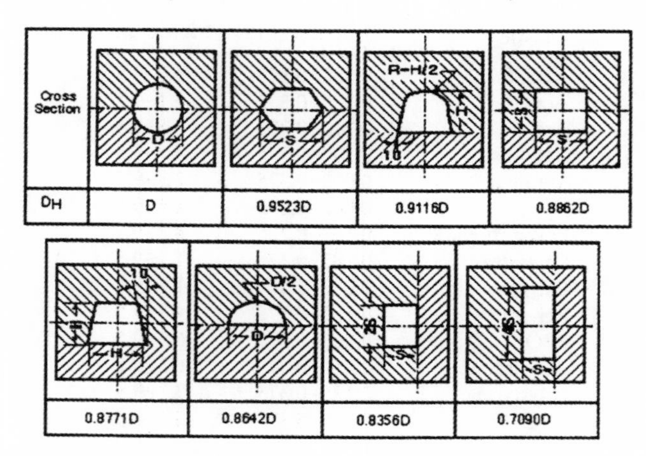

Figure 17.42 Equivalent hydraulic diameters for common runner configurations.

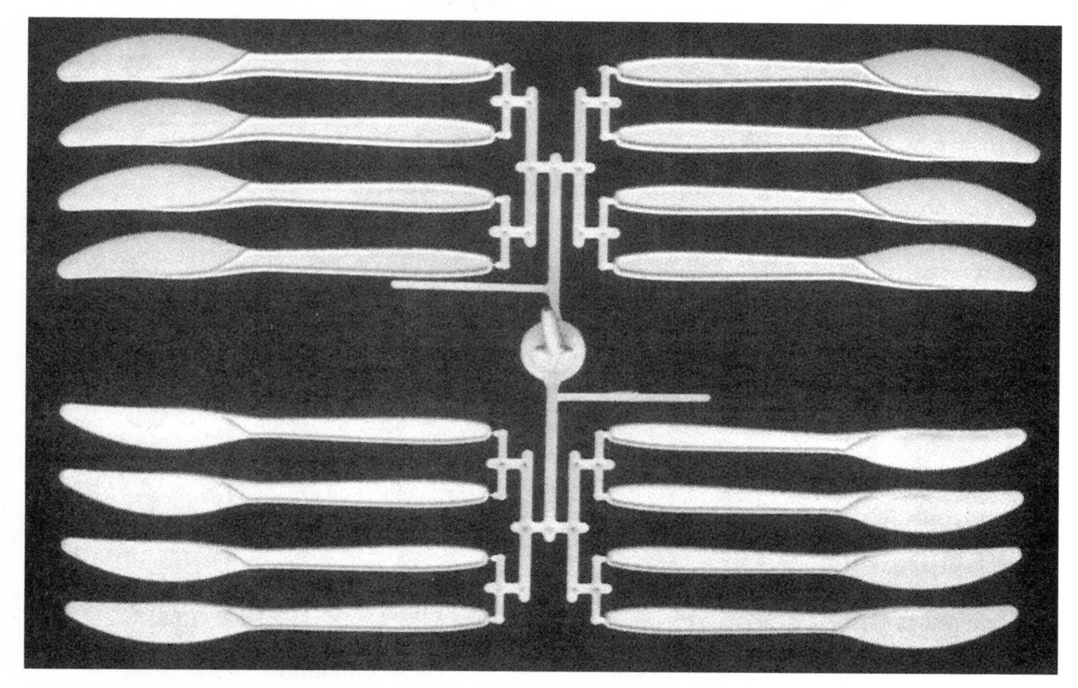

Figure 17.43 Balanced cold runner with edge gates.

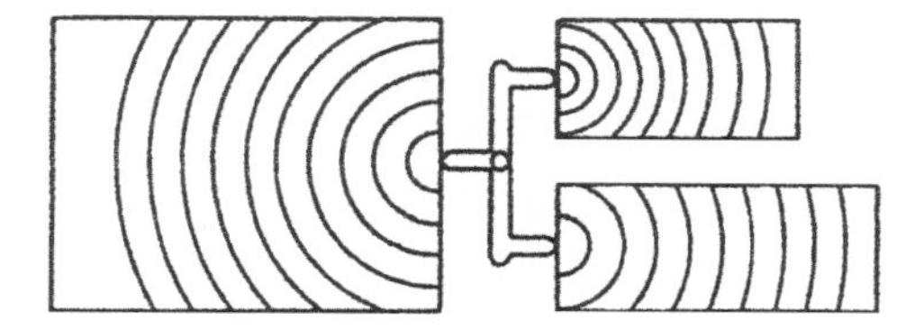

Figure 17.44 Example of dissimilar cavities in a family mold.

If flow-simulation software is used, runner dimensions will be calculated precisely and can then optionally be adjusted to a standard cutter size. If a simulation is not available, the following guidelines suggest suitable runner sizes for use with TPs, such as polypropylene (PP). These are given as a function of wall thickness and may need to be adjusted for parts that are thicker than 4 mm or thinner than 2 mm. In any case, fine-tuning may be necessary during mold trials. The easiest way to do this is to start with small runners and enlarge them if necessary (Fig. 17.45).

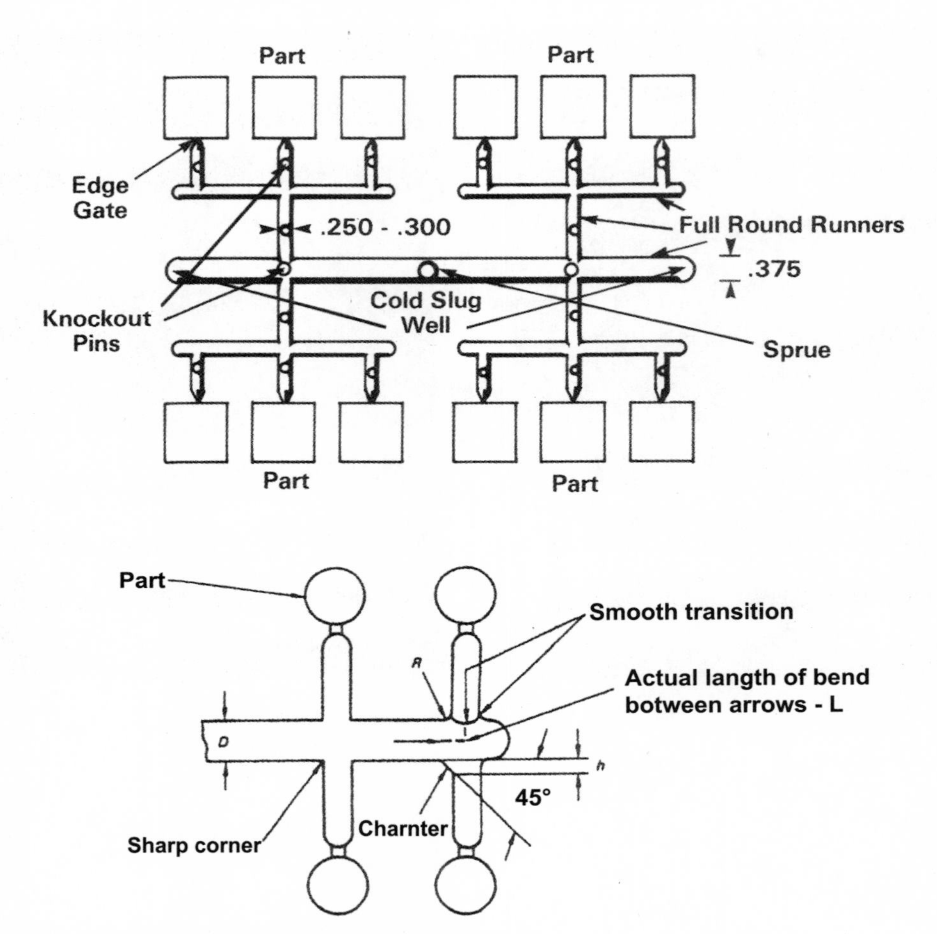

Figure 17.45 Examples of unbalanced cold runner molds.

DEVELOPING COLD RUNNER SIZE

Here is a cold runner design approach without going through detailed calculations that follow. It uses three simple rules of thumb. Take the nominal wall of the part. The runner feeding the cavity is equal to the part thickness. Assuming most runners are laid out in a modified H-pattern layout as one works one's way back toward the nozzle, the flow rate in the previous runner segment will double as one works toward the nozzle. The flow is proportional to the cube of the diameter so that each runner diameter is multiplied by the previous diameter times the cube root of 2. The procedure is repeated until one is finished with the sprue, when the sprue diameter is the average of the opening at the nozzle end and the end feeding the primary runner. The cube root of 2 = 1.26, so the calculation is $D_n = 1.26D_{nh}$. As a rule of thumb, one can tolerate 20% regrind from the sprue and runners, so at the end one calculates to see if this is the case.

There is software for computing the minimum runner size required to convey melt at the proper rate and pressure loss to achieve optimum molded-part quality. As a result, runner design has evolved from pure guesswork into an engineering discipline based on fundamental plastic-flow principles. The computations are based on the plastic's rheological property of the material to be molded (chapter 1). This property is the material's melt viscosity at the shear rate encountered in the runner at the commonly encountered melt temperature for the material. Figure 17.46 shows the melt viscosity of nylon 66 that are typical viscosity data available from plastics manufacturers. In this evaluation, nylon, which is a crystalline TP, is being molded; it also relates to other crystalline TPs though not necessarily for amorphous TPs.

This review provides information on how runner sizes are determined. Since no single calculation will do the job, it is necessary to start with a reasonable runner size, estimated on the basis of prior experience (someone will have it), that can then be refined with the aid of calculations. Start with the product weight and configuration and its performance or appearance requirements.

A study is made of the pressure drop that is controlled primarily by the volumetric flow rate or injection speed, melt viscosity, and channel dimensions. Although it is possible to reduce the melt viscosity by increasing the melt temperature that results in reducing the pressure drop, most IM materials have an ideal melt temperature that provides fast cycles and optimum product quality. Start by using this melt temperature. A resin supplier's molding manual provides this temperature. Follow with an assumption concerning the amount of pressure drop that can be tolerated.

The IMM is usually capable of delivering at least 20000 psi (138 MPa) of pressure. Since common sense rules against designing a mold to demand the absolute pressure limit of the machine, the

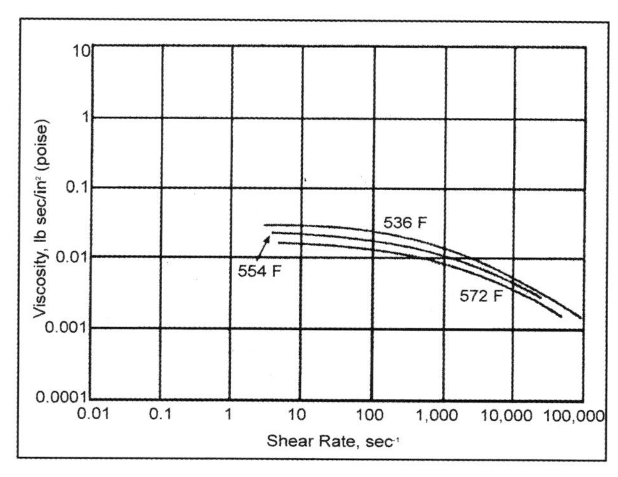

Figure 17.46 Examples of melt viscosity data.

mold should be designed so that the pressure required is somewhat less than the machine's capacity such as 25% (15000 psi [103 MPa]). Unless the product design is unusual—if it calls for long, thin parts, for example—or experience dictates otherwise, a pressure of 5000 psi (34 MPa) is usually adequate to fill and pack out most cavities. This means that the runner system can be designed for a 10000 psi pressure drop that is used in an eight-cavity, balanced runner layout. Figure 17.47 shows the melt flow from the sprue to primary, secondary, and tertiary runners to the eight cavities.

All runners are the full-round type, material specific gravity is 1.0, and product weight is 15 g. For eight-cavity molds, the total amounts to 120 g or 19 cm^3 (7.31 in^3). Lengths of the primary, secondary, and tertiary runners are shown in the figure. A typical fill or injection time for this amount of nylon is 3 s. The aim is to determine the optimum runner diameter. Start by determining the runner volume V that is calculated in the equation

$$V = \pi r^2 L,$$

where

$\pi = 3.141593$,
r = runner radius, and
L = length.

Runner sizes are as follows:

$$\text{Primary runner: } V_p = \pi\,(0.125)^2\,(10) = 0.49 \text{ in}^3,$$

$$\text{Secondary runner: } V_s = \pi\,(0.100)^2\,(12) = 0.38 \text{ in}^3,$$

$$\text{Tertiary runner: } V_t = \pi\,(0.075)^2\,(8) = 0.14 \text{ in}^3.$$

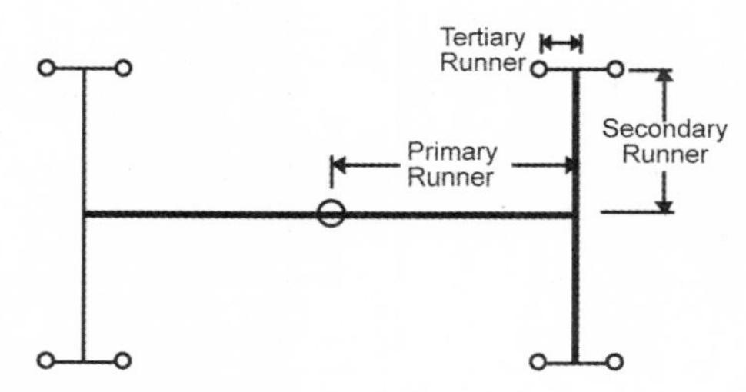

Figure 17.47 Balanced runner system in an eight-cavity mold.

In turn, the total volume of melt required from the IMM nozzle (runner + cavities) is

$$\text{Total IM shot volume} = 7.31 + 0.49 + 0.38 + 0.14 = 8.32 \text{ in}^3 \ (21.13 \text{ cm}^3).$$

With the melt flow splitting evenly at the intersection of the sprue and primary runner, it is only required to calculate the pressure loss through one half of the mold. Thus the volume of melt that is to be conducted through the primary runner in this half of the system is 10.57 cm^3 (4.16 in^3). With the 3 s fill time, the flow rate is 1.39 in^3/s (3.53 cm^3/s). This is the volumetric flow rate Q. The shear rate S_r can be calculated:

$$S_r = 4Q/\pi r^3 = 4(139)/\pi(0.125)^3 = 906 \text{ s}^{-1}.$$

The chart similar to Figure 17.46 provides the means to obtain the melt viscosity at this shear rate and at the specified melt temperature resulting in $\mu = 0.016$ lb-sec/in. Following this calculation, the shear stress S_s is determined:

$$S_s = \mu S_r = (0.016)(906) = 14.5 \text{ psi}.$$

Now the pressure drop P through that runner segment can be calculated:

$$P = S_s(2L)/r = (14.5[2][5])/0.125 = 1160 \text{ psi}.$$

The next runner segment is to be calculated, where the total volumetric flow through each secondary runner is 4.16 in3 minus the volume in the primary runner, resulting in

$$V_p = (4.16 - 0.25)/2 = 1.95 \text{ in}^3.$$

The flow splits in half again at the secondary runner, resulting in the volumetric flow rate in each secondary runner segment of 1.95/3 or 0.65 in3/s. The result is

$$S_r = 4(0.65)/\mu(0.100)^3 = 827 \text{ s}^{-1}.$$

With the melt viscosity at the shear rate of 0.017 poise, the following calculations are made:

$$S_s = (0.017)(827) = 14.0,$$

$$P = (14)(2)(3)/0.100 = 840 \text{ psi}.$$

To determine the volumetric flow through each tertiary runner, the calculation is made by subtracting the volumes of primary and secondary runners, or simply by adding the total tertiary runner volume and total product volume and dividing by eight cavities, or

$$[0.14 + 7.31]/8 = 0.93 \text{ in3 } (2.36 \text{ cm3}).$$

The result is that the volumetric flow rate is 0.93/3 or 0.31 in3/s, and

$$S_r = 4(0.31)/\mu(0.075)^3 = 936^{-1}.$$

Viscosity corresponding to this shear rate is 0.016 poise and

$$S_s = (0.016)(936) = 15.0,$$

$$P = (15)(2)(1)/0.075 = 400 \text{ psi } (2.76 \text{ MPa}).$$

Total pressure loss from the sprue to each gate is the sum of pressure losses through each segment:

$$\text{Pressure loss (total)} = 1160 + 840 + 400 = 2400 \text{ psi } (16.54 \text{ MPa}).$$

The result of these calculations shows that much smaller channels can be designed to accommodate a 10000 psi (69 MPa) pressure loss. Repeating the calculations for progressively smaller runner diameters, the desired pressure loss is obtained. The relationship between the diameters of primary, secondary, and tertiary runners can be arbitrary. However, since each successive stage of the runner system carries less melt than the previous stage, the successive runner diameters run smaller.

It may be necessary to build molds in which the number of cavities is not two, or it is not possible to balance the cavity layout for equal flow distances to all cavities. Although this type of design presents no particular problem in molding products with loose tolerances, the effect on dimensions and product quality must be considered carefully when designing runner systems with critical dimension products. Its runner system requires that all cavities fill at the same rate. This is necessary to ensure that they cool at the same rate and provide uniform shrinkage; surface gloss is not affected.

Molders will frequently try to balance the fill rates of individual cavities by changing the gate size. While this has some possibility, it is ineffective for unbalanced runner layouts. The land length of the gate is too short to make any significant difference in pressure drop from one cavity to another. Best approach is to vary the runner diameters and control fill rate.

Figure 17.48 depicts a large, unbalanced, six-cavity mold, where the sprue is offset from the center of the runner system. Since all the cavities fill at the same rate, what is required is a

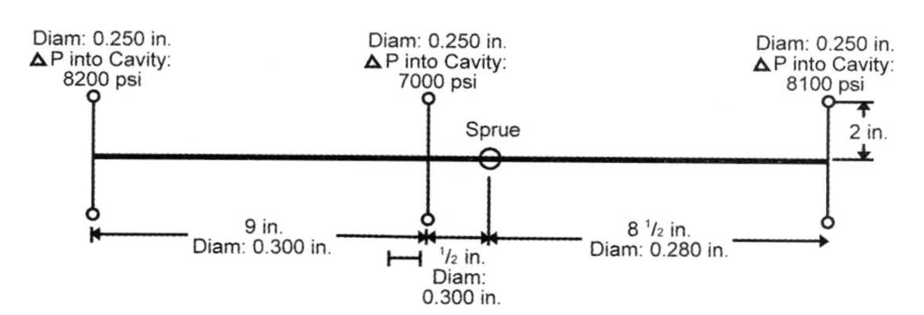

Figure 17.48 Unbalanced runner system in a six-cavity mold.

computation of the runner diameters that will provide the same pressure drop from the sprue bushing to the gate of each cavity.

With the runner lengths being different for each pair of cavities, different runner diameters will be required. Using the previous equation, pressure drop is proportional to runner length. Thus the longer runner segments will need to be slightly wider. Figure 17.48 shows the actual lengths and diameters for each segment of the runner system. The total pressure drop into the various cavities is similar though not identical for each. It is often impractical (and unnecessary) to exactly balance the pressure drop into each cavity.

In this example it was considered impractical to go smaller than 0.3 cm (0.125 in). for the diameter of the secondary runners closest to the sprue in order to raise the pressure drop there to a level closer to that of the other secondary runners. The parts all filled uniformly, despite some degree of difference in the pressure drop leading into the cavities.

The next example is a ten-cavity family mold with different cavity sizes that range from 0.64 cm (0.25 in) to 5 cm (2 in) in diameter and 1.27 cm (0.05 in) to 2.54 cm (1 in) in length. To balance melt flow and pressure, runner sizes—not gate sizes—are changed. In Figure 17.49, P_c is the pressure drop in the cavity and P_t is the total pressure drop.

The approach used in calculating the required diameter of the final runner segments of a three-plate mold with multiple drops into the cavity is the same as the pressure drop approach reviewed. However, for most three-plate molds with multiple drops, it is usually difficult to design them so that an equal volume of melt passes through each drop. However, for circular parts with tight tolerances, it is required that the cavity fill equally from each gate in order to minimize out-of-roundness. Since the pressure drops are usually tapered, the diameter is not constant. The difficulty can be circumvented by using the diameter at half the length as a basis for this calculation. Sucker ejector pins in the pressure-drop area will influence the pressure loss and can provide additional restrictions to help equalize flow into each drop. Both the length and the diameter of the sucker pin can be used to regulate the flow.

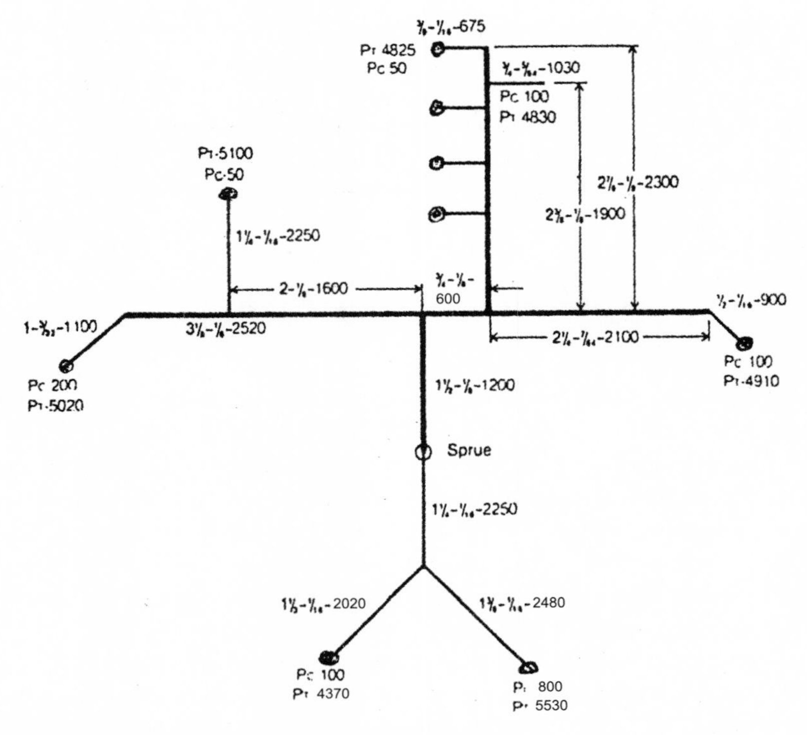

Figure 17.49 Unbalanced runner system in a ten-cavity mold.

HOT RUNNER

Hot runner systems for TPs maintain the feed channels in a permanently molten state (Fig. 17.50). They eliminate the waste associated with a solidified cold runner system. Another advantage is that the pressure drop through the feed system is less than that of a comparable cold runner arrangement. The principal requirement is the need to maintain two very different temperature regimes within the mold. The hot runner system in a mold may be operating at 260°C in close proximity to mold cavities at 20°C. This results in difficult problems of temperature control and differential expansion. The cost of a hot runner mold is also much greater than that of the cold runner equivalent. However, their important difference is not the cost of the mold but that of the molding. Hot runner molds have the big advantage.

Unless production quantities are low or frequent color changes are required, there are grounds to prefer the hot runner mold. Some processors may still have control and engineering fears associated with hot runners. There continues to be great progress in this area, and very sophisticated and reliable designs are now available as stock components from different suppliers. Most manufacturers of hot runner components now offer completely engineered systems that free the molder and

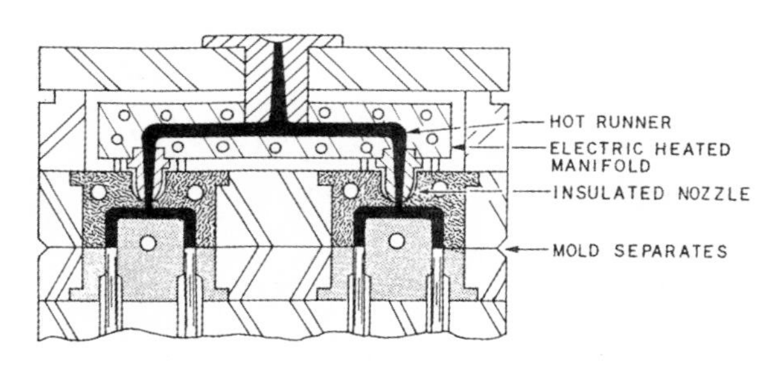

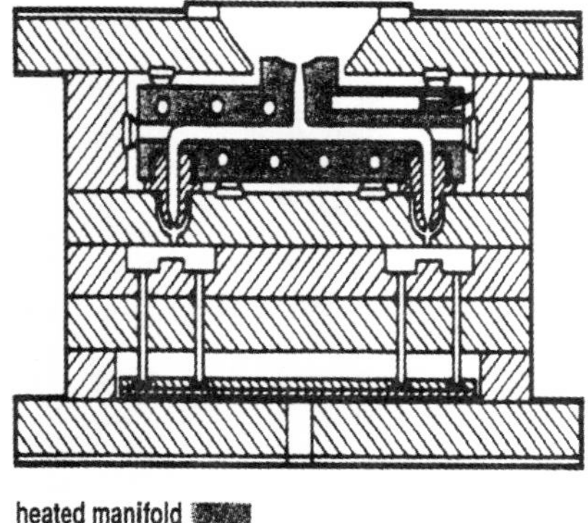

Figure 17.50 Schematics of hot runner mold systems.

mold builder from design responsibility. Traditionally, cold runner molds have been the norm and hot runners the exception, but progress has been such that the project concept really should now start from the opposite premise.

The heart of a hot runner system (Fig. 17.50) is the manifold. This is a distribution block containing flow channels maintained at controlled melt temperature. The channels distribute the melt from the single entry point at the sprue to multiple outlets at nozzles that feed individual cavities in a multicavity mold or serve as multiple gates in a single large cavity. The latter case is one of the most important applications of hot runners; multiple gates allow the flow lengths in large part to be brought within easily managed proportions. The same strategy also limits the necessary clamp force.

Contact between the hot manifold and the remainder of the relatively cool mold must be kept to a minimum to prevent heat flow. This is normally achieved by the use of air gaps that also allow for expansion, but care must be taken in the design to ensure that the mechanical strength of the mold remains adequate to deal with molding and clamping forces. The heat-transfer problem is reduced with internally heated hot runners (Fig. 17.51) but the disadvantage is the relative inefficiency of the annular flow channel. These channels have higher pressure drops than unobstructed channels and are also prone to "slow flow" areas where material may stagnate and decompose. Internally heated hot runners are not on the whole recommended for use with certain plastics, such as PP.

A primitive variant of the hot runner is known as an insulated runner (Fig. 17.52). The mold employs an unheated manifold with very large runners such as 20 mm to 35 mm (0.75 in to 1.375 in) diameter. It relies on the poor thermal conductivity of plastics to ensure that a flow channel in the center of the large runner always remains molten. Insulated runner molds, which have oversized passages, are formed in the mold plate. Their open center passages are of sufficient size that, under conditions of operation, the insulating effect of the plastic combined with the heat applied with each shot maintains an open flow path. A layer of chilled plastic that forms on the runner wall provides runner insulation. This insulated runner mold is not capable of precise and consistent control and is now rarely used.

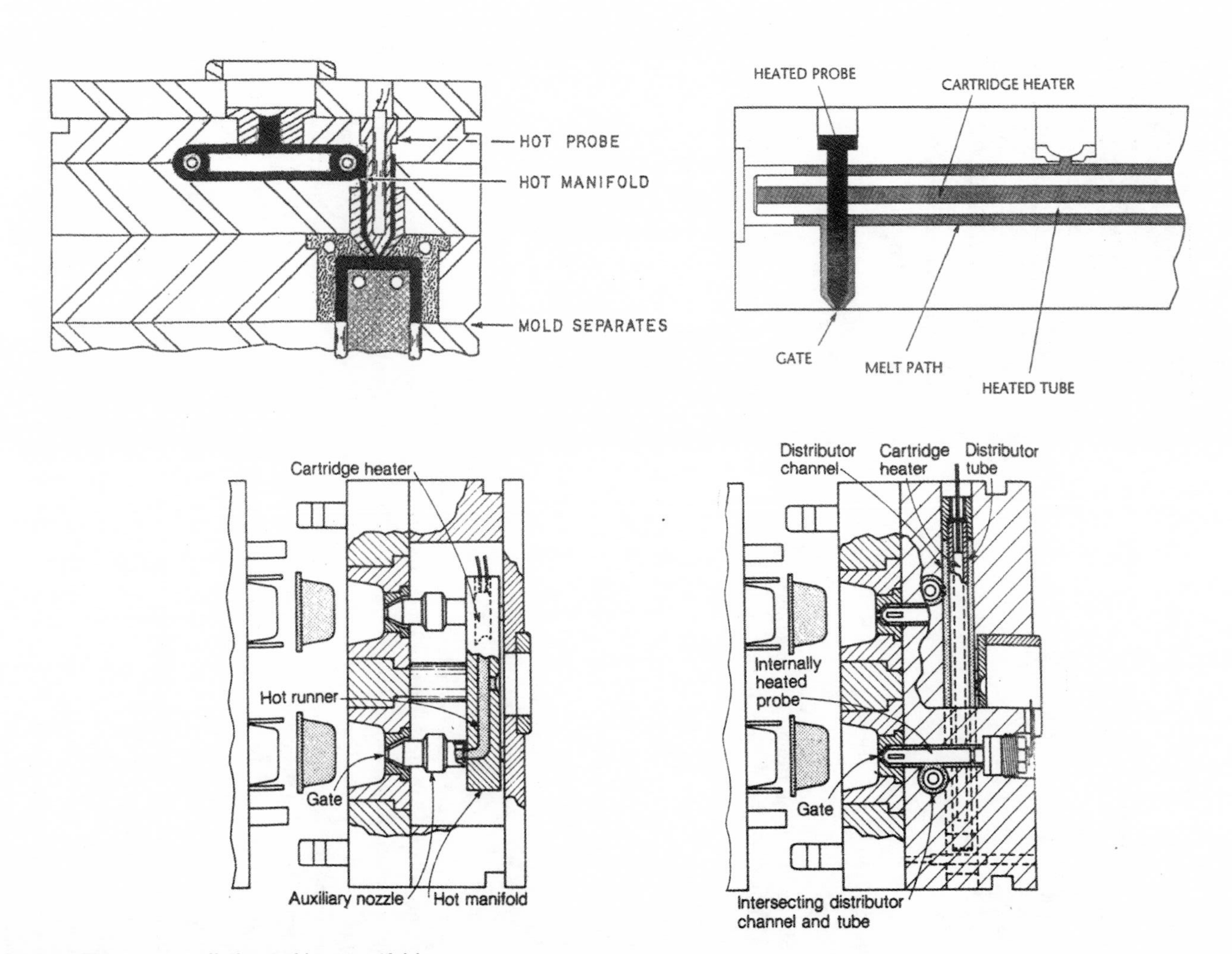

Figure 17.51 Internally heated hot manifold.

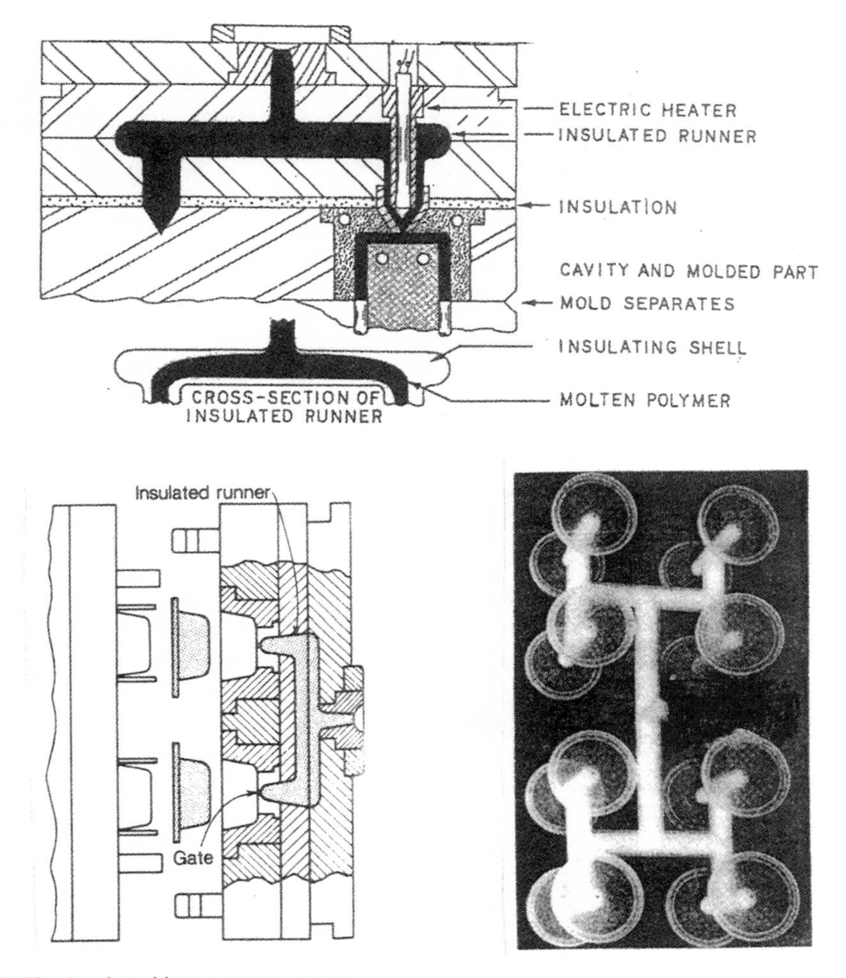

Figure 17.52 Insulated hot runner systems.

The connection between the hot runner manifold and the cavity is made by means of a hot nozzle that may operate in conjunction with a surrounding bushing. A wide variety of nozzle types allow for many gating options (Figs. 17.53). The types depicted in Figure 17.54 can be used either with or without witness marks. Hot runner nozzles may have shutoff valves for precise control of flow. Major progress continues to be made in the design of hot runner nozzles for all sizes of cavities, particularly small and closely spaced cavities. These designs allow one nozzle to feed a number of cavities, and also permit gating options that are almost undetectable on the finished part.

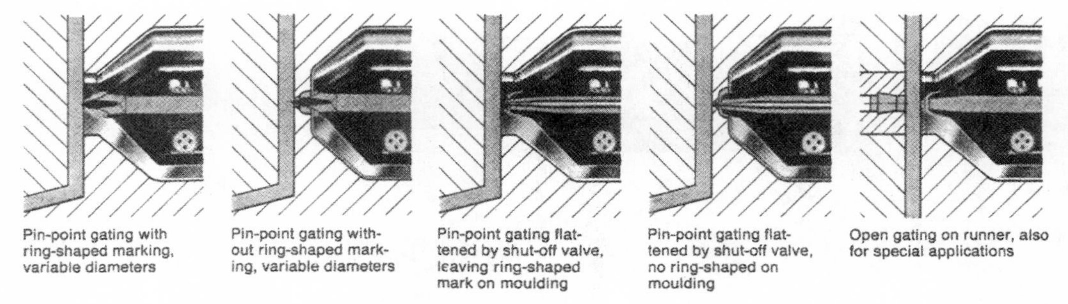

Figure 17.53 Examples of direct hot runner gates.

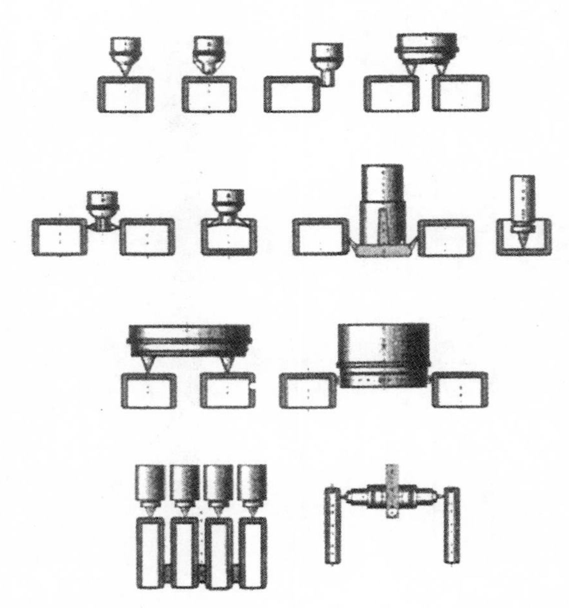

Figure 17.54 Advanced types of hot runner gates.

The conventional hot manifold (no internal heating) continues to make its operation predictable and reliable. They are designed to take all types of loads (Fig. 17.55). Figure 17.56 is an example of a manifold, hot stack mold with forty-eight cavities on each side (ninety-six cavities total), and Figure 17.57 is an example of a hot manifold stack mold with six cavities on each side (twelve cavities total).

For example, the heating power for manifolds should be at least 60 to 80 W/in^3 of steel. Heaters should be positioned to eliminate hot and cold spots. Manifold flow channels should be at least 12 mm (0.5 in) in diameter. Large manifolds and high shot volumes will need bigger channels. Flow

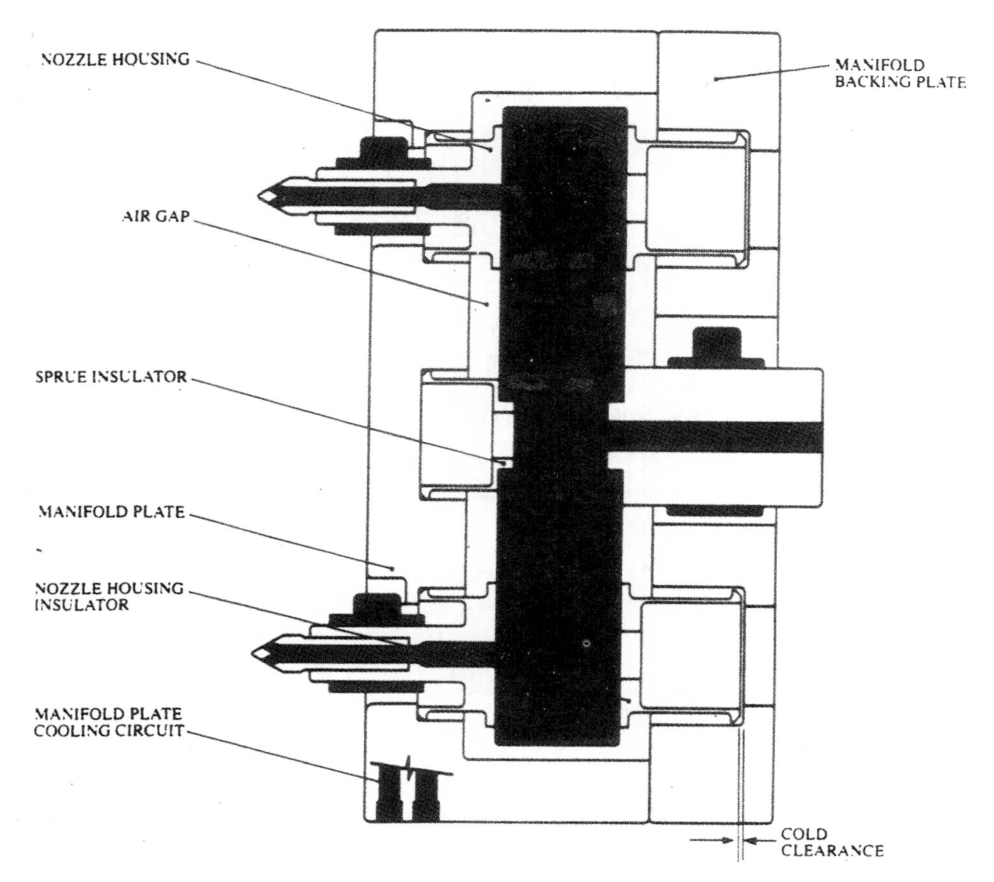

Figure 17.55 Example of a hot manifold support system.

channels should be streamlined to prevent slow or stagnant flow. Corners should be radiused by contoured end plugs. The manifold air gap should be at least 1.5 mm (0.62 in). Manifold support pads should have minimal surface contact and should be furnished in materials of relatively low thermal conductivity, such as stainless steel or titanium. Precise temperature control is necessary for consistent results. The temperature of the manifold block should be uniform throughout. Each nozzle should be provided with individual closed-loop control.

Developing hot runner size

Hot runner components used for TP molding have been in use since the 1940s, with most of the activity starting during the early 1960s. Initial developments were subjected to certain problems, such as drooling, freeze-off, leakage, and high maintenance. Gradually, new design concepts and

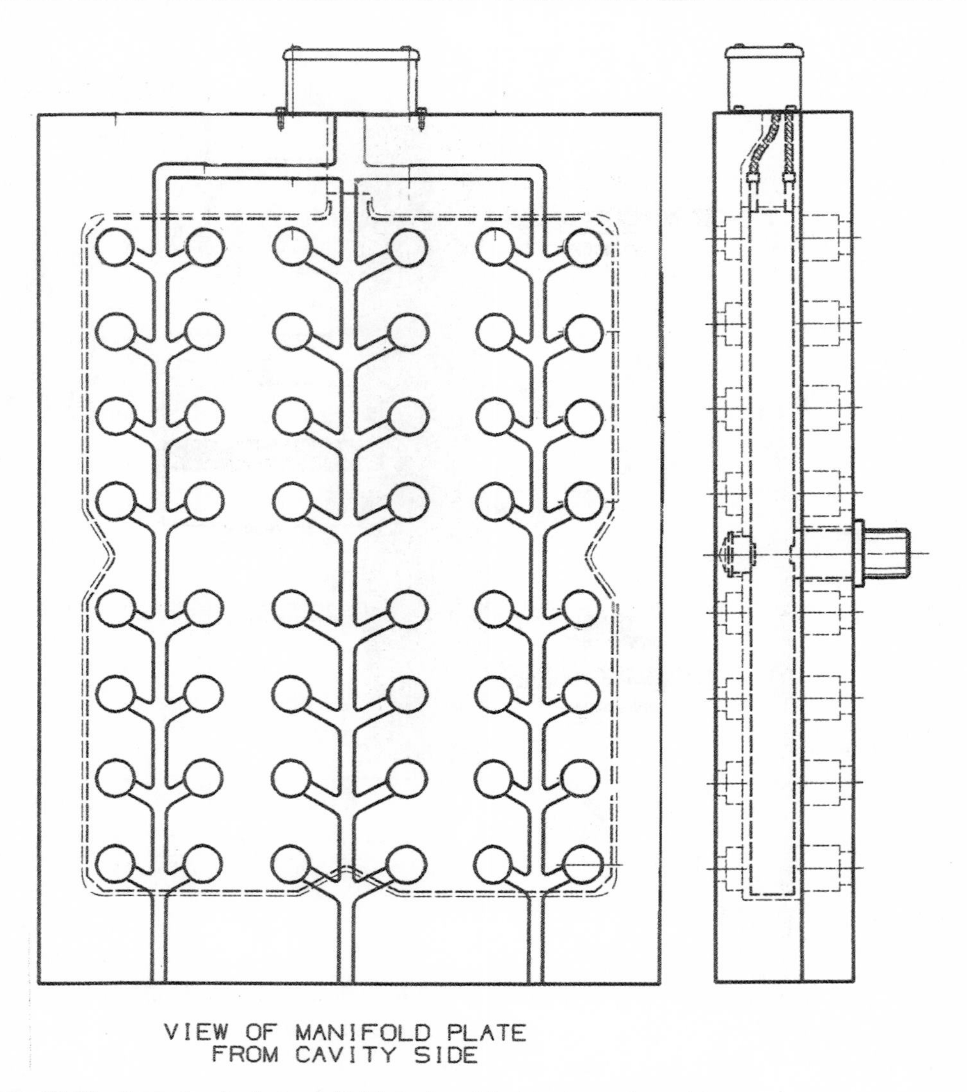

Figure 17.56 Example of a hot manifold stack mold with ninety-six cavities.

tool-building methods overcame these disadvantages, and today's tools for runnerless molding are highly efficient and relatively free of fault.

The term *runnerless* refers to the fact that the runner system in the mold maintains the plastic in a molten state. This plastic melt does not cool and solidify, as in a conventional two- or three-plate mold, and is not ejected with the molded product. It is a logical choice for any high-speed operation whose goals are to reduce cycle time, reduce plastic consumption, and eliminate runner scrap.

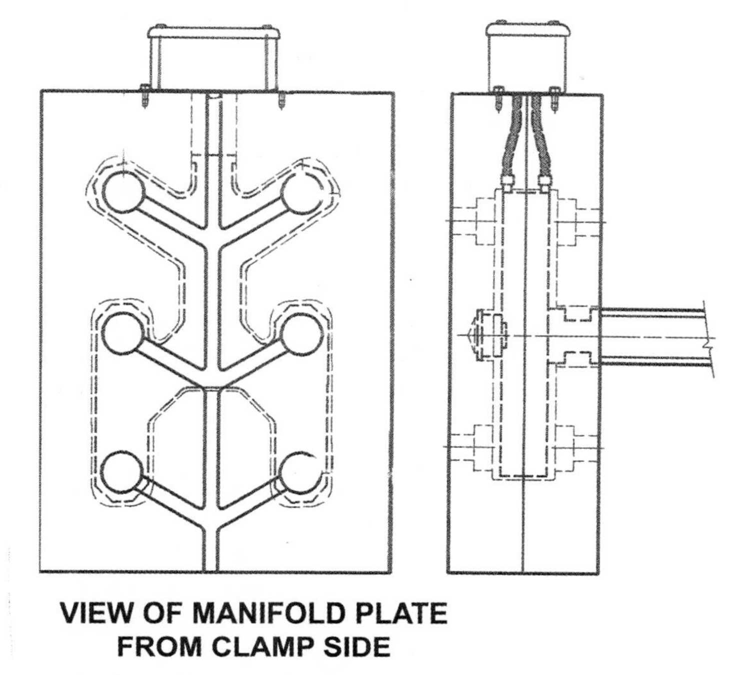

Figure 17.57 Example of a twelve-cavity hot manifold stack mold.

With the exception of a mold with a single cavity that is fed directly from the machine nozzle, all injection molds have a runner system, so describing them as runnerless can be misleading. This term originated in the use of insulated or heated runner channels, in which the plastic does not cool and solidify. No plastic is ejected from the runner channel when the mold is opened and the molded product ejected. Thus the term is indicative of the absence of scrap from the runner system.

Some plastics or products are not adaptable to runnerless molding. The major problems are encountered with heat-sensitive materials, in which the time-temperature relationship can be a problem. However, with today's technology, they can be used with proper available controls. Examples of these plastics include the acrylonitriles and polyethylene terephthalate (PET).

More popular are the hot runner molds with either an insulated manifold or a cartridge-heated manifold (Fig. 17.52). There is a type with internally heated flow passages; the heat is furnished by a probe or torpedo located in the runner passages. This system takes advantage of the insulating qualities of the plastics to avoid heat transfer to the rest of the mold. Although the insulated runner molds are generally less complicated in design and less costly to build than hot runners, they also have a number of limitations: freeze-up at the gates, fast cycles required to maintain a very even melt state, long start-up periods to stabilize melt temperature and flow, and problems in uniform fillings of molds.

Hot runner molds provide a very even and controllable distribution of heat. In addition to a complete, compact hot runner manifold, components that enhance the design and construction of hot runner molds are available. These standard components include a variety of cartridge-, band-, or coil-heated machine nozzles, sprue bushings, manifolds, and probes; heat pipes; gate-shutoff devices; and electronic controllers for various heating elements. In the compact hot runner manifold, melt flows through a tube to the individual hot-channel nozzles. Since the runner dimensions are precisely defined, the pressure loss in an externally heated system can be easily calculated using an appropriate CAD software program.

Certain considerations are to be taken in the rheological design of a runner system. For instance, pressure loss in the runner system must be as low as possible. In systematic design, the channel diameters have upper (plate-out problems) and lower (pressure loss too great) limits. For this reason, the diameter that is specified cannot always be the one that is best rheologically. To avoid dwell-time problems such as plate-out, it is advisable that a particular limiting shear rate not be exceeded. The relevant limiting values for various materials are arrived at by experience. Because the flow lengths are necessarily long, there is certainly a large loss of pressure in the hot runner system. To reduce pressure loss, the aim is to specify large diameters and short flow lengths to individual injection points. Such a design procedure results in an unsymmetrical system. In order to balance it for a particular operating point, an approach is to adjust the channel diameters so that, at the operating point, the manifold behaves like a balanced system with small pressure loss.

However, determining the corresponding channel diameters takes quite some time, since flow impedances over the various flow lengths have to be calculated. This effort can be reduced by means of a dedicated computer program for the calculation of pressure loss and balancing of hot runner systems. The program developed provides answers to questions such as volume-flow distributions, pressure losses, and channel diameters.

The design of hot runner molds takes into account the thermal expansion of various mold components. It applies to the center distances between the nozzles, supports, set bolts, and centering points. The bends in the hot runners to the nozzles should be generously radiused to prevent dead corners of melt. In the design, each nozzle contains a device such as a capillary to act as a valve to prevent plastic leakage or some equivalent problem. Heating elements positioned around the nozzles provide proper temperature control. When thick-walled products are molded, the long after-pressure time may necessitate the use of nozzles with needle valves, as capillaries tend to freeze up rather quickly.

Hot runner manifolds provide heater-loading requirements for specific types of plastics. For example, for general-purpose materials (polystyrene, polyolefins, etc.), it is 0.09 to 0.12 W/cm^3 (15 to 20 W/in^3) of manifold. For high-temperature TPs (nylon, etc.) it is 0.12 to 0.18 W/cm^3 (20 to 30 W/in^3) of manifold.

A major portion of cycle time for a plastic product is cooling time, which is the amount of time it takes the plastic to set prior to opening the mold and part ejection. In a cold runner mold, the thickest wall section is often found in the cold runner, and the molding cycle may wait until the runner is solid enough to be ejected. Whether it is freefall or by sprue picker, the elimination of the runner results in a reduction in the cooling portion of the cycle, thus reducing the overall cycle time. Cycle time can be reduced by as much as 50%.

The elimination of the cold runner means less recovery time is required since the injection unit does not have to plasticate the cold runner that in turn reduces the energy consumed per part. Less pressure is needed to fill the mold, all adding up to additional energy savings. The reduction in pressure loss during fill is achieved with the use of heated flow channels. The flow length found in a hot runner system also tends to be shorter, further reducing the pressure losses found in a cold runner system. Reduced injection pressure means less stress in the product, providing better structural quality. A reduction in pressure will result in easier filling of the cavities, which reduces the deflection in both the platens and mold, in turn reducing the amount of flash, again improving quality.

If the runner made up 30% of the shot weight, this would reduce the recovery time proportionally. If recovery time hindered the overall cycle previously, this would also reduce cycle time. The reduction of the overall shot weight also means that injection time is reduced since the same injection rate needs to be maintained for required fill rates. The plastic's flow path is much shorter. Mold-open dwell time is reduced since the system does not have to wait for the ejection of the runner, further reducing cycle time. The hot runner approach eliminates the need for a sprue picker and grinder, which also require energy and personnel to operate.

The elimination of the runner-plate movement reduces the clamp motion because the stroke is shortened and runner stripper plates controlled by shoulder bolts are not required. With shoulder-bolt ejection, the stroke needs to be profiled to ensure that the shock loading is controlled. Elimination of this action allows full clamp speed to be incorporated, again reducing cycle time.

As the melt flows through the cold runner, a solid layer sets up on the channel wall, restricts flow, and requires greater injection pressures from the machine to help overcome losses. The higher pressure at the injection end of the runner is required to achieve the needed pressure to overcome the gate restriction, flow losses, and cavity filling. Keeping the resin molten in the hot runner reduces the pressure drop to each cavity since the flow is less obstructed. The hot runner system provides an easy approach in providing balanced melt flow to each cavity, resulting in consistent product weight from cavity to cavity. Balanced flow also produces fewer rejects.

Unfortunately, it is important to understand that the hot runner manifold increases the cost of a mold and the extra expense needs to be justified by the application and/or production quantity. On average, a hot runner system adds 10% to 15% to a mold's cost, but sometimes it could double the mold's cost. Such higher costs can best be justified for high-volume production, the molding of expensive plastics, and high-quality molding, where gate vestige should be minimal. Products that weigh less than 1 g or 10 to 160 kg (22 to 350 lb) or more and are extremely large, like a big trash container, can be made with these systems. As engineering plastics continue to be more sophisticated and expensive, there will be more of a need for hot runner systems to eliminate or significantly reduce the waste of plastics or build up their residence time.

CONVERTING COLD RUNNERS

Based on past experience with long production runs, consistent tight tolerance requirements, and/or other factors, existing molds using cold runners offer opportunities to improve profitability by converting to hot runners. In some cases, existing mold design precludes complete conversion from a cold to a hot runner system. For example, a conversion to hot runners could provide a cycle savings of about 10% for a forty-machine plant that in turn would free up four machines, or it could increase the revenue from the plant by 10% without adding any new machines. The elimination of a cold runner, as reviewed, can also reduce energy consumption and mold maintenance, eliminate granulator and sprue picker, and improve part quality and the efficiency of cavities.

The hot runner conversion can be made on both two- and three-plate cold runner molds. The conversion can be either to a full hot runner or a hot/cold combination runner. The latter would have a hot runner feed a smaller cold runner, which is a system with many advantages. This approach substantially reduces the runner weight and can provide a more balanced delivery of melt. The elimination of the sprue and thick feed runners offers the advantages of smaller shot size, reduced injection pressure, and possible cycle savings. This combination runner may also require sucker pins and sucker-pin motion to eject the runner. This can be determined after the mold design is reviewed.

When considering a change there are factors to consider. The existing cavity may need to be modified to accommodate the hot runner nozzle tip. The existing material may not be reworkable; new cavities or gate inserts may be required. The hot runner system may add to the height of the mold.

The gating required by the product needs to be reviewed to ensure it can be accommodated. The existing cavity must provide space to install a hot runner probe. The location of the gate may need to be changed if insufficient space or cooling exists. The type of plastic will also play a factor in the gating style, as some are more degradable than others. The addition of a hot tip into the cavity requires a close look at the cooling in and around the gate to ensure that the desired thermal equilibrium can be achieved to produce consistent quality gates. Gate quality may vary due to the elimination of the cold sprue. The operating sequence on many existing IMMs is to inject, hold, recover, and then decompress. The act of recovering with backpressure keeps the resin in the manifold under pressure. The screw decompressing afterward tends to decompress the resin in the barrel, not that in the hot runner. The type of sequencing with the hot runner may cause a variation in gate quality.

RUNNER OVERVIEW

Traditionally, there have been a number of misconceptions about proper runner design. In the past, many molders and tool builders felt that the larger the runner, the faster the melt would be conveyed to the cavity. They also believed that the lowest possible pressure loss through the runner system to the cavity would be the most desirable. Runners were commonly machined into the mold with these objectives in mind. However, it is important to select the minimum runner size that will adequately do the job with the material being used.

Cavities should be placed so that the runner is short and, if possible, free of bends. They should also permit delivering the required amount of melt that is balanced. This means that the shape and size of the runner, its length as well as the gate size are for all practical purposes identical, providing a balanced condition in the cavities. This becomes especially important for precision parts.

A balanced supply ensures that any change made in any one of the molding parameters will affect all cavities in the same direction. It is good practice to use a runner plate of the same grade of steel as the cavities, which has a surface machined to 50 RMS.

The material processing data from hardware and material suppliers give a range of runner sizes for each material. The smaller sizes can be applied for cases in which the length of runners does not exceed 5 cm (2 in) and the volume of material is less than 38 cu cm (15 cu in). Keep the runners on the smaller end because it reduces the amount of scrap or regrind, and also accelerates the freezing of the gate, all of which reduce cycle time. It becomes a matter of proportioning runners in relation to the spacing of cavities, wall thickness of products, length of cavities, and corresponding gate sizes in order to develop a streamlined pressure drop.

The surface finish of the runner system should be as good as that in the cavity, for example, machined to 50 RMS. A good surface finish not only keeps the pressure drop low, but also prevents the tendency of the runner to adhere to either half of the mold. Such adherence would aggravate the highly stressed area of the gate portion to an even higher stress level.

It is important in TPs molding that the runners in multicavity molds must be large enough to convey the plastic melt rapidly to the gates without excessive chilling by the relatively cool molds that are used for TPs. Runner cross-sections that are too small require higher costly injection pressure and more time to fill the cavities. Large-diameter runners produce a better finish on the molded parts and minimize weld lines, flow lines, sink marks, and internal stresses.

If the runner (not hot runner or runnerless) is excessive, problems develop: increased costs caused by longer chill times, increased material subtracts from the available machine capacity, and increased scrap. In two-plate molds containing more than eight cavities, the projected area of the runner system adds significantly to the projected area of the cavities, thus reducing the effective clamping force available.

RUNNER MANIFOLD

There are more than fifty hot runner manifold system manufacturers. There are also moldmakers that have their designed systems (354). This flexibility that hot runners provide can help optimize cavity orientation, cooling, and mold simplification (Fig. 17.58).

Suppliers of these hot runner manifolds supply information and in turn the molder can request certain information about the following topics:

1. *Molding automation.* Hot runner manifolds in molds have a definite advantage when used with automation. In addition to part accuracy (dimensional consistency) and

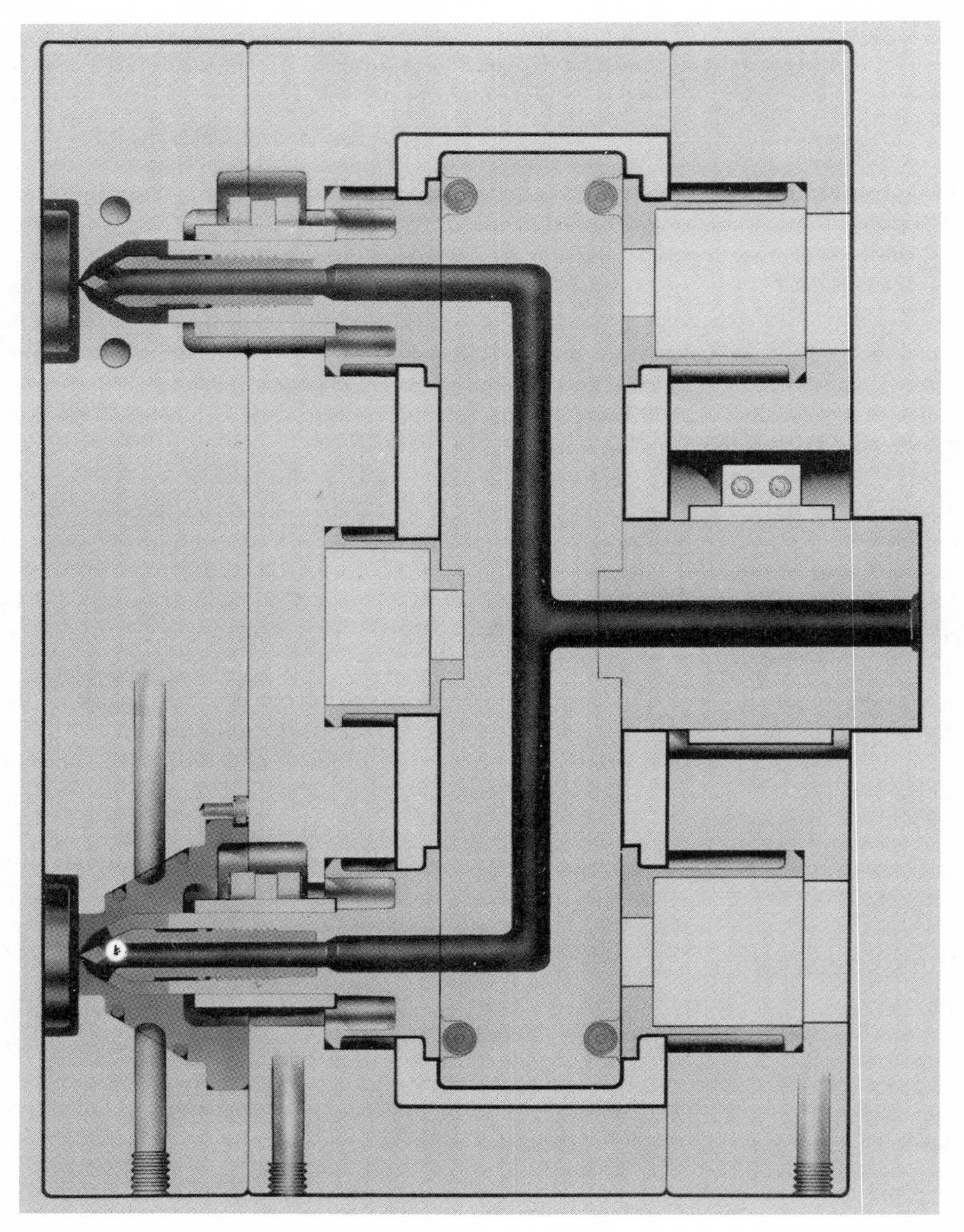

Figure 17.58 Heated manifold for TP hot runner system.

flash-free parts, there are no runners present to get tangled in the mold mechanisms, robots, conveyors, assembly machinery, and so on.

2. *Balanced melt flow.* The flow advantage that hot runners have is that the melt channels are in separate, externally heated manifolds, which are insulated from the surrounding mold plates. Different to a cold runner plate, the hot manifold can be designed to have flow channels on multiple levels to ensure that the plastic flows the same length from the molding machine nozzle to each cavity with the same channel profiles and diameters, numbers of turns, heat soaks, and pressure losses. In other words, the plastic reaches each of the cavities, whether there are two or ninety-six, with the same pressure and thermal history. This designed-in temperature and pressure control is particularly favorable for plastics that have narrow processing windows.
3. *Available manifold.* The supplier offers flow analysis, channel sizing, and other design capabilities to produce a mechanically and thermally balanced hot runner system—a system where the runners are large enough to give a relatively small pressure drop through the system without causing too much residence time.
4. *Melt channels.* Are the channels designed and manufactured to be smooth with rounded corners, and to have no sharp corners or "dead" spots? Can the manifold be cleaned should an incident leave the system full of degraded resin?
5. *Insulation.* The hot runner manifold and nozzle assemblies must be insulated from the mold plates to avoid heat loss and its resulting excessive power consumption.
6. *Plates.* Hot runner plates machined from solid blocks of prehardened stainless steel to ensure ruggedness, maximum support around the manifold, minimum deflection under high clamp tonnage and injection pressures, minimum maintenance, and hot runner longevity.
7. *Part quality.* Eliminating the cold runner will give better filling and packing conditions. When plastic is flowing through the cold runner, it loses heat to the mold plates, causing pressure drops that can result in sink marks and underfilled parts. The hot runner maintains a balanced melt flow at a constant temperature from the machine nozzle to the gate to fully fill and pack the cavities. Thus hot runners help molders take full advantage of highly accurate and interchangeable cavities to achieve plastic part dimensional accuracy and quality. Should a cavity get damaged, be out of specification, or for any reason, it is very easy to reduce the heat to its hot runner nozzle and stop the cavity from producing faulty parts.
8. *Complete hot runner assembly.* Fully assembled and tested bolt-on systems are available as full systems and their components are available separately, leaving the moldmaker to design and machine the surrounding hot runner plates, assemble the complete system and then test the completed system.

9. *Product range.* To achieve the optimum system for the molding application, does the supplier offer an extensive nozzle range of hot tips, valve gates, edge gates, hot sprues, multitips, and so on for maximum design flexibility?
10. *Experience.* The supplier should be able to offer guidance in gate location and possible part orientation for optimum filling and packing, gate/nozzle type, gate strength and gate cooling, and so on. An understanding of the relationship between the mold and the hot runner system is a critical supplier strength.
11. *Plastic testing.* Is a fully equipped resin testing or R&D facility available to assist the moldmaker in choosing the best hot runner system, nozzles, and other parameters for a new hot runner application or resin?
12. *Ease of maintenance.* Determine if the gates can be cleaned in the machine. Can wear items such as nozzle tips, thermocouples, and heater bands be replaced without removing the mold from the machine? Can the valve stems in valve-gated systems be adjusted or replaced in the molding machine?
13. *Service.* Does the supplier offer product service, training, and startup assistance if required?

ENERGY CONSUMPTION

Consider two runner systems. One weighs 50 g and another weighs 20 g. The mold produces 750,000 shots/y at an electrical cost of 10¢/kWh and energy requirement of 350 Btu/lb to plasticate nylon, and the cost of molding the extra material in the overweight runner system is about $600/yr. The latter figure assumes close to 100% mechanical and electrical efficiency. Given the actual efficiency factors typical of many molding machines, an added cost of $1000/mold/yr with a poorly designed runner is not unreasonable. Thus it is best to eliminate the waste by carefully considering runner size.

MATERIAL OF CONSTRUCTION

Injection molds are subject to rigorous requirements that have a direct influence on the materials of construction. Examples of the materials are shown in Tables 17.49 and 17.50 (details on the materials of construction were reviewed at the start of this chapter). Mold materials must withstand high injection pressures and clamp forces. They must be good thermal conductors, easy to machine, and capable of reproducing fine detail and taking a high polish. They must be resistant to corrosion, abrasion, and wear.

The traditional material has always been steel, and over the years the choice of alloyed and sintered steels available to the moldmaker has steadily expanded. Some are used only in specialist applications; others are generally used just for specific parts of the mold. The average modern injection mold will contain a number of different steel types, some in the as-machined state and others in a hardened and tempered condition.

Alloy type	Alloy UNS number	Thermal conductivity (W/m.K)	Rockwell hardness	Tensile strength (MPa)
STEEL				
Type 420 stainless steel	S42000	24.9	C27 — C52	863–1725
H–13 tool steel	T20813	24.9	C38 — C54	1421
P–20 tool steel	T51620	38.1	C28 — C50	1007
ALUMINUM				
Type 6061 T6	A96061	166.9	B60	276
Type 7075 T6	A97075	129.8	B88	462
COPPER				
Aluminum bronze	C62400	62.3	B92	725
BeCu — high hardness	C17200	104.8	C41	1311
BeCu — Moderate hardness	C17200	131.0	C30	1173
BeCu — High conductivity	C17510	233.6	B96	759
Cr — Hardened copper	C18200/18400	325.5	B60 — B80	352–483
NiSi — Hardened copper	C64700	162.6	B94	725
NiSiCr — Hardened copper	C18000	216.3	B94	690

Table 17.49 Property comparison of some mold construction materials

Type	Characteristics	Werkstoff number	DIN description	Rockwell hardness	Applications
Medium carbon steel		1.1730	C45 W3	C10	Unhardened parts Back plates Cavity and core holder plates
Pre-hardened sulfur-free tool steel	For high compressive stresses. Good for spark erosion. Not recommended for high polish.	1.2311	Cr Mn Mo7	C32	Cavity and core holder plates Cavity and core inserts
Pre-hardened sulfur-free tool steel	For high compressive stresses. Good for spark erosion.	(AISI P20)		C30 — C32	Cavity and core holder plates Cavity and core inserts
Pre-hardened tool steel	For high compressive stresses. Not recommended for high polish.	1.2312	40 Cr Mn MoS 86	C32	Cavity and core holder plates Cavity and core inserts Support plates
Through-hardening tool steel	High toughness. High resistance to wear and corrosion. Suitable for mirror-finish polish.	1.2767	X45 Ni Cr Mo4	C52 — C56	Cavity and core inserts
Pre-hardened stainless steel	High resistance to corrosion. Suitable for mirror-finish polish.	1.2316	X36 Cr Mo 17	C30	Cavity and core inserts
Through-hardening stainless steel	High resistance to corrosion. Suitable for mirror-finish polish.	1.2083	X42 Cr 13 (AISI 420)	C48 — C53	Cavity and core inserts
Through-hardening hot-work tool steel	Good dimensional stability in heat treatment	1.2344	X40 Cr Mo V–51 (AISI H–13)	C38 — C54	Cavity and core inserts

Table 17.50 Applications of principal mold steels

The choices of alternatives to steel are also growing, and the principle candidates are aluminum and copper alloys. There are two main reasons for using one of these materials in place of steel. They are generally softer and easier to machine. These characteristics speed up mold production and reduce fabricating costs. The other reason is that they have much greater thermal conductivity than steel. This property can be exploited in areas of the mold where it is difficult to engineer adequate cooling channels or when a high rate of heat transfer is required for rapid cycling. The principal disadvantage is that these materials are much softer and weaker than steel alloys. Molds using aluminum and copper alloys are far more prone to damage and are rated for a shorter production life than a steel mold. Some of the latest materials are closing the gap on steel but only to a limited extent.

Steels too can vary in thermal conductivity. In this respect, stainless steels are substantially inferior to normal alloys, and this can prove a major handicap in molds with high thermal requirements. The tables compare the main materials of mold manufacturing and outlines the uses of the leading types of mold steel.

There are many plastics that are normally noncorrosive in contact with all common mold steels. However, there are a few cases of corrosion, particularly at vents. By using cavities and cores produced in an alloy with a chromium content of 10% or more, there should be no difficulty. A copper alloy can itself affect certain plastics, such as PP. These can induce degradation and should not be used in hot runner components that are in direct contact with the melt. However, there should be no difficulty in using copper alloys for cavity and core components because the melt is rapidly cooled on contact. It is the high heat extraction demand posed by most plastics that has the most direct bearing on the choice of mold material. Stainless steels should be avoided, and if there is any difficulty in engineering adequate cooling channels, then materials with high thermal conductivity should certainly be used.

COOLING

In the past perhaps the least understood and least well-applied factor is the inclusion of cooling channels to meet proper heat transfer from the plastic melt to the cooling liquid (for TPs). Usually, insufficient space is allowed between cavities, particularly in molding the crystalline plastics. The cooling rate of melted plastic is usually the final control in the variable associated with the final plastic product performances. This variable influences factors such as melt flow rate, residual stress, and degree of orientation. Heating and cooling rates for amorphous and crystalline plastics differ. If not properly controlled, product performances are either not meeting maximum values or they are defective.

It is a fundamental requirement of the injection mold to extract heat from the molded plastic. With each cycle, the mold acts as a heat exchanger to cool the injected plastic from melt temperature to at least ejection temperature. The efficiency with which this is done has a direct bearing on the speed (cycle) of production. Heat is removed from the injection mold by circulating a fluid coolant through channels cut through the mold plates and particularly through the cavities and cores (Fig. 17.59). The coolant is usually water but may be an oil if the mold is to be cooled at

Core diameter Core width d (mm)	Description	Design
≧3	Heat elimination by air with open mould	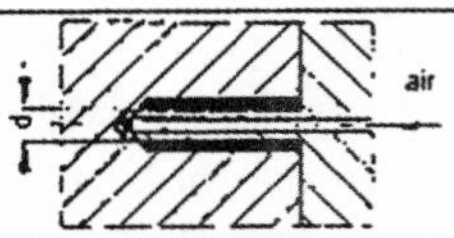 air
≧5	Copper or thermally conductive pipe to conduct heat to the heating/cooling medium	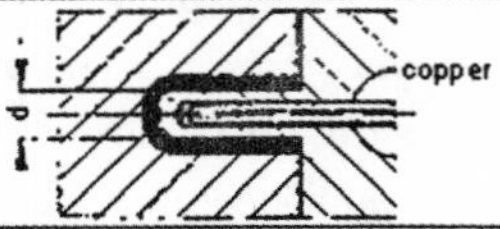 copper
≧8	Long slender heating/cooling channel (with spiral)	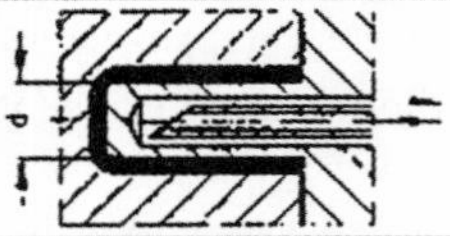
≧40	Spiral heating/cooling channel	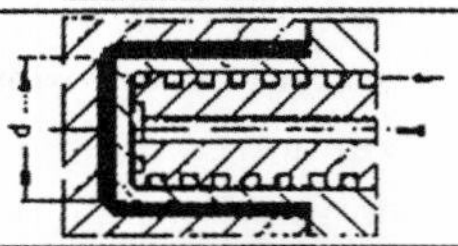
Pipe core s ≧4	Heating/cooling for pipe core with a double flighted spiral	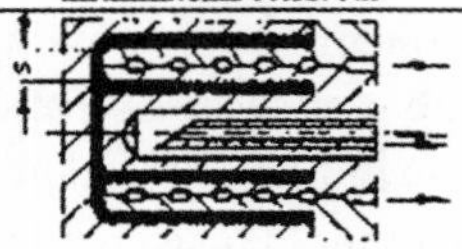

Figure 17.59 Cooling arrangements for cores of various sizes.

temperatures near or above the boiling point of water. Such a mold appears hot in relation to the ambient temperature, but it is still cold compared to the plastics melt. If the mold is to be chilled at low temperatures, it is customary to circulate a mixture of water and ethylene-glycol antifreeze. Occasionally the coolant may be air, but this is an inefficient means of heat transfer and should be regarded as a last resort.

Cooling channels represent a real difficulty in mold design. Core and cavity inserts, ejector pins, fasteners, and other essential mechanical features all act as constraints on the positioning of cooling channels, and all seem to take precedence over cooling. However, uniform and efficient cooling is crucial to the quality and economy of the molding, so channel positioning must take a high priority in the total mold design.

Cooling channel design is inevitably a compromise between what is thermally ideal, what is physically possible, and what is structurally sound (Fig. 17.60). The thermal ideal would be flood cooling over the entire area of the molding, but the pressurized mold cavity would be unsupported and mechanical details such as ejectors could not be accommodated. (Flood cooling is included as a cooling method for blow molding; chapter 6.) Interrupting the flood-cooling chamber with supporting ribs could provide support, but the mold construction is complicated by the need to

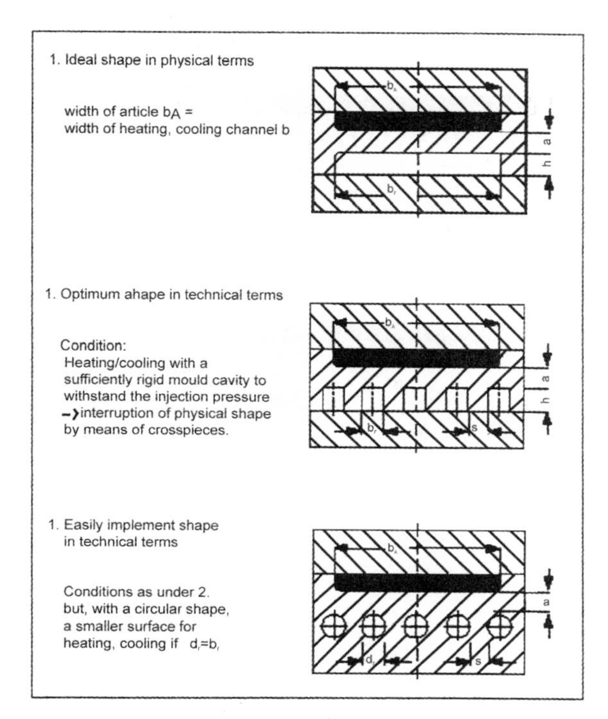

Figure 17.60 Cooling channel considerations.

fabricate and seal the cooling chamber. The reasonable and practical compromise provides the cooling channels in the form of bores that can be easily machined through that may be linked either inside or outside the mold to form a complete cooling circuit.

The correct placement of the channels is very important. If they are too widely spaced, the result is a wide temperature fluctuation over the cavity or core surface (Fig. 17.61 and Table 17.51). If the channels are too closely spaced or too close to the cavity surface, the mold becomes structurally weak.

REYNOLDS NUMBER

Another important consideration in the design of cooling channels is to ensure that the coolant circulates in turbulent rather than laminar flow (Fig. 17.62). The coefficient of heat transfer of the cooling system is drastically reduced in laminar flow.

The Reynolds number (Re or N_{Re}.) determines the condition of laminar or turbulent flow. This is a dimensionless number given by the equation

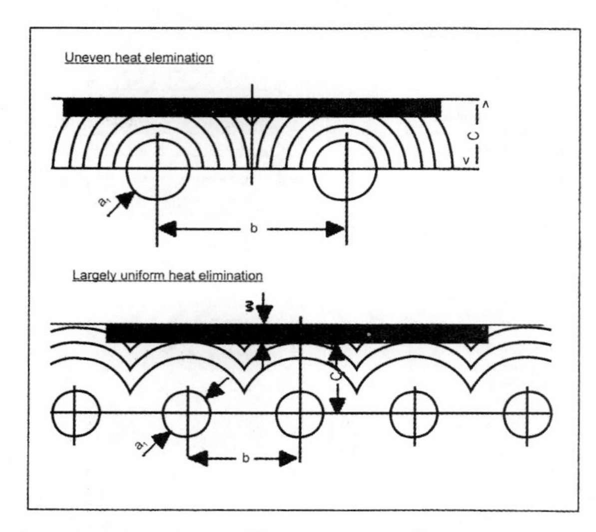

Figure 17.61 Poor and good cooling channel layouts.

Wall thickness w (mm)	Channel diameter d_1 (mm)	Channel spacing b (mm)	Channel depth c (mm)
<2	8–10	24–30	20–25
2–4	10–12	30–36	25–30
4–6	12–15	36–45	30–35

Table 17.51 Guide to cooling channel diameters for PP (see Fig. 17.61)

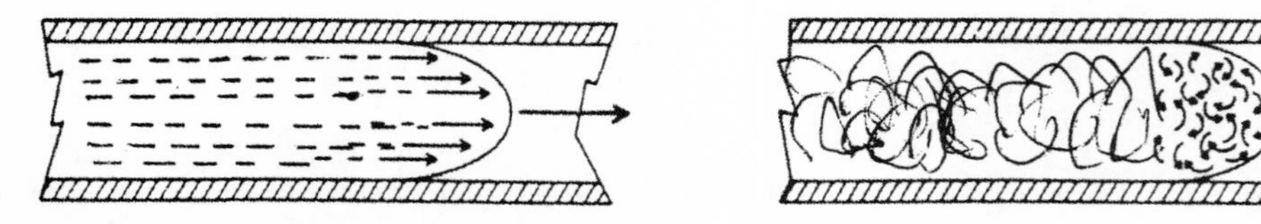

Figure 17.62 Schematic of laminar flow (left) and turbulent flow (right) in coolant channels.

$$Re = Dvp/\eta,$$

where

D = channel diameter,
v = coolant velocity,
p = coolant density, and
η = coolant viscosity.

Re is significant in the design of any system in which the effect of viscosity is important in controlling the velocities or the flow pattern of a fluid. It is equal to the density of a fluid times its velocity, times a characteristic length, divided by the fluid viscosity. This kinetic velocity is defined as viscosity divided by density. In the aforementioned formula, the proper units must be used. The coefficients 7740 or 3160 take care of the conversion factors to make each expression come out dimensionless.

For a channel of circular cross-section, turbulent flow occurs when the Reynolds number is greater than 2300. The coefficient of heat transfer of the cooling system continues to increase as turbulence increases, so the design limit of the Reynolds number for cooling channels should be at least 5000 and preferably 10000. If the volume flow rate of the coolant remains constant, then reducing the size of the channel can increase the Reynolds number. This runs counter to the natural impulse that a larger channel must always result in better cooling. It is more difficult to achieve turbulent flow with oils or antifreeze solutions because of their greater viscosities compared with water.

It is true, up to a limit, that the greater the velocity of coolant flow, the greater the heat transfer from the mold to the coolant. After a certain point, increasing cooling velocity does not appreciably improve heat transfer from the mold.

A laminar, nonturbulent flow is not desirable in a coolant system since it flows in parallel streamlines with a stagnant layer at the wall that prevents heat transfer within the coolant. With turbulence, more heat will be removed because as the fluid on the inside surface of the channel is heated, that heated fluid moves away to be replaced by cooler fluid to rapidly remove more heat (Fig. 17.62).

When good turbulence cannot be achieved (as in passages with very small diameters) baffles can be used in the mold's water channels. Baffles channel the fluid to the wall and then away again to circulate the fluid from the wall to the center of the passage and back in order to remove heat from the plastic at the required uniform rate and in the shortest time period. Other devices, such as bubblers and baffles, are used as well. When inserted into a mold cavity they allow water to flow deep inside the cavity.

LAYOUT PATTERN

Good cooling channel design is particularly important for molds to be used with plastics that have high heat content (Fig. 17.63). This means that slender cores and corners will rapidly and inevitably heat up unless special attention is paid to cooling by means of extensive coolant circulation coupled with the use of components such as heat pipes, bubblers, pipette bubblers, baffle plates, and mold materials with high thermal conductivity (Fig. 17.64). Many different patterns that meet different cooling requirements are available from manufacturers of preengineered mold components. As a general rule, it is impossible to design too much cooling into a mold.

Cooling passages above 2 cm (0.78 in) are not economical because they require tremendous volumetric flow rate. The cross-sectional area increases with the square of the diameter and for an increased area the velocity decreases for a constant flow that leads to laminar flow. Thus the flow rates for large diameters require more electricity to pump the fluid and maintain the Re at 5000 to 10000. The size of the cooling passage also determines the cavity and core plate thickness.

A rule of thumb says that the center of the cooling passage should be located 1.5 diameters below the cavity surface for mechanical strength of the plate. Hence a cooling passage with 1.3 cm (0.5 in) diameter requires a plate of 3 diameters or 3.8 cm (1.5 in) plus the cavity depth. Similarly, to maintain a uniform temperature along the cavity surface for uniform cooling and to prevent warpage of the product, the pitch or spacing between passages should be 3 to 5 diameters. Figure 17.65 is a nomogram that can be used as a guide for determining depth, width, and number of cooling channels required in a mold. In this example, the 455 gal/min value is inked with a tentative passage depth of ½ in, and a mark is ticked off on the reference line. If one end of the straightedge is held steady on the reference mark, and the other end moved up and down across the last two scales, various combinations of numbers of passages and passage depths can be explored. Two combinations are shown here: six passages, each 4 in wide; and eight passages, each 3 in wide. The nomogram can be used in reverse order. Set a tentative combination of width and number of passages, tick a mark on the reference line, then align a straightedge across this mark and the water requirement and read off depth of the passages on the second scale.

One of the major mistakes in mold construction relates to pressure drops in the cooling channel. A 0.5 in diameter cooling passage is rated at 2 gpm (gallon/minute) with a pressure drop of 10 ft of head of water. Note that 34 ft of head is one atmosphere or 14.7 psi. In most molds the cooling passage is no more than 2 or 3 ft long, hence the pressure drop should be less than 1 psi. This is not the case because of all the plumbing fittings that are used to attach the coolant hoses to the mold. The internal diameter of the passages should be the same diameter as the diameter of the cooling passage.

High pressure drops can occur in a cooling system going from lead-in lines and through the mold. A well-designed cooling passage with elbows or turns should have a passage drop of less than 5 psi. As a rule, a cooling passage with a pressure drop of more than that is designed poorly. Nevertheless it may work because commercial chillers' coolant or water supplies have discharge pressures of about 50 psi and can overpower the existing restrictions. Commercial plumbing fittings such as

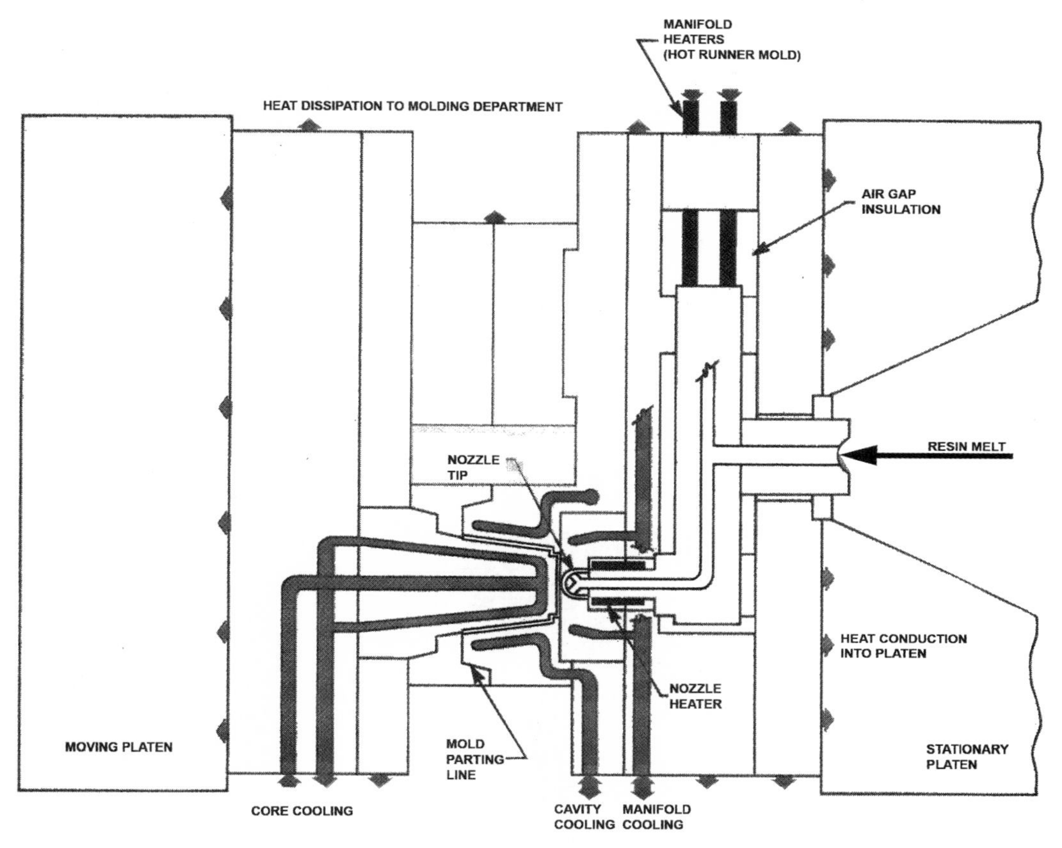

Figure 17.63 Heat-transfer characteristics in a typical hot runner mold (courtesy of Husky Injection Molding Systems Inc.).

manifolds and water bridges are now available as off-the-shelf items that make it easy to keep the pressure drop low and outside the mold hardware compact.

One can calculate the total amount of heat to be removed based on the production rate of plastic used in lb/hr. The chiller load is expressed in tons of refrigeration, where 1 ton is 12000 Btu/h of heat removed. The formula relates heat load to flow rate for a mold-temperature controller or chiller, where the chiller tons of refrigeration = (gpm ΔT)/24. The ΔT is the difference between the inlet and outlet temperatures of the coolant. Preferably it is 3°F (–16°C), and it should not exceed 5°F (–15°C). The ΔT increases as the cooling circuit adds additional loops. If a cooling tower is used, the formula is changed slightly because the efficiency is not as great, thus the tower tons of refrigeration = (gpm ΔT)/30.

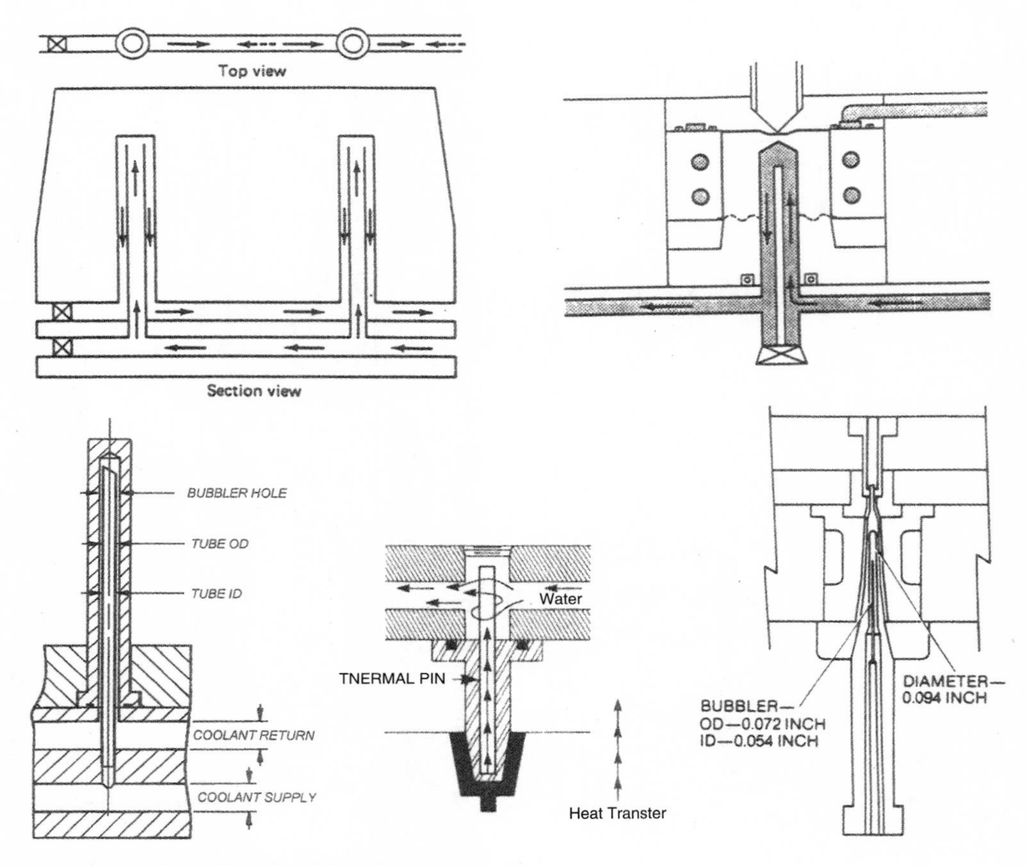

Figure 17.64 Examples of mold-cooling components.

In a cooling tower, a small part of the water evaporates into the air to cool the rest. The system must be provided with makeup water and one must purge about 1.5% of the flow with treated water to keep the chemical balance in the system and prevent mineral buildup. For untreated water, 2% of the flow should be purged.

Based on the heat content of various plastics and their processing temperatures, the following chiller guidelines in lb/h/ton of plastic can be used: 50 for high-density polyethylene (HDPE), 45 for PVC, 35 for low-density polyethylene (LDPE), 40 for polystyrene (PS), 40 for PET, and 35 for PP.

CONTROLLER

A mold temperature controller may only heat water. This type can only control the mold temperature from ambient conditions to the boiling point of the pressurized system. For water with

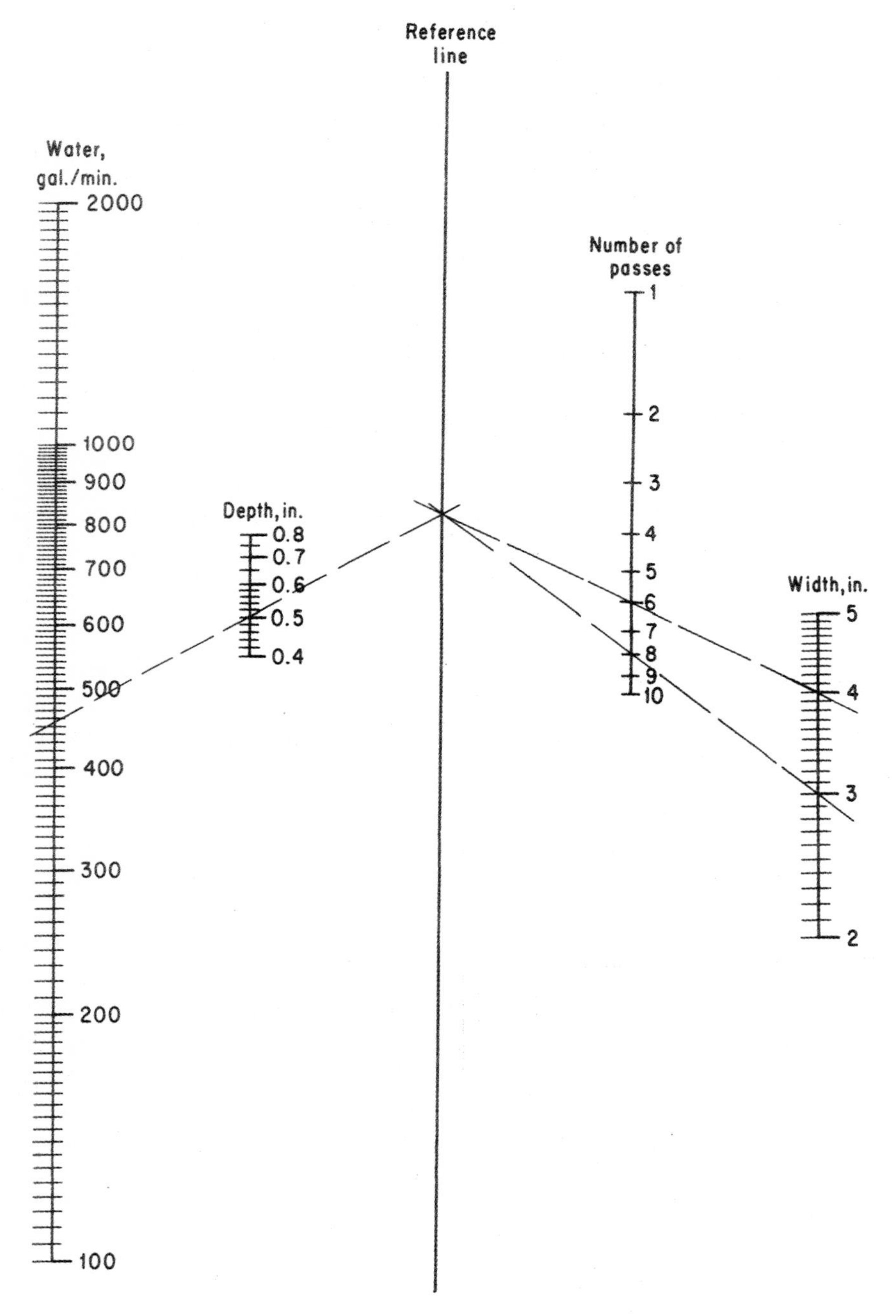

Figure 17.65 Nomogram guide for determining cooling channels.

a minimum pressure of 30 psi at the suction side of the pump, the attainable temperature is about 250°F. In case a mold has cooled below ambient temperature conditions, a chiller must be used. Here a combination of antifreeze (ethylene glycol) and water is circulated. The lowest recommended temperatures for the coolant are as follows: water at 40°F, 10% antifreeze at 25°F, 20% antifreeze at 15°F, 30% antifreeze at 0°F, 40% antifreeze at –15°F, and 50% antifreeze at –40°F. No mixture higher than 50/50 should be used.

High antifreeze concentrations result in increased viscosities, lower Reynolds numbers, and less heat transfer than pure water. One should remember that it is not how cold the mold gets but how fast the coolant will flow that is important. The cavity surface cannot be cooled below the dew point of the ambient air that causes moisture condensation and a cosmetic defect on the molded product's surface when the moisture flashes to steam when it contacts the plastic melt. This is the orange-peel effect (this is different from polishing orange peel). Since humidity tends to be high in summer, the mold cannot be run as cold and the cooling capacity is less as compared to the winter months.

One of the plastics with the most stringent cooling requirements is PET in the molding of preforms that are used to stretch injection blow mold (chapter 6). The mold has to be kept below 40°F (4.4°C) or the plastic will cool too slowly, crystallize, and turn opaque. The blown products will stick to the mold cavity.

COOLING METHOD

There are methods other than those reviewed so far. For example, there is the melt-pulse cooling that does not subject the mold to a fast cooling shock. Rather than having the coolant continually running, it only flows after the melt fills the cavity. Solenoid-controlled valves placed in the incoming coolant are used to open or shut off the coolant flow. It is a cyclic process that is supposed to shorten cycle time by 6% to 10%. It also results in less molded-in stress and reduces scrap and energy use.

With conventional continuous coolant flow, the mold-temperature control generates isotherms, which are contour/wavy lines that move around the cooling channels. Although heat travels quickly from the plastic to the metal to the coolant, the mass of steel between the channels is not used.

Pulse cooling, because it is not continuous, eliminates the isotherms that segregate a conventionally cooled mold. Heat from the part is absorbed not only by the cooling channels, but also by the large mass of steel around the mold. When the fill stage is complete, coolant circulates quickly, removing excess heat, and quickly brings the mold and part back to the minimum temperature.

There is also the flood-cooling method that is used primarily in molds for blow molding (chapter 6). It involves an internal flooding action with or without baffle plates in a confined open chest that surrounds the mold cavity rather than using holes in the blow mold. However, the typical drilled drilled-hole mold is also used, either alone or combined with the flooding action.

CAVITY VENTING

When an injection mold fills, the incoming high-velocity melt stream is resisted by air. It must displace the air in the feed system and cavities (Fig. 17.66). Molders may rely on incidental air gaps between the parting faces and between the assembled parts of the core and cavity to provide a leakage path for air. This is no substitute for properly engineered venting, which should be designed into all molds (Fig. 17.67). Provided the vent is machined to the correct depth, the injected material will not flash through the air gap. More than one vent may be used; up to 60% of the peripheral dimension can be vented.

Venting can also be used to prevent or minimize gas traps. A gas trap occurs when the flow of plastics melt surrounds and isolates an area in the cavity, trapping air that is rapidly compressed and heated. Burned or charred material is often associated with gas traps. The best remedy is to redesign flow to do away with the gas trap. If that design is not possible, venting will help if the vent is positioned in the correct spot. Most gas traps cannot be vented through the parting faces of the mold. Instead, the vent has machined inserted pins or plugs. Such vents are prone to blocking unless one of the parts is in motion. For this reason, an ejector pin with or without a spring is often used for venting. An alternative method is to use an insert of porous steel with the coolant. This method develops a negative-pressure vent (Fig. 17.68).

Venting should be "machined steel safe," where the least amount of steel is removed. One can always grind the vent depth deeper if the vents fail to remove air in initial molding trials. Ribs are difficult to fill and to vent—particularly the blind ribs. When possible, ribs should be vented via ejector pins.

To overcome the trapping of air or gas in a cavities that are difficult to vent effectively, molds may be designed such that all cavity vents feed into a space that is sealed from the outside of the mold (when closed). Sealing is accomplished by a heat-resistant O-ring, which is connected to a vacuum reservoir through a vacuum line containing a solenoid-operated valve or other device. In operation, as soon as the mold is closed and the transfer plunger enters the pot, the aforementioned solenoid valve is automatically opened, causing the cavities to vent rapidly into the vacuum reservoir before the molding compound has entered or filled the cavities.

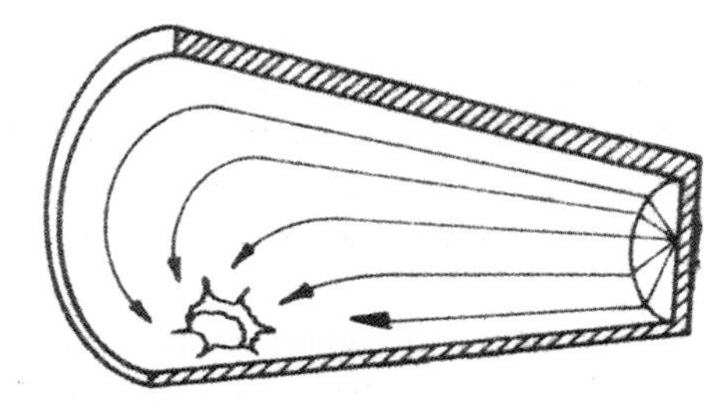

Figure 17.66 Without proper venting, air entrapment can occur in the mold cavity.

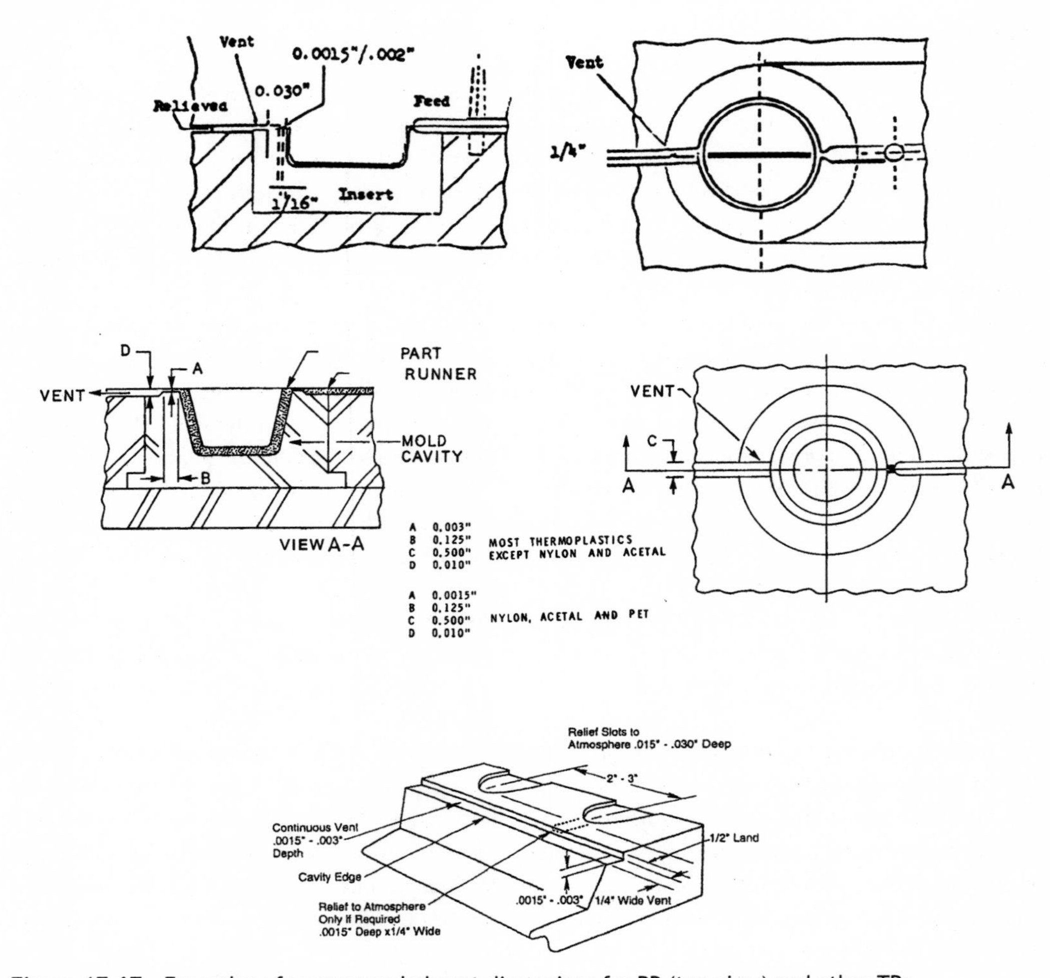

Figure 17.67 Examples of recommended vent dimensions for PP (top view) and other TPs.

An advantage with the vacuum system is that the melt does not have to force the air from the cavity through the vents and enters with a minimum of back pressure and thus fills the cavity more rapidly, leading to faster cycles. With this system, there is essentially no entrapped air, which results in no voids in the product. If a minute quantity of air is present, it is readily absorbed into the melt because of the molding pressure.

In venting products, the most minute flash may be objectionable. Although the depth of venting specified for each material is obtained after testing and/or experience, one must remember, in addition to the measured depth, to consider the peaks and valleys from the surface roughness of

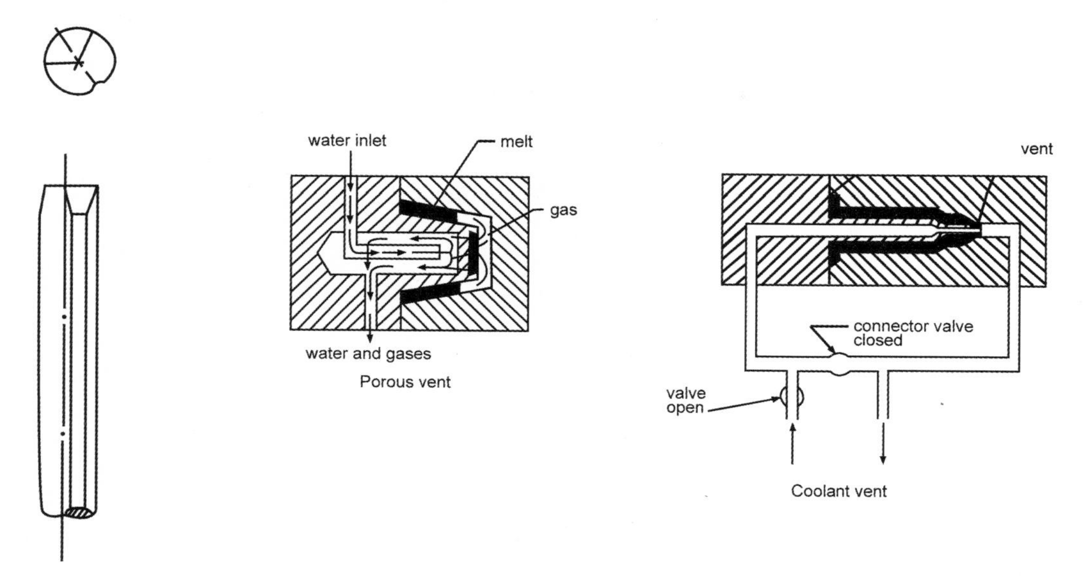

Figure 17.68 Examples of vents.

machining. This roughness measurement plus the micrometer depth should be considered as the value indicated in the tabulation. In the case of products such as gears, it is wise to vent the runner system and all ejector pins, water-blast mating surfaces at the parting line with 200-grit silicon carbide abrasive, polish the vent in the direction of flow, and polish the cavity in the direction of melt flow to eliminate problems on products such as rough surfaces and melt sticking in the mold.

Products with depth—containers, for example—tend to adhere in the cavity. After the cavity separates from the core, the atmospheric air pressure would make it difficult, if not impossible, to remove the product. This suction problem is eliminated using a vent pin that is tapered at the end, where the cavity is located (Fig. 17.69). It is held in its normally closed position by the spring during melt injection in the cavity. When the material is injected, the pressure of the material on the head of the pin forces it too tightly close. When the product is to be ejected, the pin will be released when the knockout system is activated, venting the interface between the core and plastic. Additional air can be pushed through the opening if more force is required to remove the product.

Negative pressure venting

For extremely difficult venting, one can consider a suction system (patented by Logic Devices), where the air is exhausted through the cooling line under negative pressure by using an ejector pin or porous plug. Water coolant can also be used. This water-transfer process was designed primarily to cool long, thin cores, such as those for pen barrels, that leave a hole in either end.

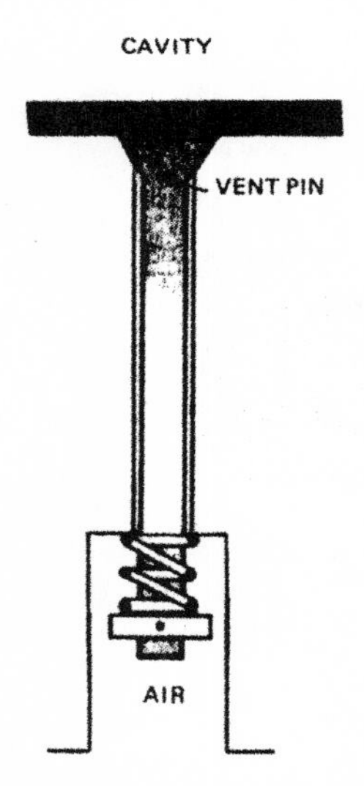

Figure 17.69 Example of a vent pin used to break the vacuum between core and plastic.

It is a technique based on he negative-pressure coolant technology (less then 14.6 psi atmosphere) pulls the coolant system, producing a negative pressure in the coolant system. The molding gases vent into the coolant rather than the atmosphere. Coolant does not leak into the cavity because it is less than atmospheric pressure.

When a suction pin is used it is kept cool because it passes right through the coolant, preventing it from overheating. It eliminates a buildup of gummy melt deposits that exude from the heated plastic. The gummy deposits in turn tend to plug up the vent and make the problem even worse:

Generally it is very difficult to cool molds that have core pins with small diameters. Usually they are so small that there is no room for conventional bubblers or baffles. However, one can also draw the coolant through this small core pin via negative pressure and at the same time do the venting. The location of water channels no longer dictates that a pin cannot be placed there. The reverse, of course, is also true: the location of a critical pin no longer means that a water channel cannot be put there.

Other benefits of venting can result in different mold constructions. An example that has been widely displayed is shown in Figure 17.68: coolant venting. A bad air trap would result at the top of the product in a conventional mold. The tool would typically be designed as a three-plate mold, with its commitment to added expense and complications. In this schematic, the coolant passes from one half of the mold to the other half, right through the molded part, when the mold is closed. When the mold is open, the supply to the mold is shut off, and both ends are subject to a vacuum pressure that evacuates the remaining coolant so that no leaks occur going into the mold. This design results in a parting-line vent at the end of the molded product opposite the gate just where it is needed.

POROUS VENTING

The porous metal approach provides another venting method. Porous metal has very poor heat conductivity. The porous vent can be located any place where it can be attached in the mold if there is a waterline nearby (Fig. 17.68). The gases exhaust into the coolant. The porous metal is directly cooled by the water behind it that keeps it from overheating and plugging with hot plastic. Water coolant does not leak out because it is held at subatmospheric pressure.

Since porous metal will leave a texture on the plastic's surface, it should be located either where it is not seen in a place where the texture on the molded part does not matter, or it can be blended into a texture on the rest of the mold. A very finely woven, porous metal that leaves only a very faint texture on the plastic is also available.

TS vents are more important to TSs than TPs because they are less viscous and will seep into the slightest crack in a parting line, acting as a wedge to force apart the mold halves. What has been reviewed will be helpful. Consider that runners should be vented prior to approaching the gate. The vents should be the full width of the runner and about 0.015 cm (0.006 in) deep. The circumference of the cavity should be vented, and the vents should be spaced about 2.54 cm (1 in) apart and be 0.64 cm (0.25 in) wide and 0.008 to 0.018 cm (0.003 to 0.007 in) deep depending on the flow characteristics of the material.

EJECTION

The ejection system involves the process of removing the molded part from the mold (Fig. 17.70), usually from the female cavity on the movable platen. Ejectors inevitably leave witness marks on the molding, and this alone may determine the ejection of the part in the mold. Once that basic choice has been made, there is usually not much further freedom to position ejectors in cosmetically acceptable positions. The choices are much constrained by other features in the mold and the need to put ejectors where they are most needed to overcome the resistance of the molded part.

Ejection may be by means of pins (Fig. 17.71) or a stripper plate (Fig. 17.72), or a combination of both. The stripper plate acts on the entire peripheral wall thickness of the part and so distributes the ejection force. This is the preferred method for thin-walled parts, but the method is difficult to apply to anything other than a circular periphery. Action to move the stripper plate can be made by hydraulic systems, chains, mechanical systems, electrical systems, and combinations of these actions (Figs. 17.73 and 17.74).

A variant on this theme is the stripper bar that works on part of the periphery. Pins, usually cylindrical in form, perform most ejections. Rectangular pins may be used in constricted areas; these are often referred to as blade ejectors (Fig. 17.75). Hollow ejector pins known as sleeve ejectors are commonly used to push the molding off small core pins. An air valve that breaks the vacuum between the molding and the mold core often assists ejection of deep-drawn or thin-walled parts. When elastomer plastics of complex shapes are molded, ejection can take advantage of just using an ejection pin (Fig. 17.76). When it is required to ensure that the molded part is ejected from the

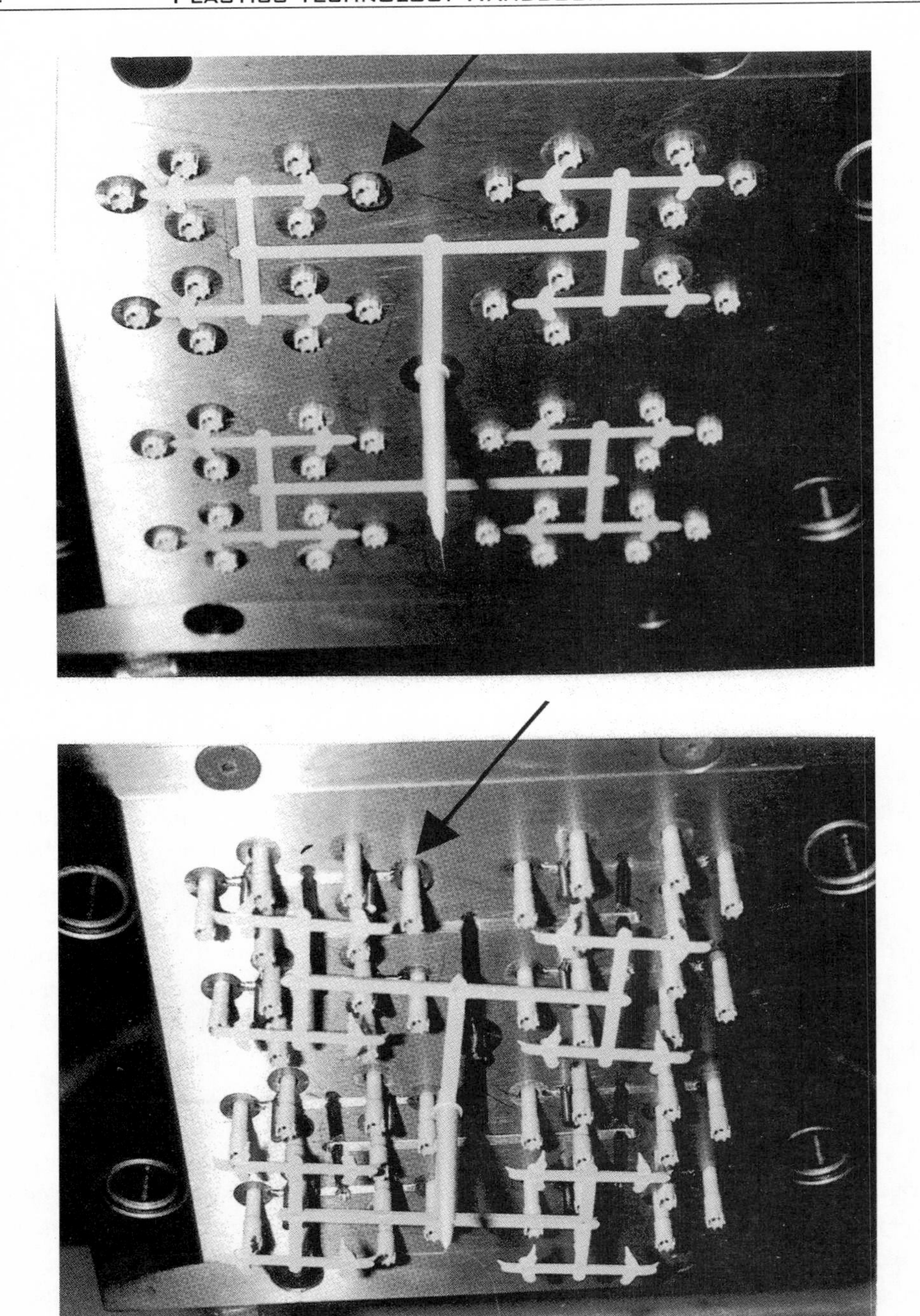

Figure 17.70 Sequence in ejection molded parts using ejection pins.

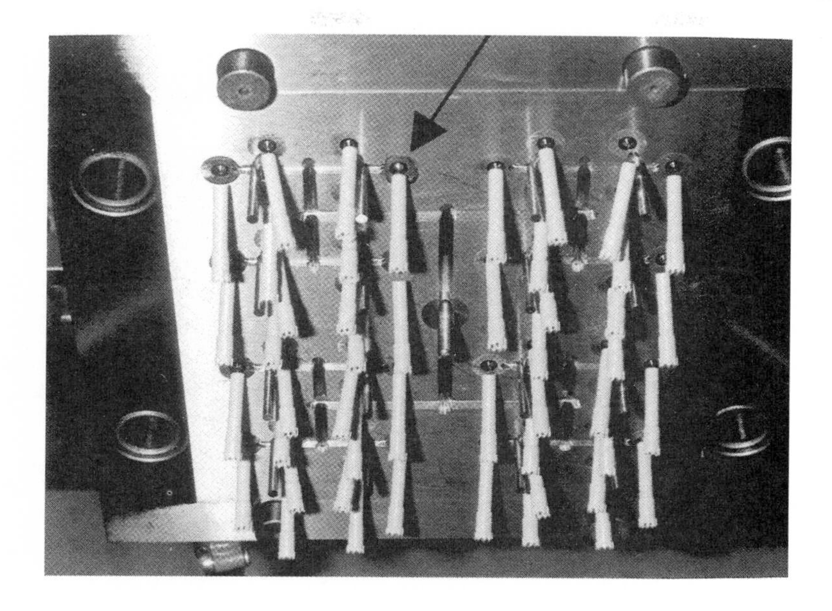

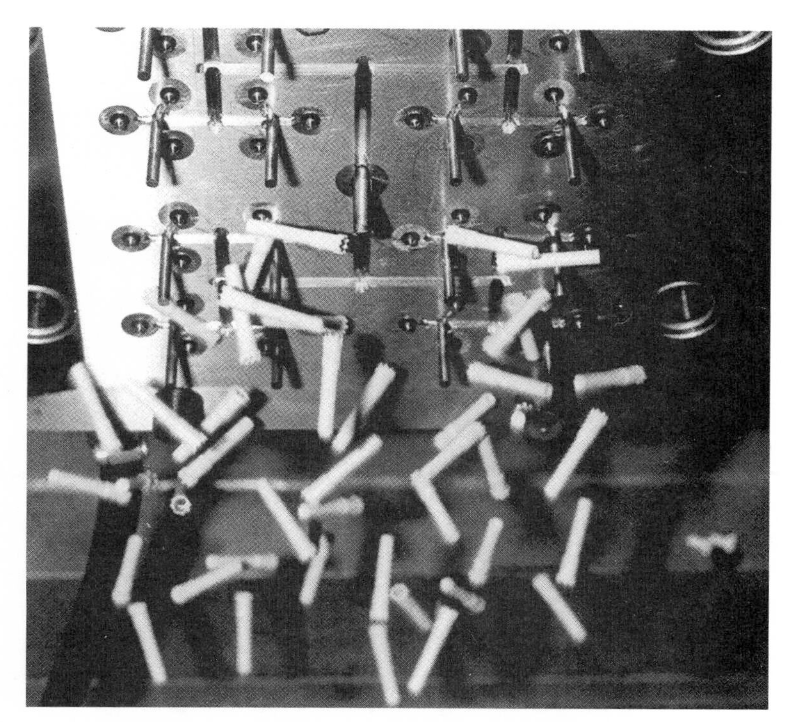

Figure 17.70 Sequence in ejection molded parts using ejection pins *(continued)*.

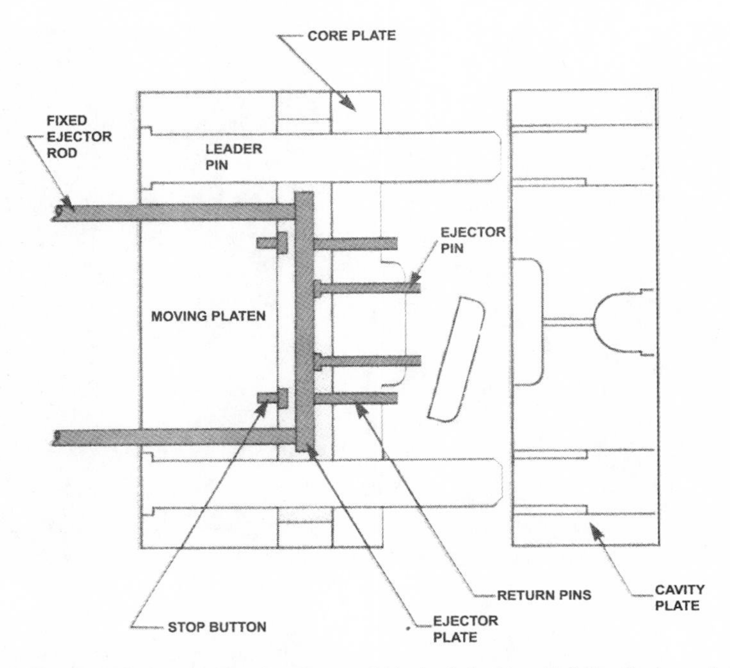

Figure 17.71 Operation of ejector pins (courtesy of Husky Injection Molding Systems Inc.).

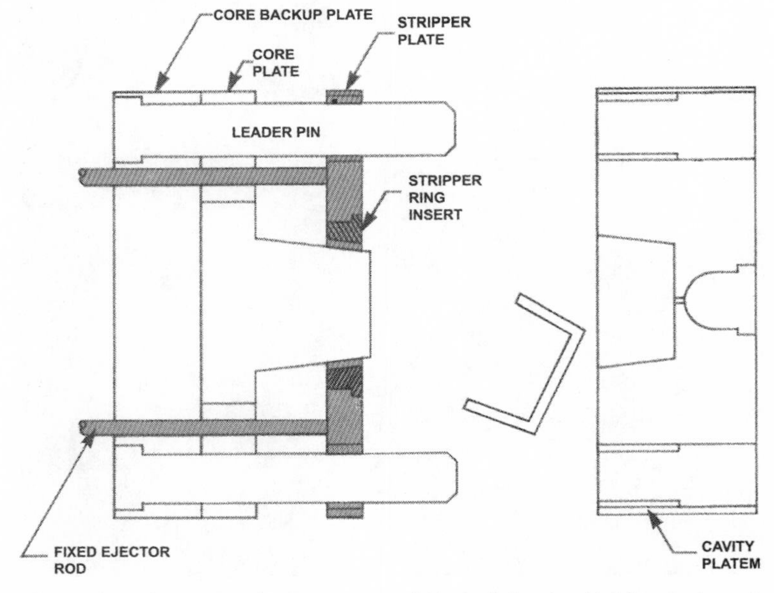

Figure 17.72 Operation of stripper plate (courtesy of Husky Injection Molding Systems Inc.).

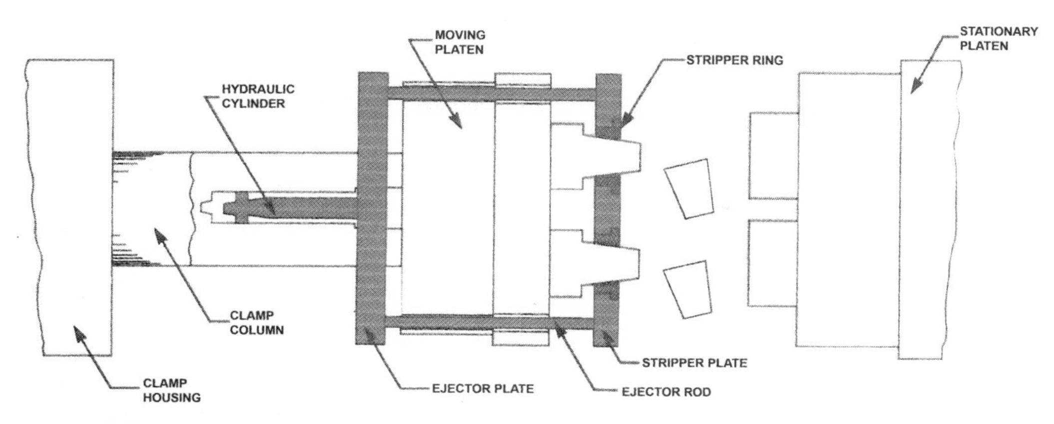

Figure 17.73 Hydraulic operation of stripper plate (courtesy of Husky Injection Molding Systems Inc.).

Figure 17.74 Chain operation of stripper plate.

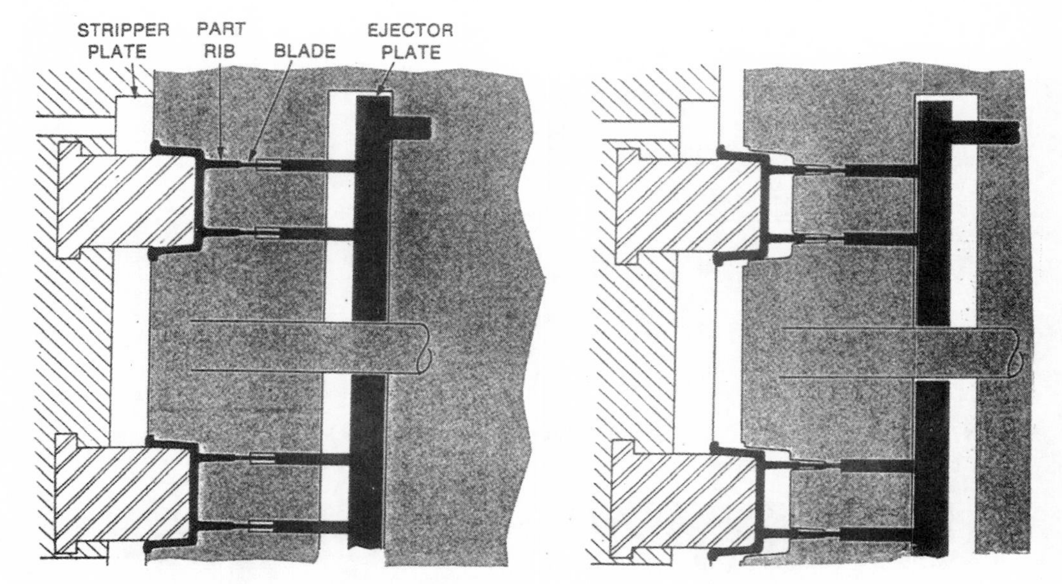

Figure 17.75 Ejection system incorporating blades.

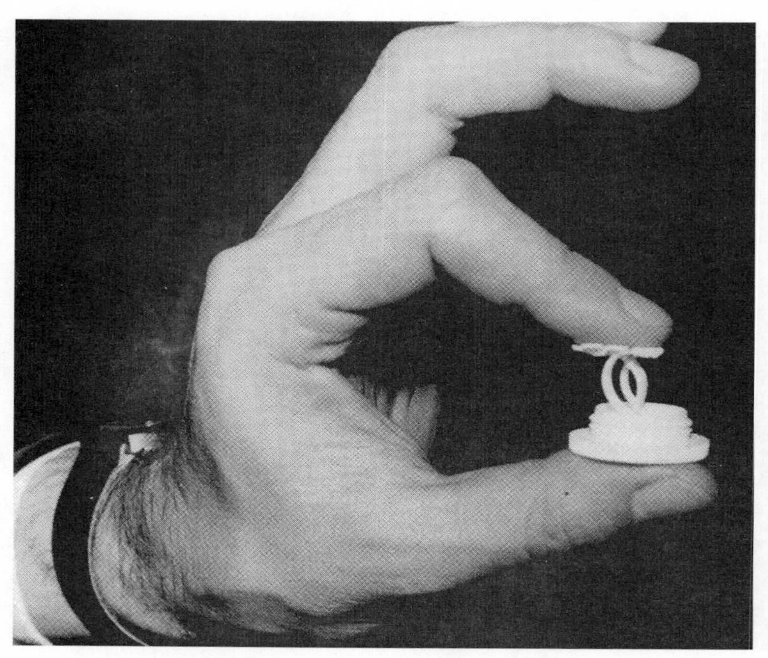

Figure 17.76 Flexible molded parts can easily be ejected from the mold cavity.

female cavity on the movable platen with its ejection system, undercut is required (Fig. 17.77). Figure 17.77 also includes a guide where maximum undercuts are strippable from a mold.

MOLD/PART SHRINKAGE

Moisture effect on a molded part's size starts from the dimensions of the part design to mold-cavity size. Plastics and other materials expand with heat. They also absorb moisture from the atmosphere, which results in an increase in part size. These factors are combined in Figure 17.78, which shows changes in length (mils/in) of a stress-free test specimen molded in nylon 6/6. These very predictable changes in the as-molded length represent steady-state values (equilibrium) with a given temperature and relative humidity.

In the typical exposure of a part to an environment of slowly varying humidity, no true moisture equilibrium is reached, but a balance is established with the average humidity. After initial moisture development has occurred, subsequent variations in relative humidity have little effect on total moisture content and dimensional changes in all but very thin sections. The time to equilibrium is highly dependent on temperature and part thickness (thin sections of nylon 6/6 absorb water very rapidly at higher temperatures). The combined effect of moisture content and thermal expansion causing dimensional changes in nylon 6/6 is easily shown. For example, assume that a part of unspecified length will be required to function at 104°F (41°C) and 50% relative humidity.

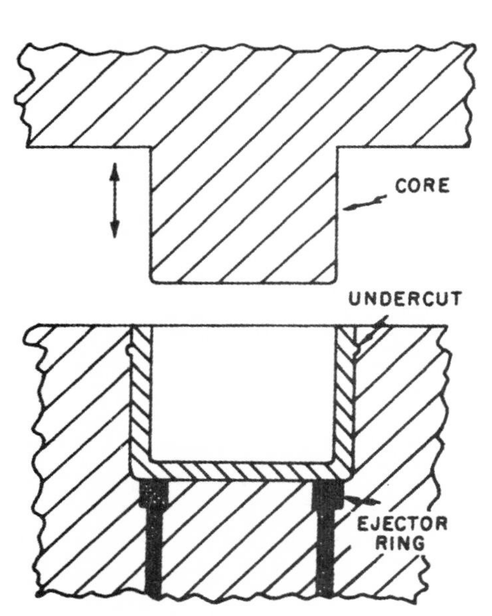

Material	Average maximum strippable undercut [mm (in)]
Acrylic	1.5 (0.060)
ABS	1.8 (0.070)
Nylon	1.5 (0.060)
Polycarbonate	1.0 (0.040)
Polyethylene	2.0 (0.080)
Polypropylene	1.5 (0.060)
Polystyrene	1.0 (0.040)
Polysulfone	1.0 (0.040)
Vinyl	2.5 (0.100)

Figure 17.77 View of undercut that ensures molded part is retained in female cavity. Data on undercuts that are strippable.

Using Figure 17.78, one can easily determine that this part will grow to be 6.8 mils/in longer in use than as molded.

ANNEALING

In addition to changes with nylon due to temperature and humidity conditions after molding is another factor that affects the size after molding is completed. Depending on part thickness and mold temperature employed during molding, dimensions can decrease with time, especially when parts are exposed to temperatures above 150°F (66°C). This is called postmolding shrinkage.

For the greatest dimensional stability at elevated end-use temperatures, annealing is sometimes employed after molding to relieve molded-in stresses and to establish a uniform level of crystallinity in the part (chapter 1). The level of molded-in stresses in most 6/6 nylons is generally low because of their high melt fluidity right up to the onset of solidification. This permits relaxation of flow stresses and orientation effects. Nucleated nylons are sometimes prone to have a higher residual stress level. Parts made in cold molds tend to be most affected because of rapid melt solidification. Flow-induced stresses in thin sections can be frozen-in, and quasiamorphous areas, often induced by cold molds, do not fully develop maximum crystallinity.

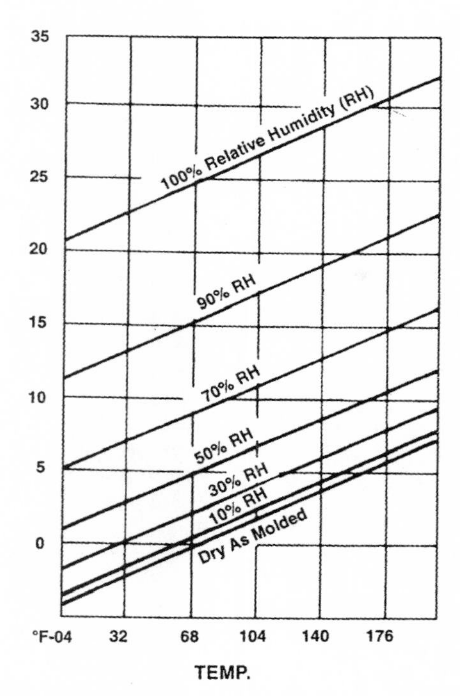

Figure 17.78 Examples of dimensional changes of annealed nylon 6/6 versus temperature at various humidities.

In general, products molded of nylon 6/6 used at temperatures less than 130°F to 150°F (54°C to 66°C) do not require annealing. Conversely, for parts exposed to higher temperatures, especially at low relative humidities in an application requiring stable dimensions, annealing is suggested. Immersion in oil at 325°F to 350°F (163°C to 177°C) for 30 minutes is typical.

Like moisture and temperature changes, the effect of annealing is very predictable. Figure 17.79 shows shrinkage during annealing of test specimens of varying thickness molded over a range of mold temperatures. These annealing changes result in contraction of the part. They often tend to negate the effect of moisture uptake at elevated temperatures that leads to expansion. In many cases, total dimensional change after molding is negligible, since the opposing expansion and contraction effects often counterbalance each other.

ESTIMATING MOLD SHRINKAGE

Molds are sized for a particular plastic, usually after the part design is finalized. It is common practice to leave metal for subsequent machining to final dimensions after trial moldings are made. This is costly, time consuming, and not always good metallurgical practice, since many tool steels should be properly heat-treated before use. Fortunately, EDM techniques allow machining of prehardened steel, which permits certain mold or cavity adjustments after trial shots are made. Nonetheless, it is desirable to size the mold as closely as possible prior to use. Parts injection molded from TPs are generally smaller than the cavity in which they were molded. The reason for this size difference is that the cavity is filled with a melt at high temperature that is less dense than the cooler solid.

The difference between the volume of the mold and that of the part is the mold shrinkage (chapter 4). Conventionally, this difference is expressed as a ratio of the original cavity dimension and is defined as mold shrinkage (MS) = $(C - P)/C$, %, or mils/in, where consistent units are C = cavity dimension and P = part dimension.

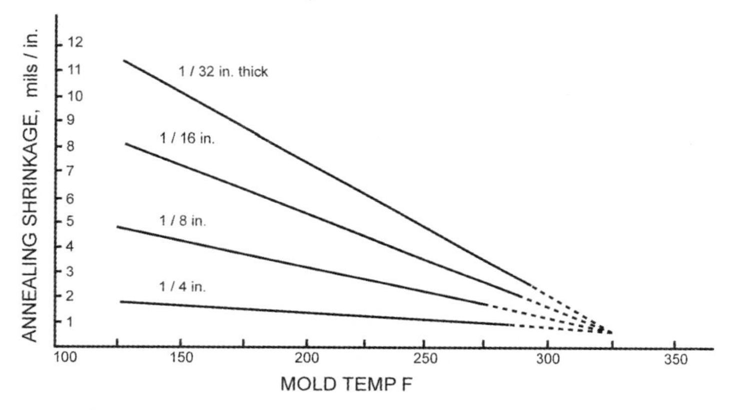

Figure 17.79 Nylon 6/6 shrinkage due to annealing versus mold temperature.

There is also a change in density during molding that is usually identified as the specific volume. The amount depends largely on the temperature of the melt and the pressure applied during processing. As melt temperature increases, specific volume increases, and as pressure increases, specific volume decreases. At the freezing point, an abrupt decrease in specific volume occurs as in an example with nylon 6/6 changes from an amorphous liquid to a semicrystalline solid (chapter 1).

With the temperature of the solid nylon continuing to decrease during cooling, the specific volume continues to decrease. Theoretically, the total volumetric change from melt to solid should approximate three times the linear mold shrinkage. In the actual molding situation, nonuniform cooling spoils this simplified approach. In practice, final mold shrinkage is determined by the temperature and pressure of the nylon melt in the cavity at the time of gate seal-off and the thickness and crystallinity of the frozen part skin (chapter 4). Since the specific volume of a solid material is considerably less than that of any melt, the greater the thickness of the solid layer, the smaller will be the size change as the part comes to room temperature. Minimum shrinkage is obtained when the part is completely solidified when the gate freezes.

With nucleation of nylon the temperature increases until solidification occurs and thereby hastens freezing of both the part and gate. The usual effect of nucleation is to reduce mold shrinkage, but it also increases the amount of frozen-in flow orientation, which can lead to nonuniform shrinkage in flow and transverse directions and, at times, part warpage. The transverse shrinkage will be greater. Pigmentation can also decrease the mold shrinkage of nylon. The greatest effect is seen with high loadings of TiO_2 and other inorganic pigments and salts that act to nucleate nylon. Organic pigments and dyes do not significantly affect shrinkage.

The shape of the part is also very important in determining the mold shrinkage of a given dimension. If the cavity contains undercuts or cores that restrain the free shrinkage of the part, the as-molded shrinkage will be less than that of an unrestrained part. The postmolding or aging shrinkage, however, will be greater for a part that is restrained from free shrinkage in the mold because of greater stresses retained within the part. A guide to mold shrinkage of unrestrained parts of simple geometry is presented in Figure 17.80, where B^1 to B^6 are different DuPont Zytel plastics. Mold shrinkage equals section A (mold dimensional variables) plus section B (process variables). If the cavity is not filled to the limit imposed by the gate seal time, then the measured shrinkage will be greater than that predicted by this graph. The injection speed and hold time (dwell under pressure) were adjusted to give maximum part weight with the injection pressure, melt temperature, and mold temperature as variables.

An example in using Figure 17.80 follows using general-purpose nylon (specifically DuPont's Zytel 101 [B^1]). Assume the following mold variables are gate width is 0.125 in (0.318 cm), gate thickness is 0.090 in (0.229 cm), and part thickness is 0.125 in (0.318 cm). Connecting scales as shown, one obtains a value of about 20 for A (sign is +). Assume the following: melt temperature is 550°F (288°C) and mold temperature is 150°F (660°C). Connect points as shown to reference line R. For example with a screw injection machine, the required melt pressure to fill is 15000 psi (103.4 MPa). Injection gauge pressure on the machine is converted to equivalent melt pressure; this factor varies

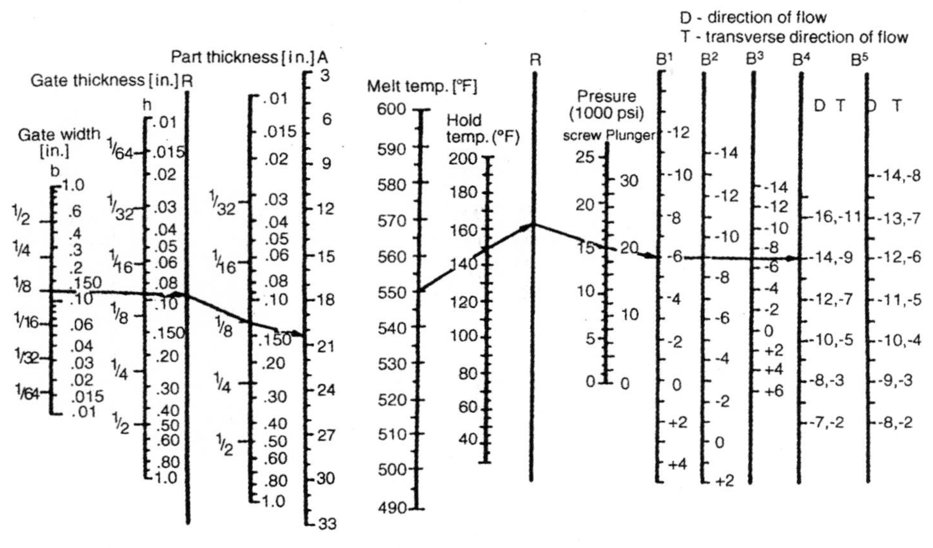

Figure 17.80 This nomograph for nylon estimates shrinkages.

with different machine manufacturers. Connect reference point (on R) with melt pressure and read –6 on scale B_1 (for Zytel 101). Mold shrinkage is A + B = 20 + (–6) = 14 mils/in.

If a plastic other than Zytel 101 nylon is molded, one would connect the point on B_1 horizontally to the specific B scale for the resin used. Note that for Zytel 131 (scale B^4) and Zytel 109 (scale B^5) both plastics are nucleated and show a different B value, depending on whether measurement is made in the direction of flow or transverse to it. Had Zytel 131 been used, the A value (+20) would be identical. However, the B value for flow direction shrinkage would be –14, and the mold shrinkage estimate would be A + B = 20 + (–14) = 6 mils/in. If transverse shrinkage were required, then A + B = 20 + (–9) = 11 mils/in. With nucleated 6/6 nylons there is a different flow and transverse shrinkage. The other nylons exhibit the same shrinkage in both directions. Surface lubrication and mold release agents do not affect mold shrinkage.

To summarize key points in this review, combine Figures 17.78 through 17.80 and estimate the mold shrinkage that would be necessary to produce a part exposed to certain environmental conditions. Assume that a part is molded in unmodified nylon. Part thickness is 3⁄16 in, gate, thickness 0.100 in, gate width 0.150 in, mold temperature 100°F, melt temperature 540°F (282°C), and injection pressure 12000 psi (82.7 MPa). The part must be used at 30% relative humidity at 175°F (80°C). The part is to be annealed after molding for maximum dimensional stability. The aim is to determine the size of the cavity to produce a part dimension (flow direction) of 1.000 in (2.54 cm) in use. Figure 17.78 shows that annealed nylon will grow 8 mils/in in use (30% relative humidity at 174°F). Figure 17.79 (interpolating) shows that for a 3⁄16 in (0.48 cm) thick part molded at 100°F (38°C), the mold will shrink in length during annealing about 4 mils/in.

General information concerning shrinkage are given in Figures 17.81 to 17.83 and Tables 17.52 to 17.55.

MOLD CONSTRUCTION

The initial aim is to simplify (Fig. 17.84) and minimize detractors that will require the manufacture of extremely complicated molds. Being involved with the product designer usually permits concessions to be made resulting in simplifying and reducing the cost of the mold. Detractors will become obvious as this review continues.

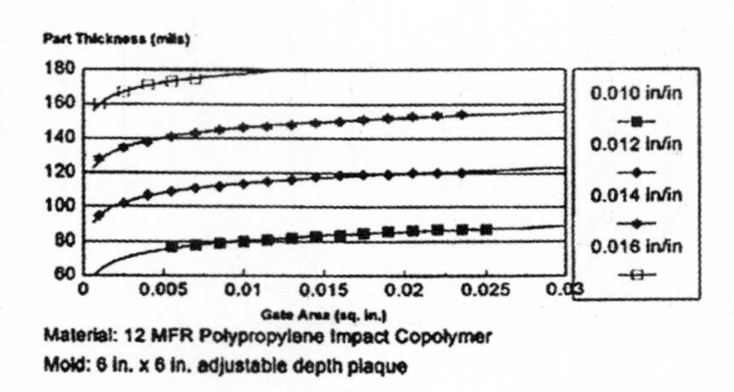

Figure 17.81 Shrinkage as a function of part thickness and gate area.

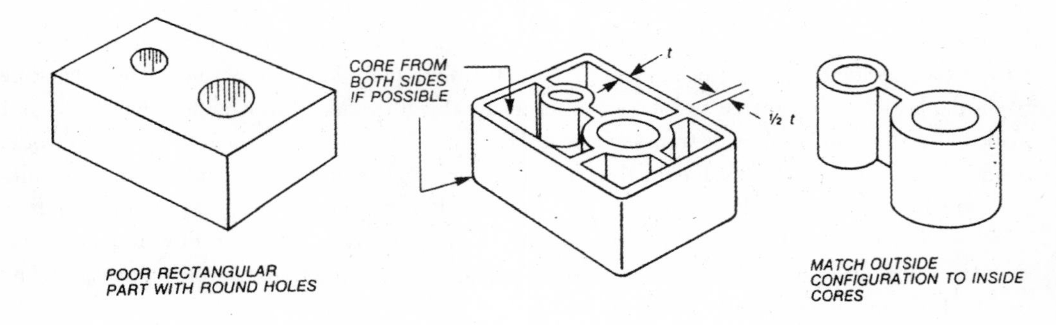

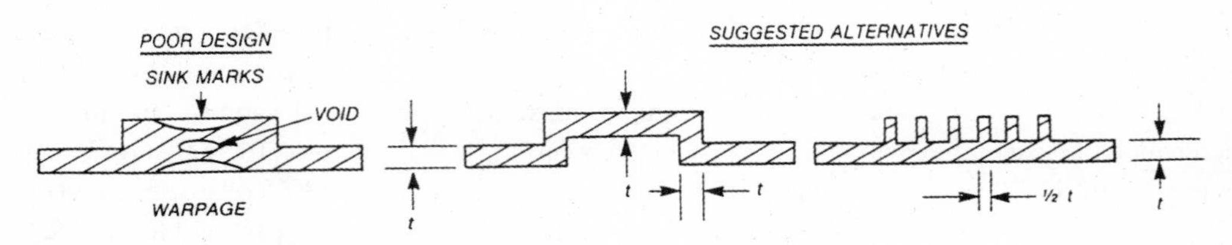

Figure 17.82 Molds can be cored to eliminate or reduce shrinkage.

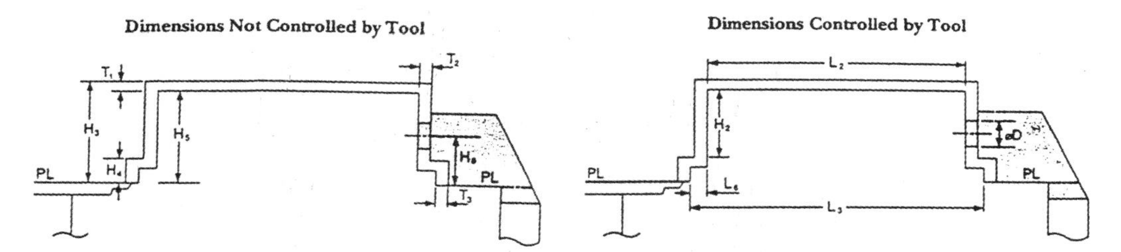

Figure 17.83 Example of shrinkage control and mold dimensions.

Figure 17.84 Example of a simplified unscrewing bottle cap mold.

Parameter	Increase	Maximum shrinkage variation (%)	Notes
Mold temperature	20°C to 90°C	+ 0.6	
Wall thickness	1mm to 6mm	+ 0.5	
Duration of holding pressure	up to 20 sec	– 0.3	min. wall thickness 2mm
Magnitude of injection/holding pressure	600 bar to 1400 bar	– 0.3	min. wall thickness 2mm
Melt temperature	220°C to 280°C	+ 0.3	
Melt flow rate	1g/10 min to 50g/10 min	– 0.3	

Table 17.52 Examples of factors that influence PP shrinkage

Material	Avg. rate in./in.	
	0.125 in. (3.18 mm)	0.250 in. (6.35 mm)
ABS		
Unreinforced	0.004	0.007
30% glass fiber	0.001	0.0015
Acetal, copolymer		
Unreinforced	0.017	0.021
30% glass fiber	0.003	NA
HDPE, homo		
Unreinforced	0.015	0.030
30% glass fiber	0.003	0.004
Nylon 6		
Unreinforced	0.013	0.016
30% glass fiber	0.0035	0.0045
Nylon 6/6		
Unreinforced	0.016	0.022
30% glass fiber	0.005	0.0055
PBT polyester		
Unreinforced	0.012	0.018
30% glass fiber	0.003	0.0045
Polycarbonate		
Unreinforced	0.005	0.007
30% glass fiber	0.001	0.002
Polyether sulfone		
Unreinforced	0.006	0.007
30% glass fiber	0.002	0.003
Polyether-etherketone		
Unreinforced	0.011	0.013
30% glass fiber	0.002	0.003
Polyetherimide		
Unreinforced	0.005	0.007
30% glass fiber	0.002	0.004
Polyphenylene oxide/PS alloy		
Unreinforced	0.005	0.008
30% glass fiber	0.001	0.002
Polyphenylene sulfides		
Unreinforced	0.011	0.004
40% glass fiber	0.002	NA
Polypropylene, homo		
Unreinforced	0.015	0.025
30% glass fiber	0.0035	0.004
Polystyrene		
Unreinforced	0.004	0.006
30% glass fiber	0.0005	0.001

Table 17.53 Guide for mold shrinkage of ¼ and ½ in thick specimens per ASTM D 955

Dimensions, in	ABS		Acetal		Acrylic		Nylon		Polycarbonate		Polyethylene, high-density		Polyethylene, low-density		Polypropylene		Polystyrene	
	Commercial	Fine	Commercial	Fine	Commercial	Fine	Commercial	Fine	Commercial	Fine	Commercial	Fine	Commercial	Fine	Commercial	Fine	Commercial	Fine
To 1.000	0.005	0.003	0.006	0.004	0.005	0.003	0.004	0.002	0.004	0.0025	0.008	0.006	0.007	0.004	0.007	0.004	0.004	0.0025
1.000–2.000	0.006	0.004	0.008	0.005	0.006	0.004	0.006	0.003	0.005	0.003	0.010	0.008	0.010	0.006	0.009	0.005	0.005	0.003
2.000–3.000	0.008	0.005	0.009	0.006	0.007	0.005	0.007	0.005	0.006	0.004	0.013	0.011	0.012	0.008	0.011	0.007	0.007	0.004
3.000–4.000	0.009	0.006	0.011	0.007	0.008	0.006	0.009	0.006	0.007	0.005	0.015	0.013	0.015	0.010	0.013	0.008	0.008	0.005
4.000–5.000	0.011	0.007	0.013	0.008	0.009	0.007	0.010	0.007	0.008	0.005	0.018	0.016	0.017	0.011	0.015	0.009	0.010	0.006
5.000–6.000	0.012	0.008	0.014	0.009	0.011	0.008	0.012	0.008	0.009	0.006	0.020	0.018	0.020	0.013	0.018	0.011	0.011	0.007
6.000–12.000, for each additional inch	0.003	0.002	0.004	0.002	0.003	0.002	0.003	0.002	0.003	0.015	0.006	0.003	0.005	0.004	0.005	0.003	0.004	0.002
	0.004	0.002	0.004	0.002	0.003	0.003	0.004	0.003	0.003	0.002	0.006	0.004	0.005	0.004	0.006	0.003	0.0055	0.003
	0.003	0.002	0.004	0.002	0.005	0.003	0.005	0.003	0.003	0.002	0.006	0.004	0.005	0.004	0.006	0.003	0.007	0.0035
0.000–0.125	0.002	0.001	0.002	0.001	0.003	0.001	0.002	0.001	0.002	0.001	0.003	0.002	0.003	0.002	0.003	0.002	0.002	0.001
0.125–0.250	0.002	0.001	0.003	0.002	0.003	0.002	0.003	0.002	0.002	0.015	0.005	0.003	0.004	0.003	0.004	0.003	0.002	0.001
0.250–0.500	0.003	0.002	0.004	0.002	0.004	0.002	0.003	0.002	0.003	0.002	0.006	0.004	0.005	0.004	0.005	0.004	0.002	0.0015
0.500 and over	0.004	0.002	0.006	0.003	0.005	0.003	0.005	0.003	0.003	0.002	0.008	0.005	0.006	0.005	0.008	0.006	0.0035	0.002
0.000–0.250	0.003	0.002	0.004	0.002	0.004	0.002	0.004	0.002	0.002	0.002	0.005	0.003	0.003	0.003	0.005	0.003	0.0035	0.002
0.250–0.500	0.004	0.002	0.005	0.003	0.004	0.002	0.004	0.003	0.003	0.002	0.007	0.004	0.004	0.004	0.006	0.004	0.004	0.002
0.500–1.000	0.005	0.003	0.006	0.004	0.006	0.003	0.005	0.004	0.004	0.003	0.009	0.006	0.006	0.005	0.009	0.006	0.005	0.003
0.000–3.000	0.015	0.010	0.011	0.006	0.010	0.007	0.010	0.004	0.005	0.003	0.023	0.015	0.020	0.015	0.021	0.014	0.007	0.004
3.000–6.000	0.030	0.020	0.020	0.010	0.015	0.010	0.015	0.007	0.007	0.004	0.037	0.022	0.030	0.020	0.035	0.021	0.013	0.005
TIR	0.009	0.005	0.010	0.006	0.010	0.006	0.010	0.006	0.005	0.003	0.027	0.010	0.010	0.008	0.016	0.013	0.010	0.008

Table 17.54 Guide for mold shrinkage for different thickness dimensions

Part Size Inches	Plastic Shrink (value in mil (0.001 in.)															
	0.004	0.008	0.012	0.016	0.020	0.030	0.040	0.050	0.060	0.070	0.080	0.090	0.100	0.200	0.300	0.400
1.0	−0.02	−0.06	−0.1	−0.3	−0.4	−0.9	−1.7	−3	−4	−5	−7	−9	−11	−50	−129	−267
3.0	−0.05	−0.19	−0.4	−0.8	−1.2	−2.8	−5.0	−8	−11	−16	−21	−27	−33	−150	−386	−800
5.0	−0.08	−0.32	−0.7	−1.3	−2.0	−4.6	−8.3	−13	−19	−26	−35	−45	−56	−250	−643	−1333
7.0	−0.11	−0.45	−1.0	−1.8	−2.9	−6.5	−11.7	−18	−27	−37	−49	−62	−78	−350	−900	−1867
9.0	−0.14	−0.58	−1.3	−2.3	−3.7	−8.4	−15.0	−24	−34	−47	−63	−80	−100	−450	−1157	−2400
11.0	−0.18	−0.71	−1.6	−2.9	−4.5	−10.2	−18.3	−29	−42	−58	−77	−98	−122	−550	−1414	−2933
13.0	−0.21	−0.84	−1.9	−3.4	−5.3	−12.1	−21.7	−34	−50	−68	−90	−116	−144	−650	−1671	−3467
15.0	−0.24	−0.97	−2.2	−3.9	−6.1	−13.9	−25.0	−39	−57	−79	−100	−134	−167	−750	−1929	−4000
17.0	−0.27	−1.10	−2.5	−4.4	−6.9	−15.8	−28.3	−45	−65	−90	−118	−151	−189	−850	−2186	−4533
19.0	−0.31	−1.23	−2.8	−4.9	−7.8	−17.6	−31.7	−50	−73	−100	−132	−169	−211	−950	−2443	−5067
21.0	−0.34	−1.35	−3.1	−5.5	−8.6	−19.5	−35.0	−55	−80	−111	−146	−187	−233	−1050	−2700	−5600
23.0	−0.37	−1.48	−3.4	−6.0	−9.4	−21.3	−38.3	−61	−88	−121	−160	−205	−256	−1150	−2957	−6133
25.0	−0.40	−1.61	−3.6	−6.5	−10.2	−23.2	−41.7	−66	−96	−132	−174	−223	−278	−1250	−3214	−6667
27.0	−0.43	−1.74	−3.9	−7.0	−11.0	−25.1	−45.0	−71	−103	−142	−188	−240	−300	−1350	−3471	−7200
29.0	−0.47	−1.87	−4.2	−7.5	−11.8	−26.9	−48.3	−76	−111	−153	−202	−258	−322	−1450	−3729	−7733
31.0	−0.50	−2.00	−4.5	−8.1	−12.7	−28.8	−51.7	−82	−119	−163	−216	−276	−344	−1550	−3986	−8267
33.0	−0.53	−2.13	−4.8	−8.6	−13.5	−30.6	−55.0	−87	−126	−174	−230	−294	−367	−1650	−4243	−8800
35.0	−0.56	−2.26	−5.1	−9.1	−14.3	−32.5	−58.3	−92	−134	−184	−243	−312	−389	−1750	−4500	−9333
37.0	−0.59	−2.39	−5.4	−9.6	−15.1	−34.3	−61.7	−97	−142	−195	−257	−329	−411	−1850	−4757	−9867
39.0	−0.63	−2.52	−5.7	−10.1	−15.9	−36.2	−65.0	−103	−149	−205	−271	−347	−433	−1950	−5014	−10400
41.0	−0.66	−2.65	−6.0	−10.7	−16.7	−38.0	−68.3	−108	−157	−216	−285	−365	−456	−2050	−5271	−10933
43.0	−0.69	−2.77	−6.3	−11.2	−17.6	−39.9	−71.7	−113	−165	−227	−299	−383	−478	−2150	−5529	−11467
45.0	−0.72	−2.90	−6.6	−11.7	−18.4	−41.8	−75.0	−118	−172	−237	−313	−401	−500	−2250	−5786	−12000
47.0	−0.76	−3.03	−6.9	−12.2	−19.2	−43.6	−78.3	−124	−180	−248	−327	−418	−522	−2350	−6043	−12533
49.0	−0.79	−3.16	−7.1	−12.7	−20.0	−45.5	−81.7	−129	−188	−258	−341	−436	−544	−2450	−6300	−13067

Table 17.55 Examples of error in mold size as a result of using incorrect shrinkage formulas

Components of the usual mold are depicted in Figures 17.22, 17.64, and other figures and in Table 17.45. Many different types of components are included in different molds to meet different performance requirements. Two examples are the sprue puller (Fig. 17.85) and the pressure transducer (Fig. 17.86).

The SPI Mold Guide provides descriptions and specifies classifications of molds. It is usually stated that two mold halves make up the mold. They may not be of equal thickness and some molds may be in three or more sections to permit ease of the molded part removal. In all cases the parting line on the molded part is formed where plates A and B meet at the cavity. Cavity plate A retains the cavity inserts and contains four leader pins. This action permits maintaining the alignment of mold halves during the closing and opening operation. The guide pins are usually mounted in the stationary mold half to ensure that the molded productwill fall out of the other mold half during ejection without hang-up. This means the core (stationary) side has the guide bushings. One of the four leader pins is offset by a small amount to eliminate the chance of improper assembly of the two halves.

Mating with the A plate is the B plate, which holds the opposite half of the cavity or the core and contains the leader pin bushings for guiding the leader pins. The core establishes the inside configuration of a part. The B plate has its own backup or support plate in the series of mold bases. Pillars below the U-shaped structure known as the ejector housing frequently support the backup B plate. The U-shaped structure, consisting of the rear clamping plate and spacer blocks, is bolted to the B plate, either as separate parts or a single unit.

The stripper stroke is the U-shaped structure that provides the space for moving the ejector plate or ejection stroke. The return pins support the ejector plate, ejector retainer, and pins. When in an inactivated position, the ejection plate rests on stop supports. When the ejection system becomes heavy because of required high injection forces, additional means of support are provided by mounting added leader pins in the rear clamping plate and bushing in the ejector plate.

The alignment of mold halves is usually accomplished using leader pins (Fig. 17.87). Many moldmakers use tolerances of +0.0008 in to ±0.0013 in (±0.0020 to ±0.0033 cm) per side pin to bushing. Tighter tolerances of ±0.0004 to ±0.0008 in (±0.0010 to ±0.0020 cm) provide more accurate alignment and less mold wear. On ejector systems, a at least four leader pins and bushings are usually used to prevent cocking of the plate, which reduces wear and prevents seizing.

With proper alignment of the two mold halves, no TP escapes from a cavity through its land. Flash occurs with TS plastic because its viscosity during the melting action resembles that of water. The land describes the area of a closed mold that come into contact with one another. The material that escapes through the clearance is called flash. The mold is designed to trap all the melt when it closes. There is very little clearance between the mating surfaces of the male and female cavities. The clearance varies between 0.004 to 0.013 cm/side (0.0015 to 0.005 in/side), depending on the size of the mold and the material to be molded. High-speed buffing, grinding, and/or tumbling usually removes flash. A raised line may be evident on the surface of a molding that is formed at the junction of the mold faces such as at the parting line after the removal of the excess flash.

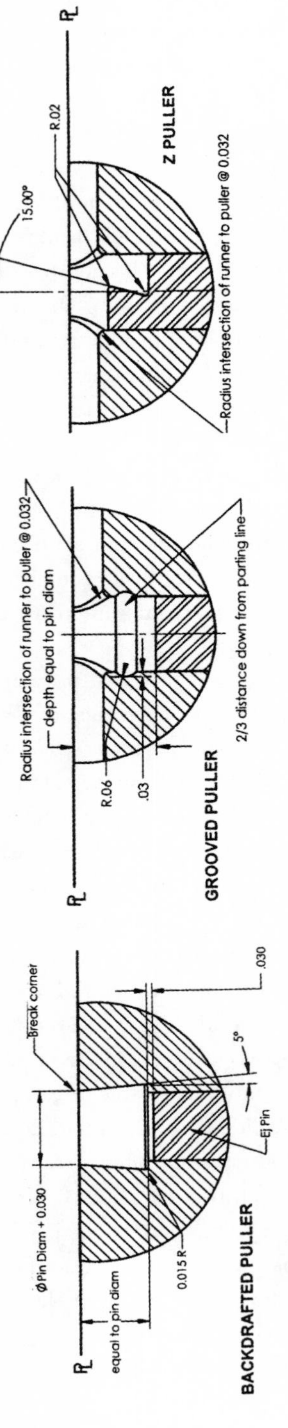

Figure 17.85 Examples of sprue pullers.

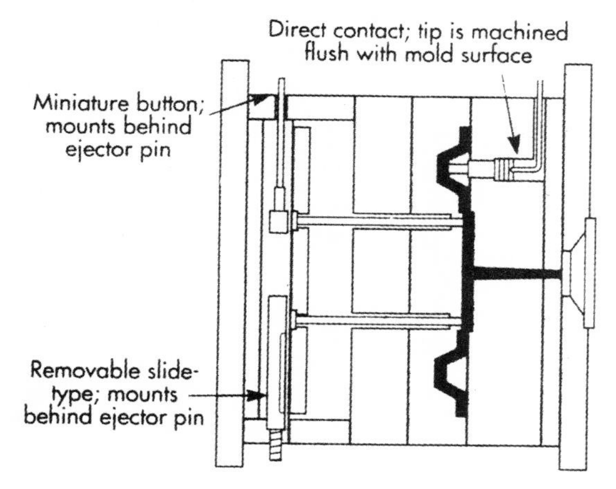

Figure 17.86 Example of the location for a mold pressure transducer sensor.

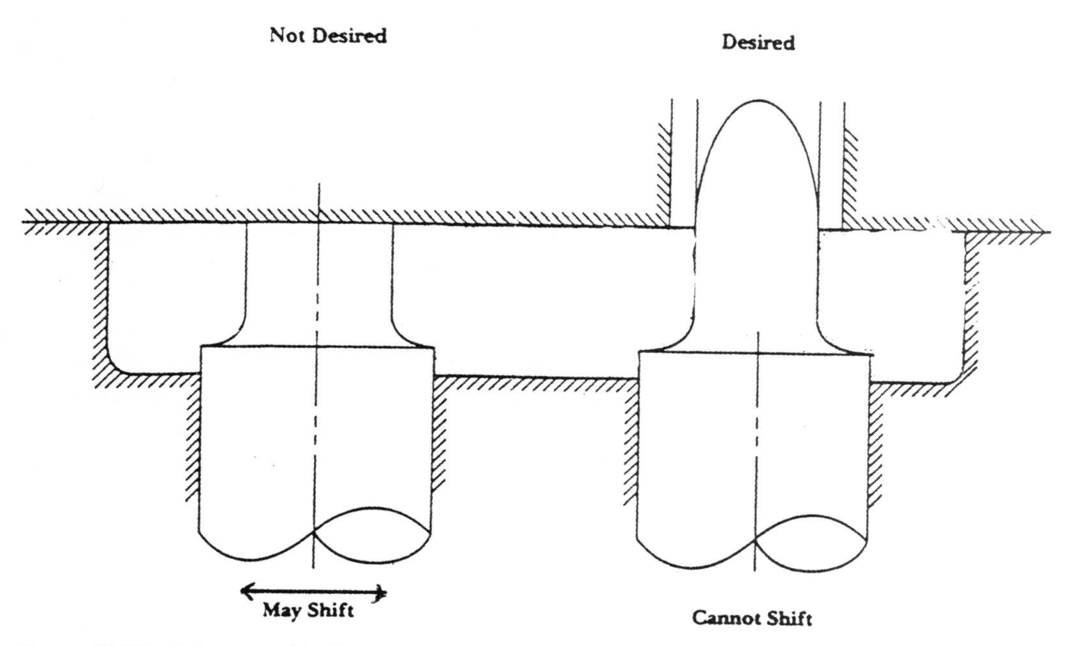

Figure 17.87 Guide to mold alignment.

There are reverse series of A and B mold bases. Molds can be ordered with pins and bushings reversed. In the A series, a separate clamping plate is used on the stationary side while a separate support is used under the core mounting plate. This makes it easier to install insert cavities and cores. In the B series, the clamping plate and cavity plate are combined into one thicker plate and on the core side the core mounting plate and support plate are combined into one thicker plate.

Proper alignment of the mold with the machine is required to permit proper melt flow and eliminate loss of melt. The mold is aligned with the injection cylinder of the machine by means of a ring in the stationary mold half, into which the cylinder nozzle seats. This locating ring surrounds the mold's sprue bushing and is used for locating the mold in the press platen concentrically with the machine nozzle. The opening into which the ring fits is usually made to a tolerance of –0.000 and +0.005 cm (–0.000 and +0.002 in). The ring itself is made 0.025 cm (0.010 in) smaller than the opening, providing a clearance of 0.013 cm (0.005 in) per side. SPI standards provide details.

Mold plates (excluding the ejector parts) and spacer blocks are ground to a tight thickness tolerance such as ±0.001 in. Conceivably, a combination of tolerances could build up to cause unevenness at its four corners. If great enough, such a condition would damage a platen when under full ram pressure. It is advisable to check the uniformity of all four corners prior to preparing the base to receive cavities.

Both mold halves are provided with cooling channels for a coolant to carry away the heat delivered to the mold by the hot TP plastic melt. With TSs, electric heaters are located in the mold to provide the additional heat required.

The quality of machine cutting tools used in manufacturing molds is absolutely critical to the operating efficiency of a mold. A significant factor has been the CNC systems for machining all kinds of tools. Although the skill and experience of the toolmaker remain an essential factor, the CNC program gives one precise control over all machining operations. Many of these are lengthy and repetitive, which makes them ideal for computerization. The computerized control of machine tools allows the complete moldmaking process to be integrated into the product and mold design. Thus, the actual computer tape or disk containing the design data can subsequently be fed into the machine-tool control system to give instructions for the detailed work to be done.

The combination of CAD, computer-aided manufacturing (CAM), computer-aided engineering (CAE), and CNC has had a further benefit in that it brings together all the expertise contributing to the manufacture of the mold (chapter 25). In the past, such experts performing each of these tasks made their contributions individually. This could produce an undesirable situation in which most of the work of current attempts to compensate for the errors/mistakes of previous ones. In regard to the CAD, CAM, CAE, and CNC disciplines, all such experts make their contributions together, virtually at the same time and certainly prior to finalized design.

SPACER BLOCK CONTACT

It is important to study the contact area of the spacer blocks in the mold. The stress on these areas should be such as to prevent the embedding of these blocks into the plates, thus causing a change of the space for the ejection system. The safe tonnage that a mold base will take as far as spacer blocks are concerned can be calculated. An example involves a 25.1 × 30.2 cm (9⅞ in × 11⅞ in) standard mold base made of low-carbon steel. The weakest section of the spacer bar is at the clamping slot area. For this size mold base, the width of the block is 1 7/16 in and the width of the clamping slot ⅝ in or 13/16 × 11⅞ × 2 in, since there are two blocks:

An important calculation is area × allowable stress = compressive force. This allowable stress for low-carbon steel is 25000 psi. Thus 13/16 in × 11⅞ in × 2 × 25000 psi = 482,000 lb or 241 tons. Higher-strength steel throughout the base can double or even triple the ability to absorb the compressive force. The addition of supporting pillars will also increase the allowed compressive force in proportion to the area that they provide. Thus two 2 in diameter pillars would add the following force:

$$\text{Area of a 2 in diameter pillar} = 3.14\ \text{in}^2 \times 2 = 6.28\ \text{in}^2,$$
$$6.28\ \text{in}^2 \times 25000\ \text{psi} = 157{,}000\ \text{lb or } 78.5\ \text{tons}.$$

BLIND HOLE

It is important to ensure that sufficient material surrounds the holes and that melt flows properly in the cavity. A core pin forming blind holes is subjected to the bending forces that exist in the cavity due to the high melt pressures. Calculations can be made for each case by establishing the core pin's diameter, its length, and the anticipated pressure conditions in the cavity.

From engineering handbooks it is known that a pin supported on one end only will deflect up to fifty times as much as one supported on both ends. This suggests that the depth of hole in relation to diameter should be small, in order to maintain a straight hole. Sometimes a deep, small-diameter hole is needed, as in pen and pencil bodies. In this case the plastic flow is arranged to contact the free end of the core, for example, from four to six evenly spaced gates. This design will cause a centering action, and the plastic will continue flowing over the diameter in an umbrellalike pattern to balance the pressure forces on the core.

When this type of flow pattern is impractical, an alternative may be a through hole or tube formation combined with a postmolding sealing or closing operation by spinning or ultrasonic welding. At the other extreme, consider a 0.6 cm (0.25 in) diameter core exposed to a pressure of 28 MPa (4000 psi) with an allowance for deflection of 0.00025 cm (0.0001 in) and see how deep a blind hole can be molded under these conditions.

RELEASE AGENT

A release agent is also called a parting agent. It is a substance put on an interior mold-cavity surface and/or added to a molding compound, such as silicone, to facilitate removal of the molded product from the mold. Using certain agents such as silicone can cause bonding problems if parts are to be bonded or decorated in a secondary operation; it may also interfere with electrical circuits.

FASTER/LOWER-COST MOLD INSERT APPROACH

Synventive Molding Solutions reports there are always new ways to build molds faster and at lower cost that result in high-quality products. The "smart" approach means using the latest technologies and computer systems that enable molders to address both quality and profitability by making only a modest capital investment in family tooling with a return on investment in four months. What this means for moldmakers is that mold-insert components are used in lieu of complete molds for production applications. This approach offers the precision and ease of start-up of single-cavity molding in multicavity tools. Moldmakers and molders can make rapid changes to a product, develop mold inserts to suit the application and ship these components anywhere in the world. By building inserts that are stocked on the shelf, lead and tooling times are reduced. The moldmaker can concentrate on building intricate core and cavity works, add value, and optimize profits.

Although family tools and modular tooling solutions are not new to the industry, it is the unique capability of smart molding to independently profile each cavity that, when coupled with family molds or multicavity systems facilitates timely and cost-effective implementation. Field tests have shown that savings in tooling costs alone can be as much as 40%.

This approach permits the molder to control the packing rate right at the nozzle; moldmakers no longer face the lengthy process of balancing molds before manufacturing commences. The molder can balance and qualify molds and rapidly produce high-quality products. Molders simply adjust the pressures at the tip of the nozzle to get optimum fill and packing rates. It is achieved with a real-time, closed-loop pressure control system that allows molders independent control of injection and packing pressure at each gate in the mold. The ability to individually create, fill, pack, and hold profiles for each cavity in the mold eliminates the time-consuming process of capability studies to qualify individual cavities.

This approach utilizes rapid tooling methods that are combined with smart molding technology. Utilizing flexible inserts, tool-build times can be reduced to 2 weeks or less, as compared with conventional methods that take up to 10 weeks. Moreover, different parts can be made in the same mold base, significantly reducing manufacturing costs while maintaining the ability to respond to changes in market demand.

Inserts can be run together as a family mold, saving unit costs. Because there is less time to build inserts and reduced time required to qualify the molds with a smart molding system, moldmakers and molders can realize a 40% to 50% savings in time required to bring products to market with a concomitant 50% reduction in overall tooling costs (255).

MANUFACTURING MOLD CAVITY

Mold cavity and core inserts are fabricated by a wide variety of methods. However, conventional machining of the cavity and core inserts is the most widely employed method of fabrication. The term *machining* is used here to denote the different mechanical and electrical milling, drilling, boring, turning, grinding, and cutting equipment. After selection of the cavity and core material, the next step is to cut and/or machine the raw material to the approximate size. This is basically accomplished with either vertical or horizontal power-sawing equipment.

The next operation is to square, true, and size the inserts. Normally, this stage of fabrication is carried out in the material's soft or annealed stage if hardened steels are being utilized for the final product. Round inserts generally are trued and sized on turning equipment, which includes lathes and cylindrical grinders. Square and rectangular blocks are milled when the hardness of the raw material allows them to be. Final sizing of prehardened or hardened blocks is completed on surface grinders

In this preliminary stage of fabrication, the usual heels required for retention in through pockets are established, or screw holes are installed for inserts that will fit into blind pockets. Although in these preliminary stages it appears that not much progress is being made on the cavities and cores, squaring is one of the most important steps in mold manufacture. Mold suppliers offer oversized tool steel inserts for square or rectangular blocks, with round inserts furnished with heels. Considerable savings of in-house labor are possible by utilizing these inserts, which are generally available in P-20, H-13, and 420 stainless steel.

The next operations to be carried out on the inserts are installation of the mold temperature-control circuits. Square and rectangular blocks normally have water, steam, or oil channels drilled directly into them with other lines connecting to form internal loops. Internal water channels are blocked with threadless-brass pressure plugs to direct the flow of the temperature-control fluid. Generally, only one inlet and one outlet are used per insert with the remaining channels blocked, using brass or steel pipe plugs. Conventional drilling equipment is normally utilized for this operation, with gun drilling used for deep holes, where accuracy of location is required. Great care must be taken in this operation, as nothing dampens the enthusiasm of the moldmaker more than hitting a waterline with a screw hole.

As modern molds require greater sophistication in temperature control to meet the challenges presented by today's design engineers, programs have been developed to accurately predict the amount and placement of temperature-control channels. Generally, the core will be required to remove approximately 67% of the BTUs generated in the IM process. This requirement presents one of the greatest challenges to the moldmaker, as less space usually is available in the core than the cavity obstructing heat-transfer capability.

In the past the standard of the industry was the reliable Bridgeport with the skill of the journeyman moldmaker coaxing accuracy from the equipment. Today, the standard is to use numerical control (NC), CNC or digital numerical control (DNC) milling, drilling, duplicating, EDM, and grinding machines. Often, the mold designer generates tapes or softwares, a practice made possible by the wide acceptance of CAD/CAM equipment in even the smallest of operations. The addition of CAD/CAM equipment has made the skilled craftsman more productive and more valuable (chapter 25).

Accuracy of modern equipment or the skill of the journeyman moldmaker is required for the machining operations that now involve the cavity and core inserts. Depending on subsequent machining operations, the knockouts for ejector pin holes may be located and established. Perhaps one of the greatest values in mold-making today is the off-the-shelf availability of high-precision ejector pins. Nitrided hotwork ejector pins, constructed from superior-quality, thermal stock H-13 steel, are available in the shoulder or straight type with an outside diameter hardness in the 65 to 74 Rc range. Ejector pins, both imperial and metric, are available in every popular fraction, letter, and millimeter size required by the moldmaker. Should requirements dictate nonstandard sizes, the component supplier also can build those pins. It is worth noting that, for purposes of future replacement, the use of standard, off-the-shelf components is highly recommended and frequently demanded by most tool engineers since they are very involved in obtaining metals conforming to very tight tolerances.

CASTING

The advantage of casting cavities is that large amounts of excess stock are not removed when compared to other methods. Various cavity materials are suitable for casting (steel, beryllium copper, Kirksite, aluminum, and others). In the casting process, a pattern must be constructed to not only incorporate the features of the desired cavity configuration, but also the shrinkage of the casting and plastic materials. Casting is usually less expensive than to machining a cavity. Fabrication economics favor this process, particularly for the larger cavities. Because the melting point of Kirksite is relatively low, patterns can be built from wood or plaster. Another advantage is that in some materials, water lines can be cast internally.

HOBBING

In the past, hobbing was very popular for mold cavities. Hobbing involves forcing a hardened negative pattern, usually steel, into relatively soft stock steel or other material under extremely high pressure. This process has a limiting factor of obtainable cavity depth to a certain ratio of hob diameter. The master hob contains the polish and precise detail desired in the finished cavity. The next operation after the hobbing itself is to cut the outside of the cavity blank to the desired size, install any required water passages and gate details, harden the product, and then apply a final polish.

In the past decade, many hobbing applications have been replaced by EDM. EDM can be used on all steels and puts much less stress onto the metal. EDM machining does not care whether the steel is hard or soft. EDM does leave a damaged surface layer that must be removed before polishing or texturizing.

POLISHING

As reviewed in the beginning of this chapter plastic molds usually require a high polish. Though the operation seemingly is gentle, polishing can still damage the steel unless it is properly done (orange peel, etc.).

PREENGINEERING

Efficient, interchangeable, standardized, and complete preengineered molds (and dies)—as well as their components—have been available for over a half century. They save costs and can be delivered quickly. They conform either to ASTM/DIN/ISO or internal work standards (chapter 22). These high-quality manufactured parts result in consistent quality and reduced mold costs. A major advantage to the molder is the cost and time savings should a component ever need replacement. They reduce cost risks in most mold shops because they are bought at fixed prices and are delivered from stock.

Included are precision components for gating, hot runners, demolding, cooling, and so on. They can provide special machining of components such as ready-to-install cavity plates with milled-out pockets for inserts, slides, different types of cooling lines, guide holes, and other special machinings according to the customer requirements.

CAD standard components libraries offer considerable support in producing molds based on the requirements provided. Parts can be displayed immediately on a computer screen. It is possible to arrange and generate the molded product on a computer screen with the complete mold around it. The components can be automatically combined into a purchasing mold or parts list.

Mold-making has been moving toward more specialization. Special devices simplify or permit molding complex shapes. Various machine operators on milling, lathe, EDM, or jig grinding machines manufacture parts of the finished mold separately. At the final stage, all these parts are brought together and assembled by the moldmaker. Standard elements are designed exactly for this type of manufacturing process.

In principle, almost all molds consist of the same basic elements: mold plates for the inserts, intermediate plates for supporting the cores and inserts in the mold plate on the ejector side, risers to limit the working distance of the ejector plates, and clamping plates to clamp the mold to the machine.

Studies have shown that the total hours required for mold production can be reduced using standard components. In addition, the manufacturer of standard elements can also handle special machining requirements. In total, these add up to some 40% of the capacity required for the production of the usual mold.

A mold-making shop that replaces rough machining capacity by specialized machining capacity for the production of contour parts utilizes manpower and machines more efficiently. Calculations can be made with lower hourly rates, since the rate of utilization is better. Mass production of standard elements for molds on large, highly efficient special machinery guarantees high quality and purchase without risk because of fixed prices and reliable delivery dates.

As in many major industries, various commercial and administrative practices have developed over the years that play an important role in the conduct of day-to-day business. These arrangements, generally expressed in the proposal, acknowledgment, and contract forms of individual companies, have been viewed as constituting "customs of the trade."

SPECIAL DEVICE

Specialty devices are those that have been engineered to simplify or improve the performance of particular mold functions. These functions can be as straightforward as returning the ejector assembly early or as sophisticated as a runnerless molding system. These devices have been standardized for installation in a variety of molding applications, and as a result, they do not have to be designed and built by individual mold designers and moldmakers.

COLLAPSIBLE CORES

These components help the molding of complex internal shapes. A collapsible core provides a means to mold internal threads, undercuts (as in snap fits, hamper-proof bottle caps, bottle or container cap, etc.), protrusions, cutouts, and so on. Many of these designs would never see the market due to the complexity of the parts or high tooling costs. Different patented designs of the collapsible core exist, and they meet different requirements that include shortening molding cycle time and special products such as the closure market. The so-called "standard" collapsible core is the oldest and is very popular, dating back to the 1950s.

It can mold circular parts with many undercuts. Its assembly consists of a center pin, collapsible core, and sleeve. The center pin, which is made up of tool steel hardened to 60 to 62 Rc, is a precision-ground shaft with a taper on one end and flange at the other. The collapsible core is a hollow cylinder with matching slots parallel to the cylinder axis, changing part of the cylinder into matching segments. These vertical segments are the flexing segments that form the undercut. It is made of tool steel that is hardened to 56 to 58 Rc. The center pin expands the flexing segments of the core and provides cooling of the molding length. The collapsible core forms the undercut with the expanded flexing segments and releases the product for ejection with segments in a collapsed position. The sleeve functions as a backup unit to collapse the core segments if segments fail to collapse on their own.

EXPANDABLE CAVITY

The patented expandable cavity device was designed to simplify molding external details. In certain mold applications it eliminates space-consuming slides, external unscrewing thread devices, grooves, or any other type of surface impression. It provides closer cavity-to-cavity locations. Instead of collapsing radially inward, the core's metal segments flower outward and move the cavity segments away from the center axis.

With reduced mold construction, different advantages develop, such as mold and product cost reductions and improved product performances. When compared with a conventionally sized mold, the expandable device permits the incorporation of more cavities. A major advantage is the reduction in molding cycle time since the mold space for heat transfer is reduced and easier to predict.

OTHER COMPONENTS AND DEVICES

There are literally hundreds of components and devices to simplify and/or reduce costs that are stock items from many companies worldwide. For example accelerated ejectors use a rack and pinion mechanism to provide additional ejector stroke. Their simple, linear movement can be used to increase the speed and stroke of ejector pins, ejector sleeves, or entire ejector assemblies. The flanges and rounded corners on these units facilitate installation within the ejector assembly. The rectangular cross-section of the racks prevents them from rotating. Included with each unit is a bumper stud that assures the positive return of the racks when the ejector assembly is fully returned. Figure 17.88 shows examples of these components and devices, many of which many are from DME of Milacron. Included are the preengineered companies that have complete assembled molds and can provide cavities to meet requirements. Special molds such as spiral flow test molds are also available (Fig. 17.89).

Figure 17.88 Examples of only a few of the many preengineered mold component parts and devices.

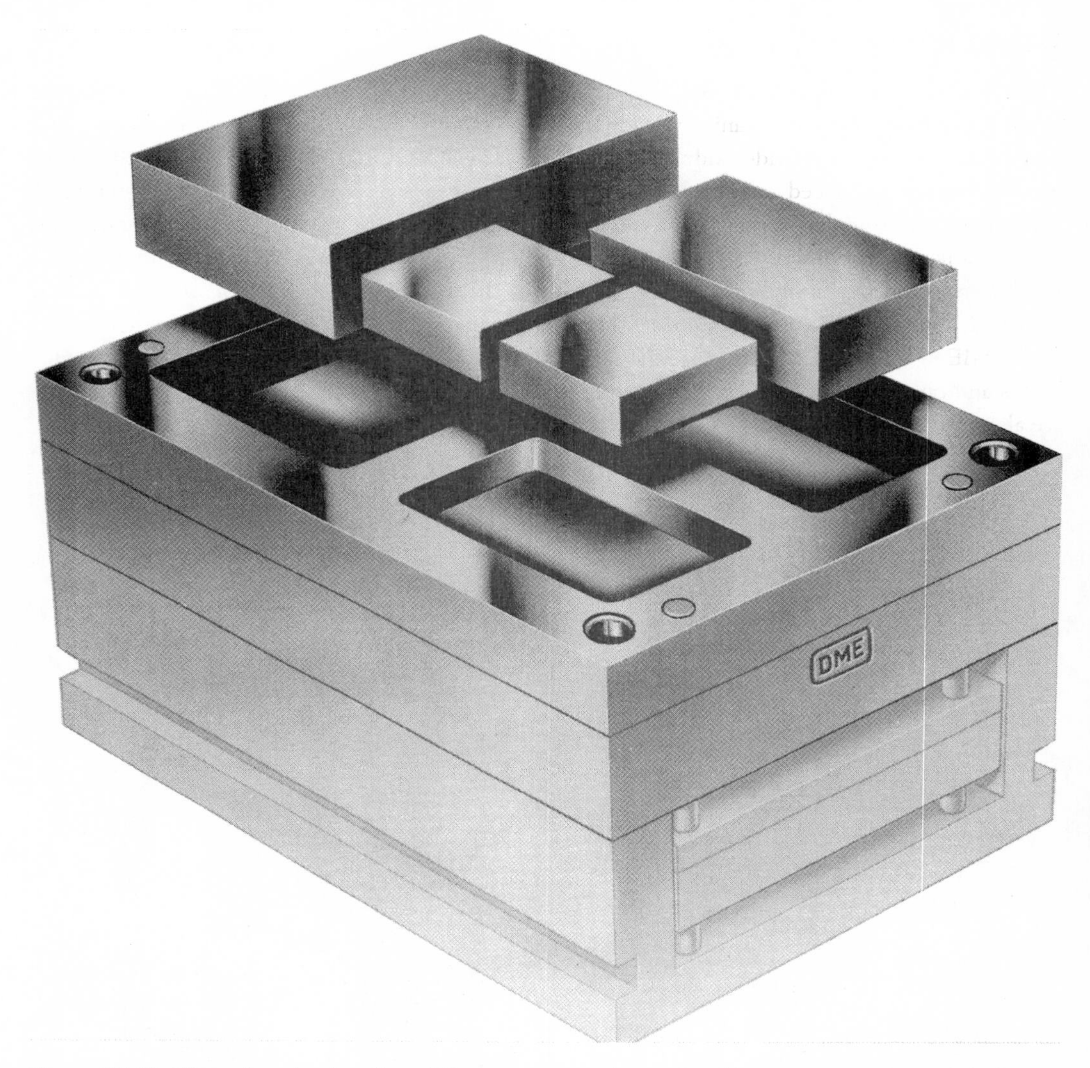

Figure 17.88 Examples of only a few of the many preengineered mold component parts and devices *(continued)*.

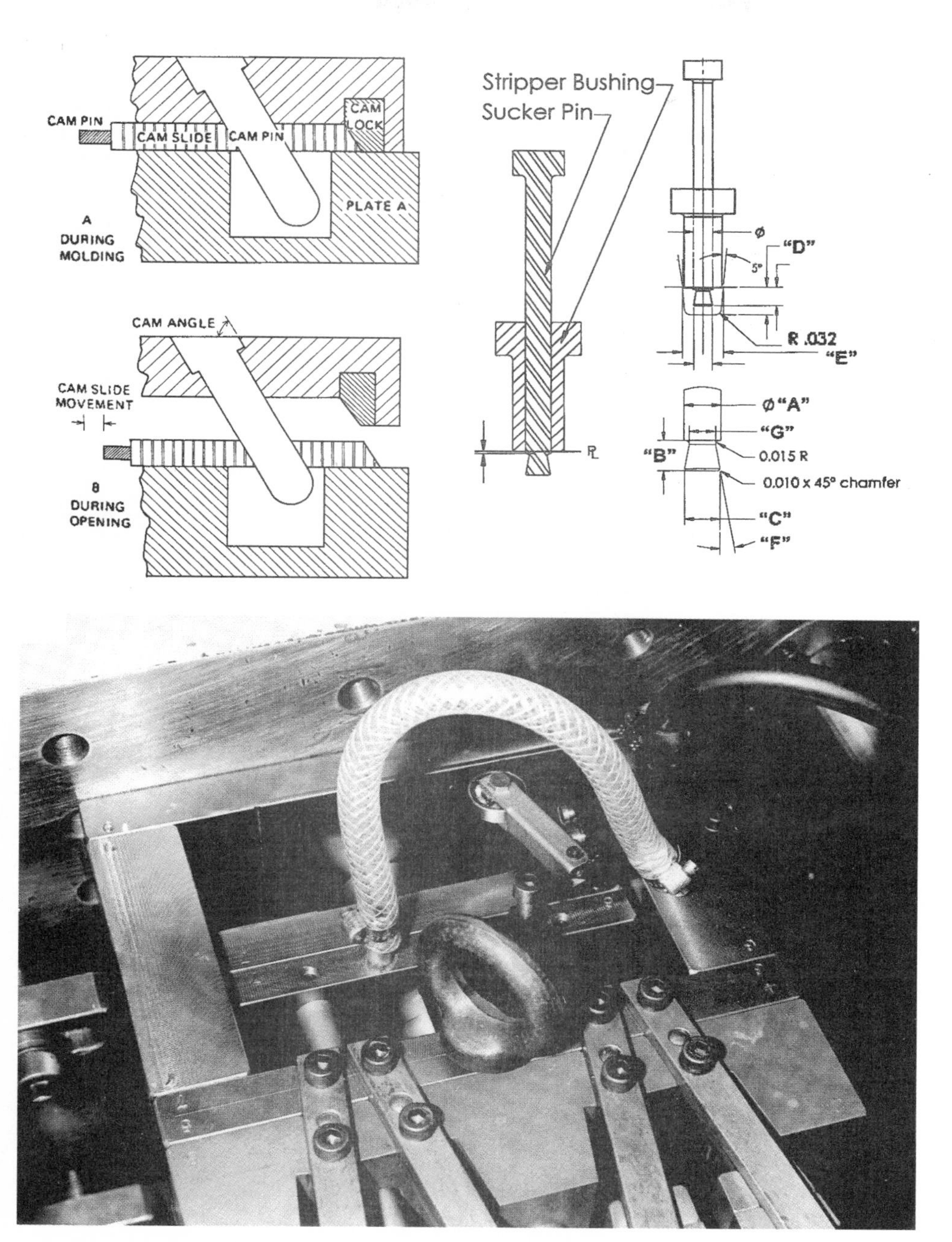

Figure 17.88 Examples of only a few of the many preengineered mold component parts and devices *(continued)*.

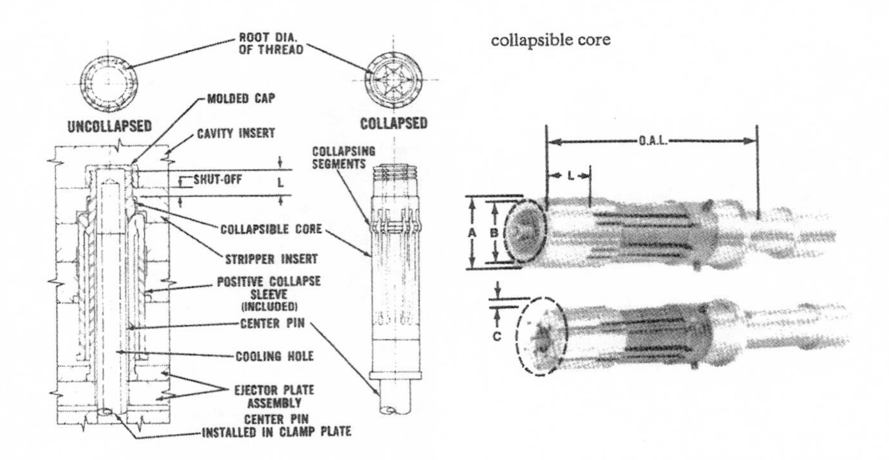

CATALOG NUMBER	A Max. O.D. of Thread or Configuration		B Min. I.D. of Thread or Configuration		Center Pin Dia. (At Top of Collapsible Core)		L Max Molded Length (Incl. Mold Shut-Off				C Collapse per Side at Tip of Core††				O.A.L. Overall Length of Collapsible Core (Only)	
	inch	mm	inch	mm	inch	mm	inch	mm	*inch	*mm	inch	mm	*inch	*mm	inch	mm
CC-200-PC	1.270	32.25	.910	23.11	.785	19.93	.975	24.76	1.150	29.21	.043	1.09	.048	1.21	7.315	185.80
†CC-250-PC	1.270	32.25	.910	23.11	.785	19.93	.975	24.76	1.150	29.21	.043	1.09	.048	1.21	5.440	138.17
CC-202-PC	1.390	35.30	1.010	25.65	.885	22.47	.975	24.76	1.150	29.21	.055	1.39	.064	1.62	7.315	185.80
†CC-252-PC	1.390	35.30	1.010	25.65	.885	22.47	.975	24.76	1.150	29.21	.055	1.39	.064	1.62	5.440	138.17
CC-302-PC	1.740	44.19	1.270	32.25	1.105	28.06	1.225	31.11	1.400	35.58	.068	1.72	.083	2.10	7.315	185.80
†CC-352-PC	1.740	44.19	1.270	32.25	1.105	28.06	1.225	31.11	1.400	35.56	.068	1.72	.083	2.10	6.065	154.05
CC-402-PC	2.182	55.42	1.593	40.46	1.388	35.25	1.535	38.98	1.700	43.18	.090	2.28	.103	2.61	7.815	198.50
CC-502-PC	2.800	71.12	2.060	52.32	1.750	44.45	1.750	44.45	1.900	48.26	.115	2.92	.125	3.17	9.625	244.47
CC-602-PC	3.535	89.78	2.610	66.29	2.175	55.24	2.125	53.97	2.400	60.96	.140	3.55	.148	3.75	11.250	285.75

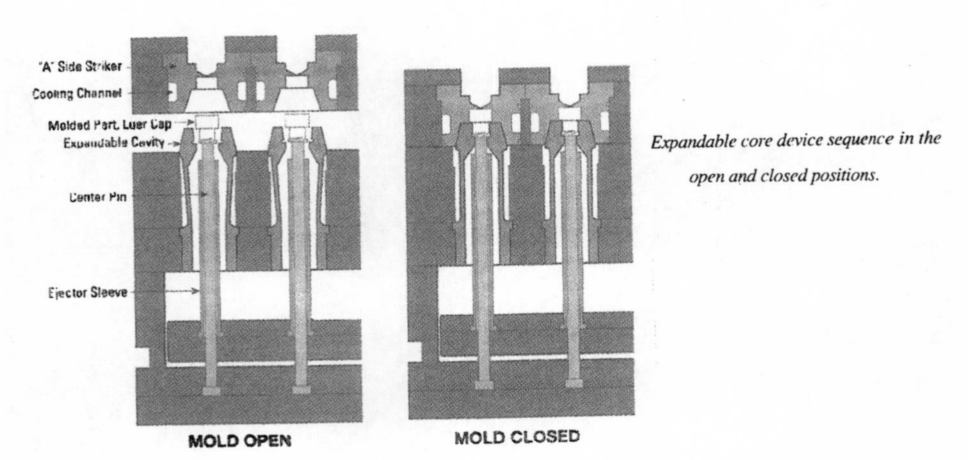

Expandable core device sequence in the open and closed positions.

Figure 17.88 Examples of only a few of the many preengineered mold component parts and devices *(continued)*.

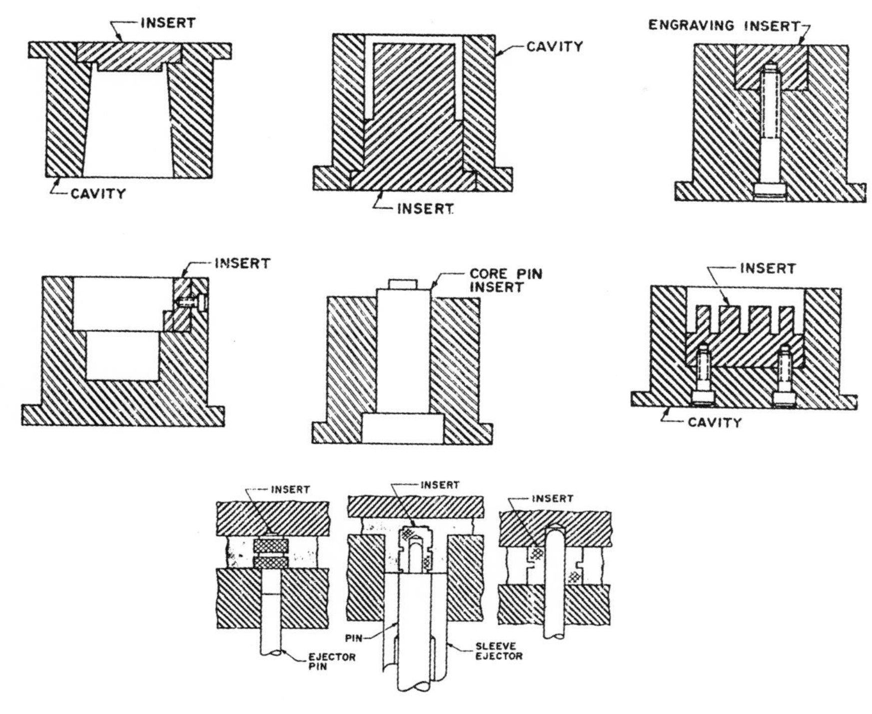

Figure 17.88 Examples of only a few of the many preengineered mold component parts and devices *(continued)*

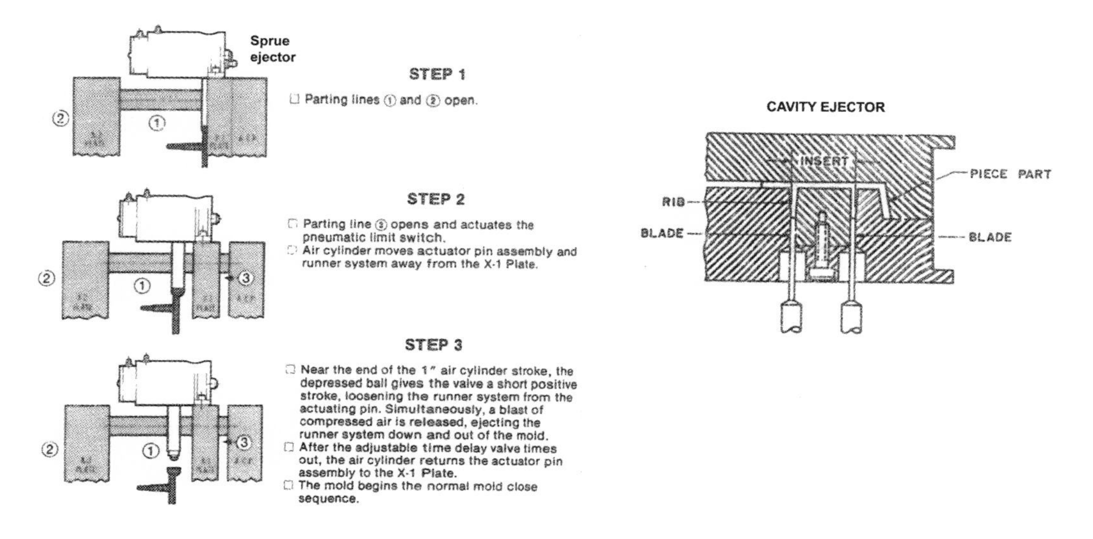

Figure 17.88 Examples of only a few of the many preengineered mold component parts and devices *(continued)*.

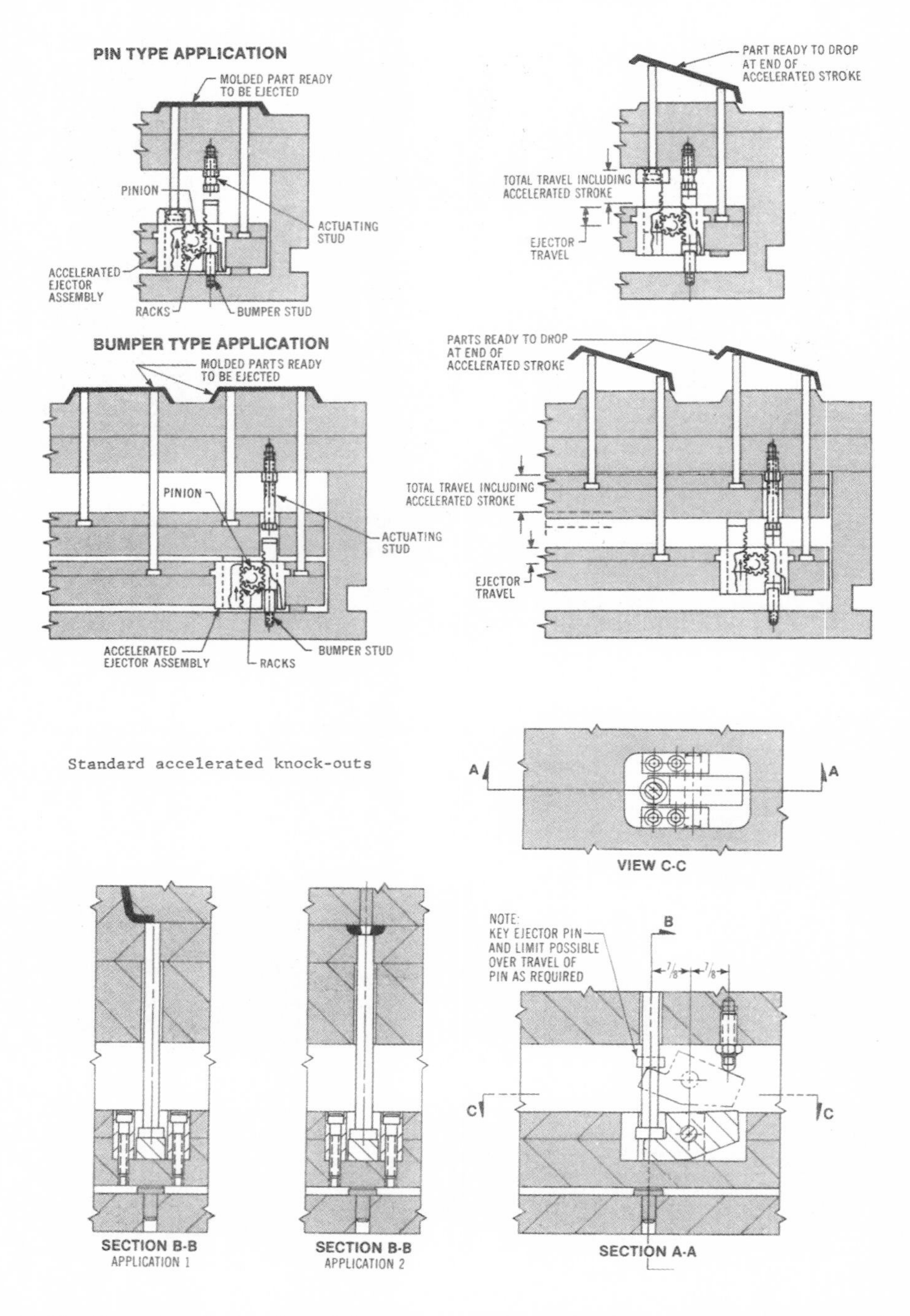

Figure 17.88 Examples of only a few of the many preengineered mold component parts and devices *(continued)*.

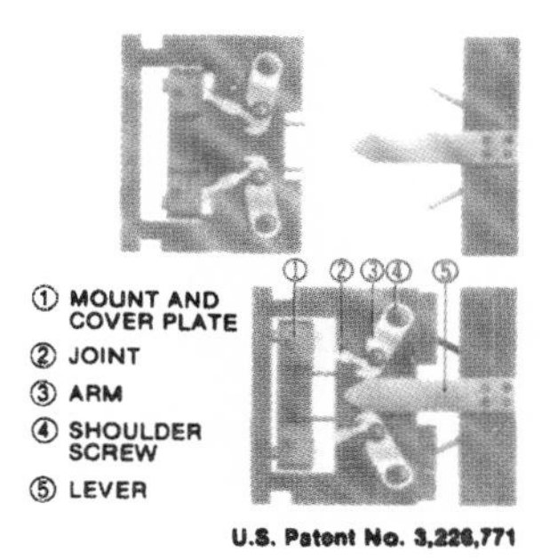

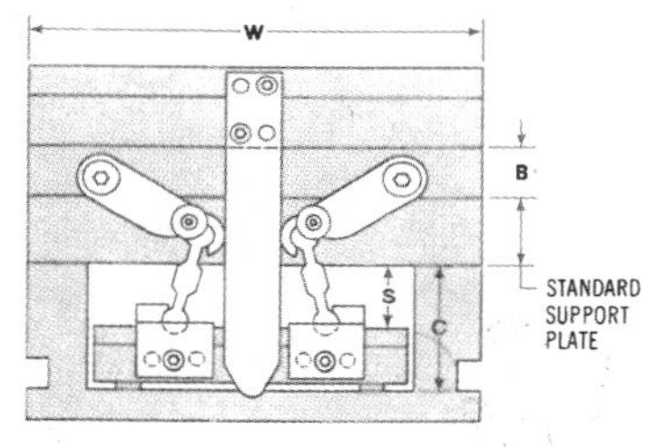

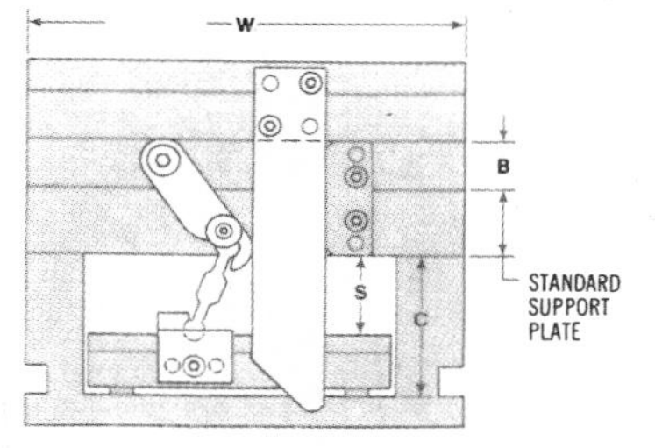

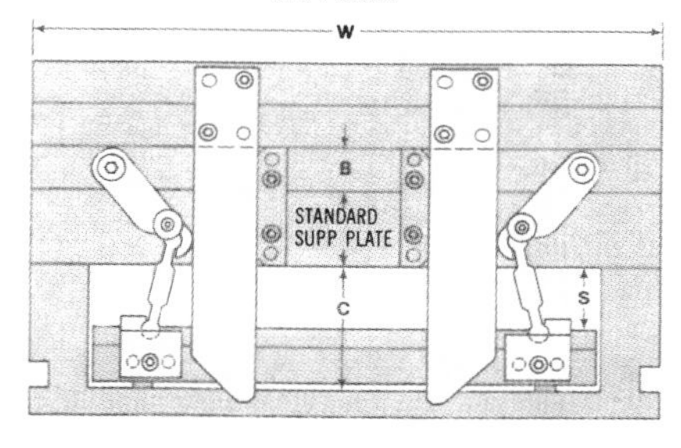

Figure 17.88 Examples of only a few of the many preengineered mold component parts and devices *(continued)*.

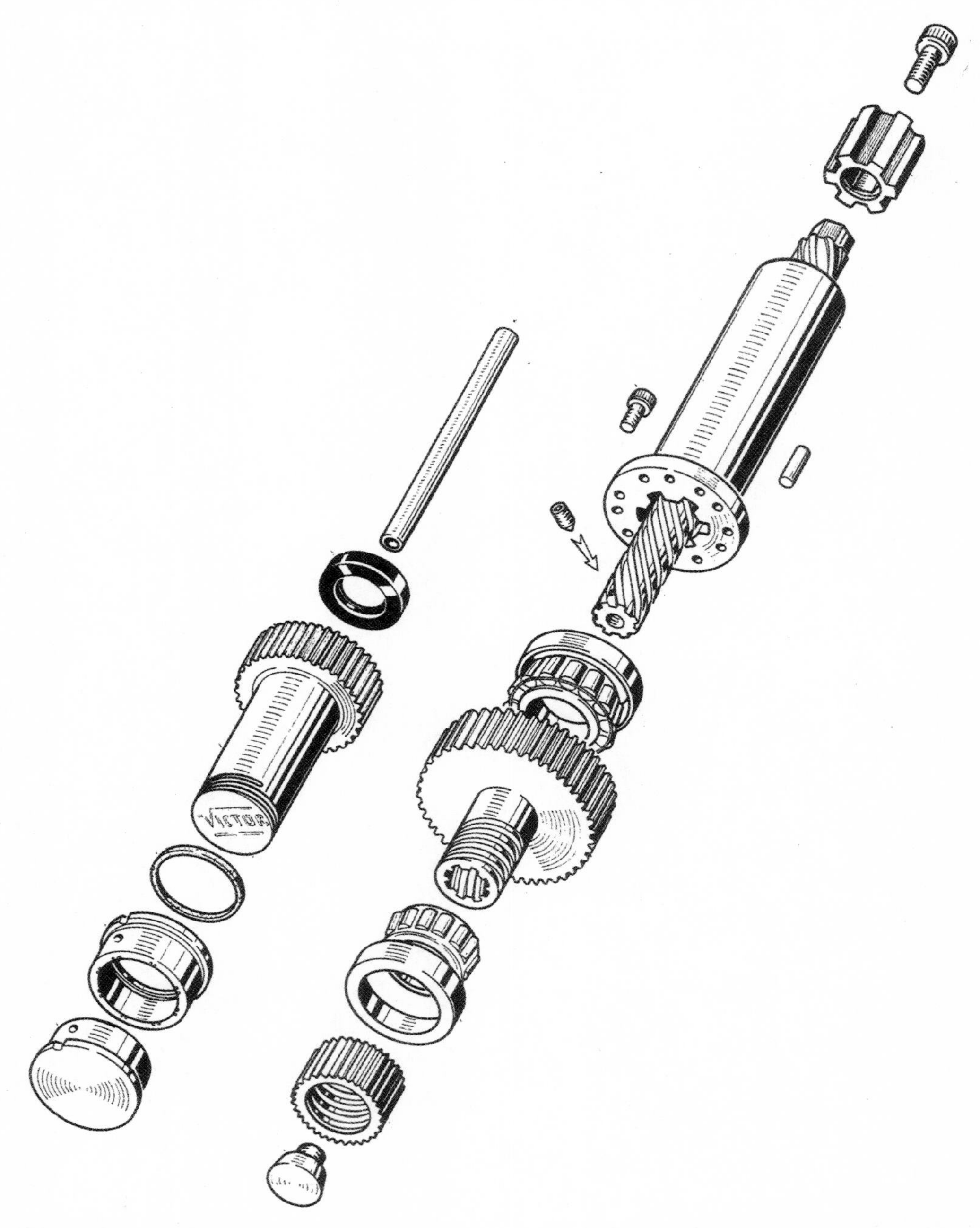

Figure 17.88 Examples of only a few of the many preengineered mold component parts and devices *(continued)*.

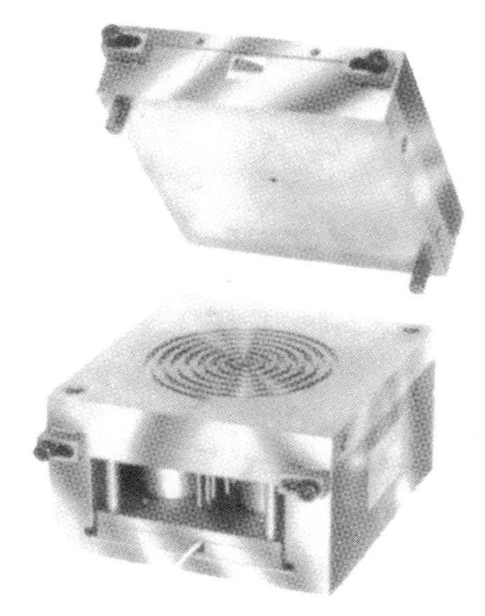

Figure 17.89 Preengineered spiral flow test mold.

SAFETY

Moldmakers and/or molding machinery manufacturers can provide information concerning safety. The following checklist from the American Society for Metals (ASM) is helpful (Table 17.56). It involves the start-up and shutdown of an IMM.

The subcommittees of the Mold Safety Committee of the Society of the Plastics Industryensure that molds meet certain guidelines for safety and good electrical practices. These groups, organized under the SPI Moldmakers Division, address different issues.

ELECTRICAL SUBCOMMITTEE

This group creates mold guidelines for the following:

1. Recommended wire size within the mold
2. Current capacity of the wire to be used in molds
3. Temperature rating of the wire to be used in molds
4. Name of the wire (e.g., appliance wire, UL listed, CSA approved)
5. Wire capacity of the wire ways (i.e., wire channels) in the molds for the thermocouples and the heater wires

1. Install mold connecting strap (or its transporting system) properly
2. Position injection fitting correctly
3. Set preliminary barrel heater temperature profile
4. Turn on the barrel feed throat cooling water
5. Start the machine
6. Open clamp to accept the mold
7. Connect chain fall
8. Position mold up against the Stationary platen
9. Carefully slip mold locating ring into platen locating hole
10. Clamp mold in place
11. Properly align and fasten "A" half of the mold to the stationary platen
12. Insert ejector rods if required
13. Adjust limit switches or settings and fully close the mold
14. Shut off machine
15. Fasten "B" half of the mold to the moving platen
16. Remove chain fall
17. Tighten mold clamps
18. Start the machine, watch for seizing
19. Turn off the machine and check mountings
20. Restart and disengage "B" half from "A" half to check for visible damage
21. Adjust settings for proper ejection
22. Lubricate moving components, clean cavity surfaces
23. Close mold and shut off machine
24. Attach hose lines from the mold temperature control unit and blow clear wi' compressed air
25. Adjust temperature settings
26. Recheck clamps, barrel temperature and heater bands
27. Close hopper feed gate, check hopper magnet position
28. Load hopper with fresh material and open feed gate
29. Purge the machine
30. Set injection and cycle limits
31. Position so nozzle seats against sprue of closed mold and lock in place
32. Open mold, bring clamp unit to full open
33. Set cycle indicator
34. Close safety gate, start first cycle
35. Adjust pressure and time settings as needed to mold the required part

Table 17.56 Checklist and guideline for operating a mold

6. Assurance that the thermocouples are put in wire ways (channels) separate from the heater wires
7. Proper ways to ground the mold
8. Setting up the proper ways to wire molds that would include such things as how wires are to be spliced within the mold (e.g., by wire splices, butt splices, wire nuts, etc.)

MOLD INTERFACE SUBCOMMITTEE

This group creates guidelines for (1) how the mold is to be wired; (2) the location of the first zone (cavity); (3) whether the manifolds, the bushings, or the sprue will be wired first; (4) how the cavities will be numbered from the parting line of the mold; (5) the top nonoperator side looking at the cavities toward the stationary side of the molding machine; (6) the numbering of the cavities from top to bottom, then left to right; (7) the numbering of the cavities in columns (Top nonoperator looking into the cavity top to bottom appears to be the best way, as there are some issues that have to do with the wiring channels within the mold. By numbering the cavities from top to bottom, the heater and the thermocouple wires will not have to be routed across or under the manifold to the wiring channels.); (8) the manifold will always have a bushing under it; (9) the pin of the connector will always be wired to a bushing; (10) where connectors or junction boxes are to be mounted to the mold; and (11) the ratings that junction boxes should have.

UNIVERSAL GUIDELINES

SPI guidelines include definitions of terms: connector, plug, receptacle, parting line of the mold, operator side of the mold, manifold, bushing, nozzle, drop, cavity, wiring channel, and junction box. It is the aim of this subcommittees to make the guidelines universal. That is, wherever a mold is built using SPI guidelines, users could be sure of the first zones being the bushing or manifolds; the location of the first zone, or cavity; the numbering of the cavity; and how the first zone is wired and to what pin on what connector.

MOLDMAKERS

Most of the plants that produce molds worldwide are small or medium in size. They may employ only a few individuals up to a maximum of about 200 employees. The exact number of shops and their sizes is not known, since many are not listed as moldmakers because they operate within another company or are classified under another part of an industry.

Moldmakers usually are experts in one type of mold manufacture. Most have their specialty fields. The design departments of these companies are usually small; a scheduling department is often found only in the earliest stage of existence or not at all. A consequence of this arrangement is that, even for more complex molds, the design is often documented only in assembly drawings. Detailing and complete dimensioning of all individual components are often dispensed initially for reasons of available time and capacity. Costs can depend on the moldmaker's available manufacturing time. As in all industries costs relate to supply and demand.

Mold prices and delivery deadlines are always the critical points in production planning and sales negotiation. These requirements put the moldmaker under constant pressure. The moldmaker is forced to limit activities to the essential—that is to the machining of cores and cavities. Everything else is necessary and must be achieved in the simplest manner possible. In practically all applications the most economical way is to use standard preengineered elements.

Even during the stages of mold calculation and planning, standard elements at fixed prices should be considered. Today, standard elements are used to a large extent as bases for molds and dies, as well as for connections to machines and other equipment and special purposes. The use of standard elements helps to keep one within the bounds of production capacities, minimize calculation and production risks, and simplifying the procurement of spare parts. The possibility of buying standardized elements at short notice considerably reduces stock keeping and shutdown times of production capacities.

The Moldmakers Division of SPI issues a bulletin on moldmaking as part of its continuing effort to improve service to molders. It is intended to assist buyers seeking guidance in mold procurement. It points out the various difficulties that can result, unless thorough understanding and communication are established between the mold buyer (molder) and moldmaker. Table 17.57 is the SPI Moldmakers Division quotation guide. Tables 17.19 to 17.23 provide additional guides for mold quotation.

Imports

Worldwide, according to Agostino von Hassell (as reported by The Repton Group, New York), the cost of molds imported into the United States during the first six months of 2000 was about $550 million. The cost breakdown by country was as follows: Canada 58.1%, Japan 14.9%, Germany 3.8%, Portugal 3.7%, Taiwan 2.7%, Italy 1.9%, China 1.4%, UK 1.3%, South Korea 1.2%, France 1.0%, and all others 10.0%. These statistics are based on US Department of Commerce (DOC) data. The DOC definition of the types of molds included in these figures includes injection and compression molds for rubber or plastic, including shoe molds and semiconductor molds, and molds not included elsewhere.

Directories

In this age of specialization, the purchasing community has found it increasingly difficult to locate the right source for the right job. The Moldmakers of the SPI provides the industry with an updated directory of its members and their special capabilities. The SPI Moldmaker members are in constant contact with the plastics industry and its ever-changing technology. The directory lists moldmakers as contract or custom services and in turn by type of mold, such as injection molds and blow molds.

There are also publications that provide buyer's guides, such as *Plastics News*. *Moldmaking Technology* magazine issues an annual buyer's guide that features directories on (1) moldmaking equipment, supplies, and accessories; (2) mold components; (3) mold-design and engineering equipment; (4) mold material; (5) machining equipment; (6) EDM equipment and supplies; (7) machining tools and accessories; (8) hot runner systems and supplies; and (9) mold polishing and repair equipment and supplies.

THE MOLDMAKERS DIVISION
THE SOCIETY OF THE PLASTICS INDUSTRY, INC.
3150 Des Plaines Avenue (River Road), Des Plaines, Ill. 60018, Telephone: 312/297-6150

TO ________________ FROM ________________ QUOTE NO. ________
DATE ________
DELIVERY REQ ________

Gentlemen:
Please submit your quotation for a mold as per following specifications and drawings:

COMPANY NAME ________________

Name of Part/s
1. ________ B/P No. ________ Rev. No. ______ No. Cav. ______
2. ________ B/P No. ________ Rev. No. ______ No. Cav. ______
3. ________ B/P No. ________ Rev. No. ______ No. Cav. ______

No. of Cavities: ________ **Design Charges:** ________ **Price:** ________ **Delivery:** ________

Type of Mold: ☐ Injection ☐ Compression ☐ Transfer ☐ Other (specify) ______

Mold Construction
☐ Standard
☐ 3 Plate
☐ Stripper
☐ Hot Runner
☐ Insulated Runner
☐ Other (Specify) ______

Mold Base Steel
☐ #1
☐ #2
☐ #3

Special Features
☐ Leader Pins & Bushings in K.O. Bar
☐ Spring Loaded K.O. Bar
☐ Inserts Molded in Place
☐ Spring Loaded Plate
☐ Knockout Bar on Stationary Side
☐ Accelerated K.O.
☐ Positive K.O. Return
☐ Hyd. Operated K.O. Bar
☐ Parting Line Locks
☐ Double Ejection
☐ Other (Specify) ______

Material

Cavities	Cores
☐ Tool Steel	☐
☐ Beryl. Copper	☐
☐ Steel Sinkings	☐
☐ Other (Specify) ______	

Press
Clamp Tons ______
Make/Model ______

Hardness

Cavities		Cores
☐	Hardened	☐
☐	Pre-Hard	☐
☐	Other (Specify) ______	

Finish

Cavities		Cores
☐	SPE/SPI	☐
☐	Mach. Finish	☐
☐	Chrome Plate	☐
☐	Texture	☐
☐	Other (Specify) ______	

Cooling

Cavities:		Core
☐	Inserts	☐
☐	Retainer Plates	☐
☐	Other Plates	☐
☐	Bubblers	☐
☐	Other (Specify) ______	

Ejection

Cavities		Cores
☐	K.O. Pins	☐
☐	Blade K.O.	☐
☐	Sleeve	☐
☐	Stripper	☐
☐	Air	☐
☐	Special Lifts	☐
☐	Unscrewing (Auto)	☐
☐	Removable Inserts (Hand)	☐
☐	Other (Specify) ______	

Side Action

Cavities		Cores
☐	Angle Pin	☐
☐	Hydraulic Cyl.	☐
☐	Air Cyl.	☐
☐	Positive Lock	☐
☐	Cam	☐
☐	K.O. Activated Spring Ld.	☐
☐	Other (Specify) ______	☐

Type of Gate
☐ Edge
☐ Center Sprue
☐ Sub-Gate
☐ Pin Point
☐ Other (Specify) ______

Design by: ☐ Moldmaker ☐ Customer
Type of Design: ☐ Detailed Design ☐ Layout Only
Limit Switches: ☐ Supplied by ________ ☐ Mounted by Moldmaker
Engraving: ☐ Yes ☐ No
Approximate Mold Size: ________
Heaters Supplied By: ☐ Moldmaker ☐ Customer
Duplicating Casts By: ☐ Moldmaker ☐ Customer
Mold Function Try-Out By: ☐ Moldmaker ☐ Customer
Tooling Model/s or Master/s By: ☐ Moldmaker ☐ Customer
Try-Out Material Supplied By: ☐ Moldmaker ☐ Customer

Terms subject to Purchase Agreement. This quotation holds for 30 days.

Special Instructions: ________________

The prices quoted are on the basis of piece part print, models or designs submitted or supplied. Should there be any change in the final design, prices are subject to change.

By ________________ Title ________________

Distribution: Use of this 3 part form is recommended as follows: 1) White and yellow - sent with request to quote. Pink - maintained in active file. 2) White original - returned with quotation. Yellow - retained in Moldmaker's active file.

Table 17.57 SPI Moldmakers Division quotations guide

SUMMARY

Different approaches are used in producing molds. One could design a mold that will fit into a specific fabricating machine. Modifying a mold is another approach. However, the best approach is to start with the product that requires a mold. Design the mold around the product. Then use a fabricating machine that meets the requirements of the mold, such as the amount of melted plastic required in the mold (shot size), sufficient clamping pressure, and that the mold fits in the machine without wasting energy (chapter 4). Figure 1.37 explains this FALLO approach.

Optimizing the mold operation to reach higher productivity requires careful examination of individual mold components. Compromises in the performance of any one of these can adversely affect productivity. Specifically, overall performance is related to designing the mold for maximum productivity and specifying the machine to obtain maximum output. It comprises many parts requiring usually the use of high-quality steels for long production runs. It also includes cooling channels and possibly hot runner channels for the hot feed of molten plastic. In many cases, it will also contain a number of moving parts, such as sliding and/or moving cores (462–467).

All kinds of actions can be used to operate the mold such as sliders and unscrewing mechanisms. Examples are shown in Figure 17.88. CAD and CAE programs are available that can aid in mold design and in setting up the complete fabricating process (chapter 25). These programs include melt flow to part solidification and the meeting of performance requirements.

Mold cooling is vital for faster cycles and uniform shrinkage. Coolant is distributed to and from each core and cavity at a uniform temperature and pressure to ensure consistent product filling, weight shrinkage, and strength. Optimum cooling is achieved through the turbulent flow of liquid in the channels located as close to the molding surface as possible. Reynolds numbers of 5000 to 10000 are recommended to assure good turbulence.

Just as the placement of cooling lines is important, so is mold material selection. For example, hardened tool steels such as H13, S7, A2, and SS420 can be used to ensure good wear and toughness characteristics. Beryllium copper (BeCu) is used in areas where improved heat transfer would reduce overall cycle time, such as on gate inserts and core caps. Stainless steel mold plates prevent corrosion and fouling of both water and air lines, thereby improving cooling, reducing maintenance, and extending mold life. Rust, mineral deposit, or algae cause fouling in the cooling system.

The ejection system can be a major source of wear in a mold. Air ejection of products is a method often overlooked; a variety of products ranging from small medicine cups to large industrial containers have been air-ejected. Because there is no mechanical contact, products can be ejected in the warm condition and the cycle time is reduced. With fewer moving parts, less wear results.

The alignment of individual cores and cavities is necessary because leader pins do not have tolerances that are tight enough. For example, thin-wall containers requiring concentricity accuracy of ±0.0005 in (±0.0013 cm) and technical parts with stepped parting lines must be protected. When the mold closes, both require individual alignment. These are called interlocks.

Wear in a mold can be minimized, but it cannot be avoided. Maintenance personnel should be consulted during mold design to ensure easy service and accessibility while the mold is operating.

Downtime and expenses due to wear can be reduced by making wear items inexpensive and easily replaceable. Also, sliding parts should have a minimum hardness difference of 5 Rockwell C to prevent galling.

In addition to the hot runner, components such as nozzle bands and tips can be replaced when the mold is in the machine by pulling the cavity plate away from the core plate. Using long-life heater bands and manifold heaters can also minimize downtime for hot runners. The hot runner system must be reliable and easy to control. Core and cavity alignment is important. All locations in the mold plates for cores and cavities are usually held to ±0.0002 in (±0.00051 cm). This accuracy ensures the interchangeability of cores and cavities within the plates. Interchangeability can also reduce spare parts inventory.

DIES

This review primarily concerns extruder dies. They are devices, usually of steel, having an orifice (opening) with a specific shape or design geometry that it imparts to a plastic melt extrudate pumped form an extruder under pressure (chapter 5). The die-opening settings influence properties of the extruded plastic. Dies have a specific orifice (opening) with a specific shape so that different products can be produced, such as sheets, films, pipes, tubings, profiles, wire coatings, filaments, and so on. These steel, precision works of art have at least a mirror finish on the melt flow-channel orifice surfaces. In addition to information presented here, there is information on dies in chapter 5. The performances of dies also relate to the downstream equipment. An example is shown in Figure 17.90.

The function of a die is to accept and control the available melt (extrudate) from an extruder and deliver it to downstream takeoff equipment as a shaped product (profile, film, sheet, pipe, filament, etc.). The aim is to minimize deviation in cross-sectional dimensions, smooth surfaces, and a uniform output by weight at the fastest possible rate. In order to do this, the extruder must deliver melted plastic to the die targeted to be a so-called ideal mix at a constant rate, temperature, and pressure. Measurement of these variables is required and usually careful performed.

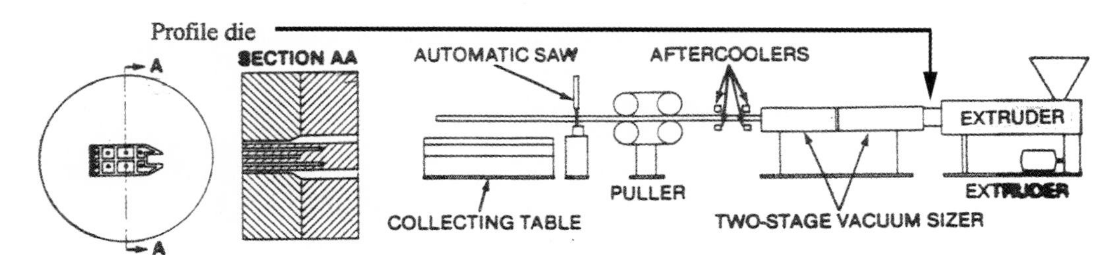

Figure 17.90 Example of an extrusion line that includes a die and downstream equipment.

The die has substantial influences on the plastic due to the melt flow orientation of the molecules, such as different properties parallel and perpendicular to the flow direction. These differences have a significant effect on the performance of the product. The die designs with melt condition (pressure, temperature, rate of travel, etc.) and its downstream equipment can provide the required unidirectional or bidirectional properties.

For the die to operate at maximum performance and lowest cost it requires that the most uniform melt be available. Details on obtaining this melt are reviewed in chapter 3. Pumping pressure and temperature on the melt entering the different designed die heads differ to meet their melt flow patterns within the die cavities. The pressure usually ranges as follows: (1) blown and layflat films at 14 to 40 MPa (2000 to 5800 psi); (2) cast film, sheet, and pipe at 3.5 to 27.6 MPa (500 to 4000 psi); (3) wire coating at 10 to 55 MPa (1450 to 8000 psi); and (4) monofilament at 7 to 21 MPa 1000 to 3000 psi).

MATERIAL OF CONSTRUCTION

Metals are used for flat film and sheet dies and are usually constructed of medium-carbon alloy steels. The flow surfaces of the die usually have protective coatings such as chrome plating to provide corrosion resistance. With properly chrome-plated surfaces, microcracks that may exist on the steels are usually covered. The exterior of the die is usually flash chrome-plated to prevent rusting. Where chemical attack can be a severe problem (such as in processing PVC), various grades of stainless steels are used with special coatings. Coatings will eventually wear, so it is important that a reliable plater properly recoat the tool, usually the original tool manufacturer.

Die materials are almost exclusively steel because of the many parameters that must be satisfied. The nonalloy steels such as AISI 1040 and other common steels can be used for simple dies—tubing and profile dies, for instance—where the ease of machining and low costs are suitable for the relatively small sizes and unsophisticated applications.

Alloy steels such as AISI 4140 and other similar alloys are used for the majority of die applications because they meet most requirements along with an inherent high quality and lack of inclusions, pits, voids, and hard/soft spots. The lack of corrosion and rust resistance of alloy steel is a potential weakness but is easily overcome by plating normally with chrome.

High-nickel alloy steels are used when certain plastics, such as PVC and PVDC, can degrade with temperature and time to produce acids that will corrode plated alloy steels. High-nickel alloy steels provide good corrosion resistance without plating and simplify manufacturing, cleaning, and repair. Stainless steel is also used with degradable materials.

Profile, pipe, blown film, and wire coating dies are examples of dies generally constructed of hot-rolled steel for low-pressure melt applications. The high-pressure dies can be made of certain steels, such as 4140 steel. Chrome plating is generally applied to the flow surfaces, particularly when processing certain plastics, such as ethylene-vinyl acetate (EVA). The steel used in the manufacture of a die varies depending on the requirements of the plastic and its application. The available spectrums of modern tool steels offer properties in numerous combinations and to widely differing degrees. Materials of construction were reviewed at the start of this chapter.

These steel precision works of art have at least a mirror finish on the melt flow channel surfaces. The slightest minute scratches can produce flaws in the extruded products. Great care must be taken during their installation, operation, removal, cleaning, and storage. When designing them, the aim is to use as few parts as possible. The dies should be easily lifted for installation or maintenance, easily disassembled, easily cleaned, and easily reassembled.

The initial aim is to simplify and minimize detractors. The major detractor is to understand melt flow behavior within the die and on exiting the die. Being involved with the product designer usually permits concessions to be made, resulting in simplifying them and reducing their cost.

TERMINOLOGY

This section includes different terms applicable to dies. This section has been prepared so that the reader will have an introduction to the terms. Figure 17.91 depicts different dies and includes a few terms; other terms will be given in the following text. (Consider reviewing the glossary at the end of this chapter for additional information.)

ADAPTER

This is also called an offset adapter, a die feedblock, or a crosshead adapter. It is the part of an extrusion die that holds the die block and can be used to reduce complex flow patterns of the melt by providing a streamlining shape. It moves the plastic melt from the plasticator (extruder) to the die.

ADAPTER, FISH TAIL

This term describes the transition from a round opening block of melt as it exits from the extruder barrel and is then flattened and spread before entering the die where a channel (tear shape, etc.) proceeds to provide a more controlled melt flow.

ADJUST DIE

Different die adjustments, such as sizes and shapes of an orifice opening, are made.

AIR GAP

The distance from the die opening to the first downstream equipment such as nip rolls (pressure roll and cooling roll).

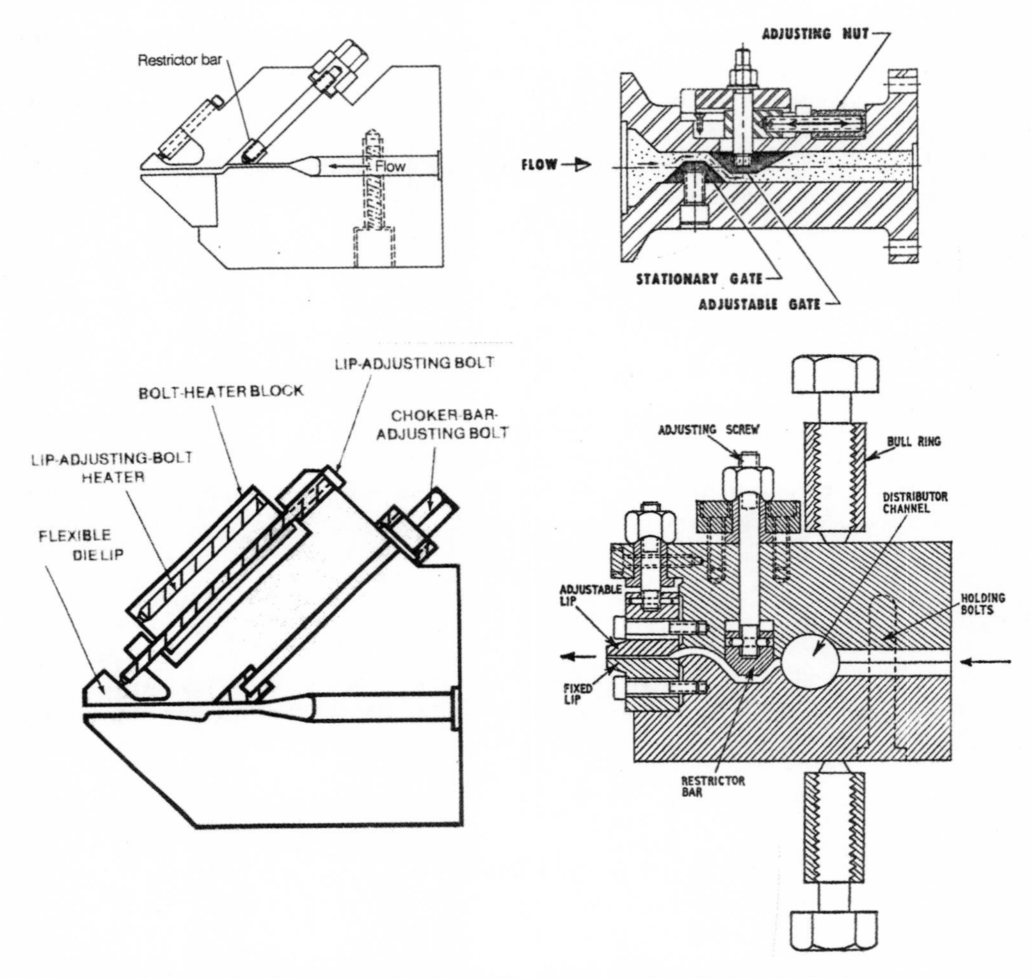

Figure 17.91 Some identifying terms for dies; other terms are described in the text.

Air Vent

This is a passage in hollow pipe, tube, or profile die to permit passage of air into the interior of the hollow extrudate.

Autoflex

This is a sheet die in which the lip opening is controlled by the expansion or contraction of thermal bolts that, in turn, respond to sheet-thickness sensors.

BLADE

This is a deformable member attached to a die body, which determine the orifice slot opening and are adjusted to produce uniform thickness across an extruded film or sheet.

BLOCK

This is the part of an extrusion die that holds the forming bushing and die core.

BLOWN FILM

Dies are designed to produce circular films that have layflat dimensions from inches (cm) to hundreds of feet (m).

BODY

The main structure of the die head excluding the bushing and pin.

BOLT HEATER

expands die bolts usually in sheet extrusion dies to control transverse thickness. It provides a very important process-control procedure.

BUSHING

This refers to a tubing die section that forms the outside diameter of the tube. It is usually an adjustable bushing, outside the ring-forming wall, at the bottom of the die body that determines the final orifice outer diameter or surface and shape.

BUSHING ADJUSTMENT

Means of moving the die bushing relative to the die pin.

CART

A trolley that supports the die and holds it at the correct height for machine connection.

CASTING

Refers to the shaping of metal products by forcing a molten metal or alloy under high pressure into a female or negative cavity by means of a hydraulic ram.

Casting alloy

There are alloys with melting points sufficiently low so that they can be cast into reusable dies; commonly used are zinc-based or aluminum-based alloys.

Cavity

Prior to the melt entering the die orifice, melt flows through a cavity to provide laminar flow of the melt. Different shapes (tear shape, for example) are used with the streamline type preferred.

Cavity deposit

It is material build-up on a cavity's surface due to plate-out of the plastic melt; usually attributed to the use of certain additives or alloys.

Choke plate

A single whole unit used between the end of an extruder barrel and the die holder to produce a controlled pressure drop in the plastic melt flowing through the die.

Coating

To protect materials of construction in metal dies against damage, such as corrosion and abrasion, special metals are used and/or coatings are applied to their surfaces.

Convergent

A die in which the internal channels leading to the orifice converge; applicable to dies for hollow bodies.

Crosshead

A device designed to extrude the plastic melt at an any angle, normally 90° to the longitudinal axis of the extruder screw.

Deckle system

A deckle rod, plate, or dam is attached to cover each end of a film- or coating-die orifice. It can be used to reduce the die slot width and/or to control the edge bead on the web. The plastic to be extruded and the extrusion process determine whether the die can be deckled. Degradable plastics are not used since deckles dam the melt flow, causing stagnated areas where degradation will occur. The most common deckle is an external type that can be bolted to a die half or permitted to slide.

External deckles can be designed to adjust manually or to be driven by power screws or rack and pinion gears.

The disrupted melt flow from behind a conventional external deckle usually causes heavy edges; they prevent good roll contact over the center of the sheet by the polishing rolls and in turn can result in lower sheet gloss. To reduce this undesirable condition, edge beads or internal deckles can be used. The edge bead rod is typically a shaped rod that is inserted into the die stream of the final land. By varying its depth of insertion, the heavy end melt flow can be controlled.

Die and pin set

Also called die and shaping. A matching bushing and pin dimension to form the extrudate.

Divergent

a die for hollow products in which the internal channels leading to the die orifice are diverging.

Diverter valve

A valve used to divert melt from one extruder die passage to another. applicable in coextrusion, blow molding and other dies.

Draw-down ratio

The ratio of the thickness of the die opening to the final thickness of the extruded product.

Drip

carbonized plastic drool formed on the face of an extrusion die face during the production cycle. If the die face is not kept clean, it can solidify, break off, and contaminate the virgin plastic.

Dry sleeve calibration

This system is for precision profile extrusions using a well-designed streamline die (with no dead spots or eddy melt flow patterns occurring), a series of vacuum calipers (similar to a sizing sleeve in a cooling vacuum tank to supply specifically metered vacuum and cooling water to the calipers), air-cooling and radiant heating stands to keep the profile straight, a puller meeting the required pulling force at an accurately controlled operating speed, and a reliable cutoff saw and stacking system.

ENTRANCE ANGLE

the maximum angle at which the melt enters the die land area measured from the center line of the mandrel.

ENTRY ANGLE

The angle of convergence of melt entering the extrusion-die lips.

EXTRUDATE

The plastic hot melt as it emerges/discharges/exits from the extruder die's orifice into a desired product form such as film, sheet, coating, pipe, and so on.

EYEBOLT AND HOLE

These holes, properly located in a die, are used to easily lift and move dies and other tools. Balanced lifting occurs with proper location and use of an eyebolt; safety devices are incorporated to ensure no accidental dropping occurs to protect personnel and the die. The different bolts have different safety load-carrying capacities.

FACE DAM

The use of brass plates at the end of the die opening to reduce and/or adjust the width of the extrudate exiting the orifice.

FAN TAIL

An extrusion die of diverging form.

FEEDBLOCK

This directs melt from the extruder to its die.

FILM AND SHEET THICKNESS

General classification can be helpful as a guide to film and sheet thickness selection. Film dies are generally applicable for thicknesses of 0.003 mm (0.010 in) or less. Thin-gauge sheet dies are normally designed for thicknesses up to 0.015 mm (0.060 in). Intermediate sheet dies may cover a thickness range of 0.01 to 0.06 mm (0.040 to 0.250 in). Heavy-gauge sheet dies extrude thicknesses of 0.02 to 0.13 mm (0.080 to 0.500 in). Different groups within and outside the plastics industry may have their own thickness definitions to meet their buyer/customer needs.

FLAT DIE

Also called a slot die. A die with its straight-slit opening used in an extrusion film or sheet line.

GAP

The distance between the metal faces forming the die opening/orifice.

GROOVING

Long, narrow grooves or depressions in the cavity surface parallel to its length. It is usually caused by die fouling or by a spot of plastic buildup on the die surface, effectively changing the shape of its cross-section and extrudate.

HANG-UP

A problem associated with irregular or uneven flow through a die; can be caused by dirt, uneven temperature, or by poor design of the die (for example, lack of streamlining).

HEAD

The entire structure used to manufacture the die.

HEAD, ADJUSTABLE

A die head whose orifice opening is adjustable.

HEAD MANDREL

This is also called a torpedo. It is the center section of the die head about which the melt flows. If a die pin is used, the mandrel supports the die pin. It describes the sizing element either in a die or in a seizer that controls one dimension of the extrudate, usually the inside diameter.

HEAD MANDREL COOLING, INTERNAL

Tube-sizing system using crosshead extrusion with a cooled mandrel to size the inside diameter of the tubing.

HEAD MANDREL DIVERTER

A die head in which the melt is fed into the side of the die head and diverted around the mandrel by flow guides.

Head mandrel movement

This is the movement of a die head whose mandrel or pin can be moved so as to vary the orifice opening. This structure contains the flow passage that carries the melt to multiple heads.

Head, programmed

A die head with provisions for varying the orifice opening at specific points during melt extrusion, such as to program the extrudate of a blow molding parison. As the parison drops, its thickness will vary based on the program setting.

Heat pipe

Also called thermal pins, heat transfer devices, or heat conductors, heat pipes are a means of heat transfer; either to remove or to add heat in all kinds of fabricating equipment.

Heater-adapter

That part of the die around which a heating element is located.

Laddering

Defective surface finish caused by melt fracture in the die.

Land

This is also called a lip. It is the parallel section just before the exit of the die head in the direction of the melt flow (Fig. 17.92). It is vital to shaping the extrudate and providing thickness-dimensional control.

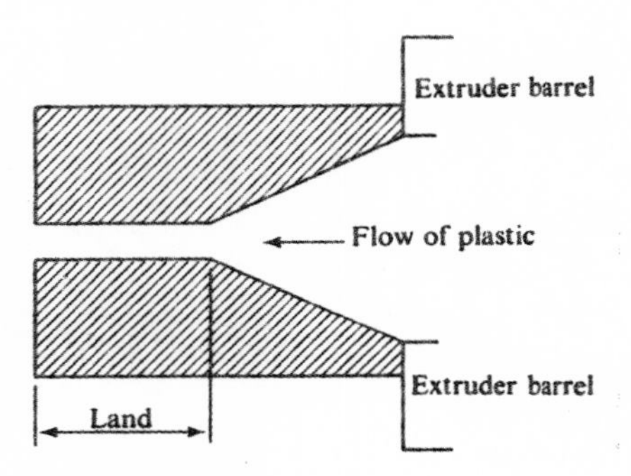

Figure 17.92 Location of the extrusion die land.

Land length

Land length is usually expressed as the ratio between the length of the opening in the flow direction and the die opening. For example, it could be expressed as 10:1.

Manifold

Also called feedblock. It directs melt from the extruder to its die.

Manifold channel

See Cavity.

Orifice

It is the opening in the die that forms the extrudate. There are other processing devices with orifices, such as a fiber spinneret that has many openings for melt flows through and drawn out into a plastic extruded shape.

Parting line

This is also called a spew line. It is the lengthwise depression or flash that can occur on the surface of an extruded product. The line can occur where separate metal parts of the die join to form the die orifice.

Preengineered

Standardized die components are available. They provide for exceptional quality control on materials used, quick delivery, interchangeability, and low costs.

Pressure transducer

Instrument mounted in different parts of a mold (cavity, knockout pin, etc.) to measure melt pressure.

Prototype

A 3-D model used in the preliminary evaluation of form, design, performance, and material processing of a die.

Restrictor bar

This is also called a choker bar or gate. It provides a means of controlling melt pressure within the die so that melt exiting the die is rather uniform.

Shrinkage

Shrinkage is the difference in dimensions between a plastic in a die and the plastic after it has been extruded and solidified. Shrinkage follows the usual expansion that occurs when the melt leaves the die and gradually solidifies; however, certain plastics may take up to 24 hours at room temperature to complete their shrinkage. In designing a product and its die, it is extremely important to make allowances for shrinkage. Each plastic material has its own shrinkage factor, and plastic materials cannot usually be changed once the die is built, since the shrinkage will probably differ.

Sink mark

Also called shrink mark or inverted blister, a sink mark is a shallow depression or dimple on the surface of part due to the collapse of the surface following local internal shrinkage after the melt solidifies. Frequently occurs on the part face opposite to a face in which the section thickness increases, as in a rib.

Spiders

Spiders are basically columns to support the mandrel, or torpedo, section in the melt stream that forms the interior of a hollow section. Melt flows over the mandrel. When the die is fed on the side, the mandrel is anchored at the rear of die and does not require spider supports.

Stock

Refers to the plastic being processed.

Tear shape

See Cavity.

Tool

A tool identifies a die. It also identifies a mold or cutting die that are virtually synonymous in the sense that they have female or negative cavity through which a molten plastic moves, usually under heat and pressure, or they are used in other operations such as cutting dies or stamping plastic sheet dies. This is the term that identifies all these devices; it is particularly used to identify dies and molds.

VACUUM BOX

Also called vacuum sizing fixture or vacuum tank, a vacuum box is a shape-holding fixture used to set the dimensions of processed plastics by holding the material against a roll or mandrel by means of a vacuum.

WEB

The extruded product.

WELD LINE

A weld line is also called weld mark, flow line, or striae. It is a mark or line when any two melt flow fronts meet after passing by a restriction in a die, such as a spider support. Under poor melt flow conditions, it can be a source of a weak strength bond.

DESIGN

Dies are expensive, delicate instruments that require careful treatment and maintenance in order to operate efficiently. Thus their design requires that they can be handled and used with little to no damage.

An important characteristic of a die is that the orifices shape and affect melt flow patterns. The effects of the orifice are related to the die design (land length, etc.) and melt condition. For example, using the popular "coat hanger" die for fabricating flat sheets, cooling is more rapid at the corners; in fact, a hot center section could cause a product to blow outward and/or include visible or invisible vacuum bubbles (Fig. 17.93). With proper orifice shape and melt control (temperature, pressure, and rate of flow) the sheet exits the die without these problems (Fig. 17.94).

MELT FLOW

The non-Newtonian behavior of plastic melt makes its flow through a die complicated but controllable within certain limits (chapter 3). Simplified flow equations are available to account for the non-Newtonian melt behavior. They provide an excellent foundation using an empirical approach that pertains to extrusion-die channels of different shapes.

The melt in the die is under pressure. Upon the melt exiting the die, the pressure is released and the melt expands in all directions. The amount of expansion can be reduced based on the die design. such as its land length, also melt temperature and rate of flow through the die and rate of pull from the die. Figure 17.95 provides an introduction to this melt behavior. By maximizing the performance of plastic melts, die designs, and takeoff equipment (rate of travel, cooling rate, etc.), dimensional tolerances can usually be held to within at least ± 3% to 5%. Tighter tolerance is achieved using suitable takeoff equipment (chapter 5).

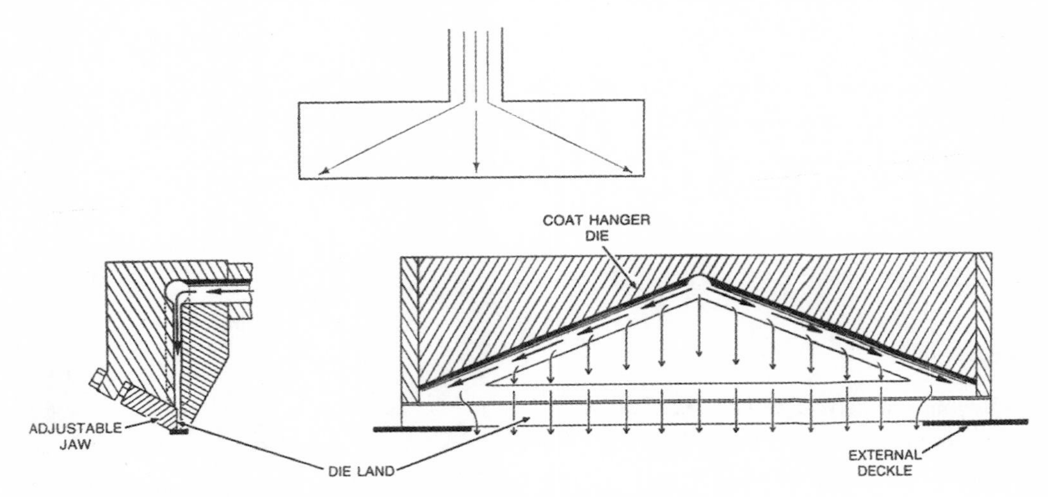

Figure 17.93 Examples of melt flow patterns in a coat hanger die.

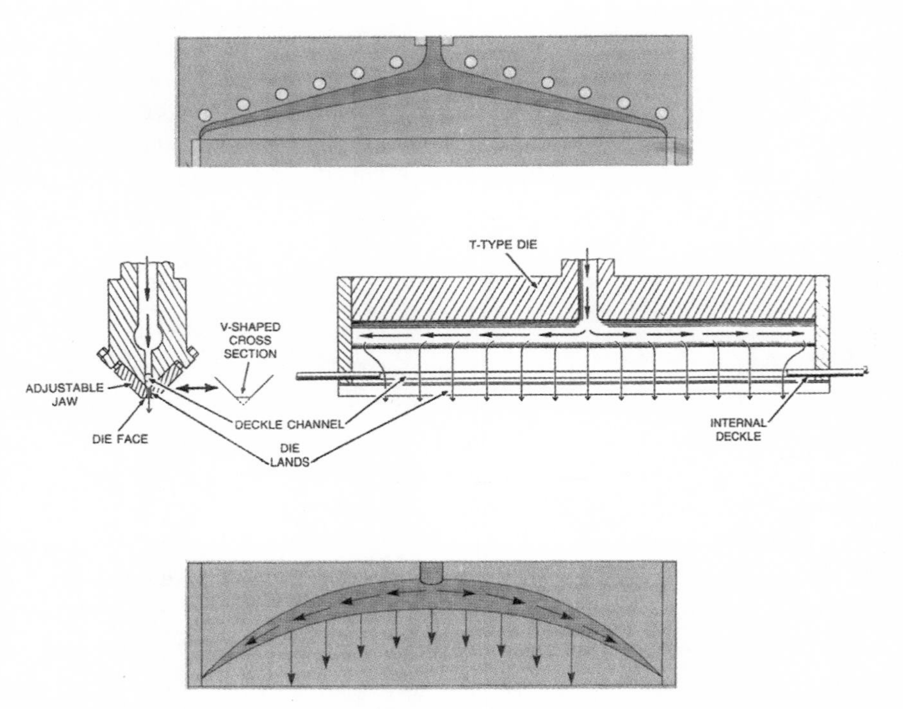

Figure 17.94 Examples of melt distribution with die geometry via their manifold channels. Each die has limitations for certain types of melts.

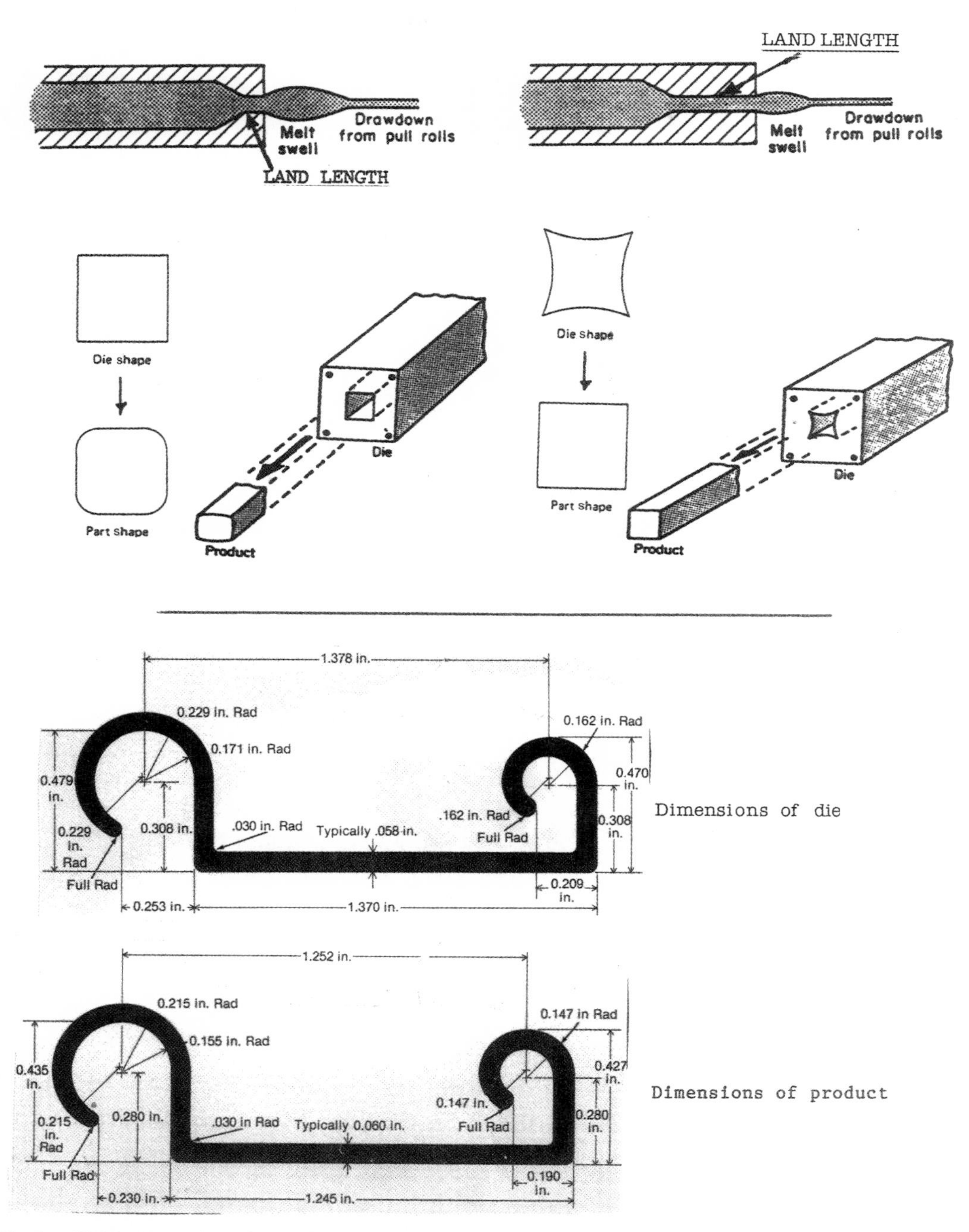

Figure 17.95 Examples of melt flow patterns based on minimum die and process control.

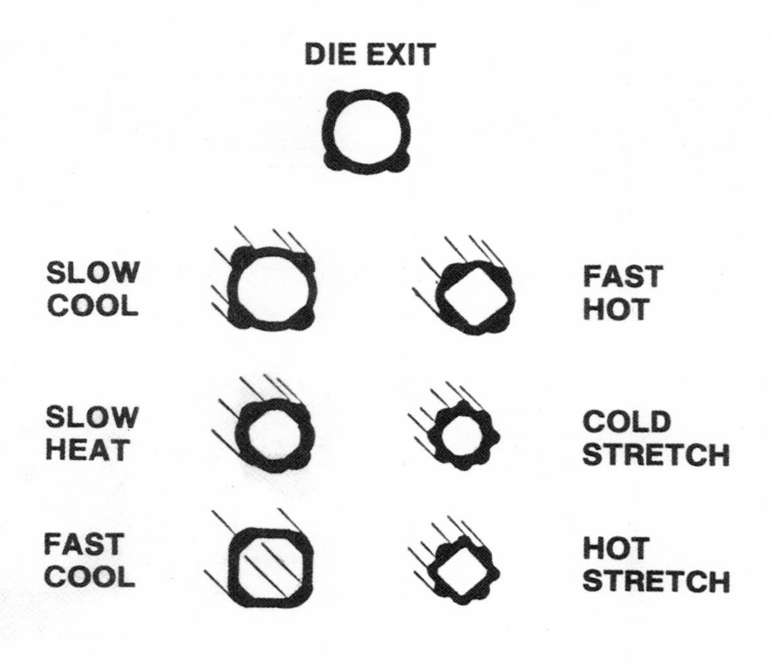

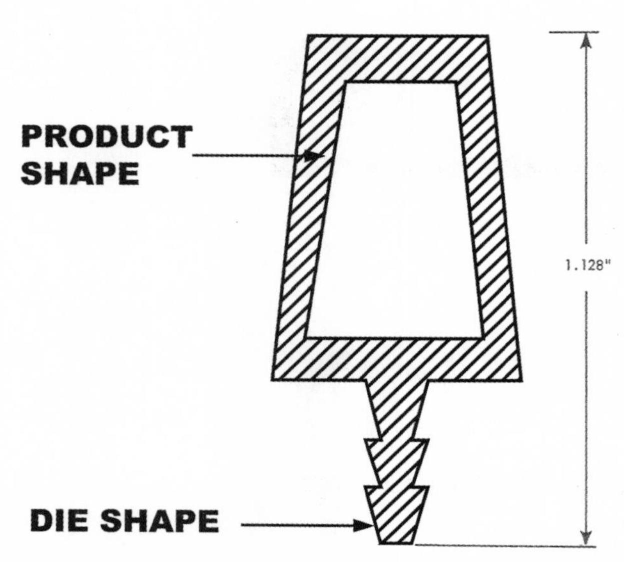

Figure 17.95 Examples of melt flow patterns based on minimum die and process control *(continued)*.

Performances of dies for other processes can relate to extrusion die performances. For example, Figure 17.96 shows the extrusion of a tubular plastic melt (called a parison used in extrusion blow molding; chapter 6) that changes its wall thickness by moving its tapered core. This type of action relates to the extrusion die's land-lip movement and in particular, to be reviewed, extruding through a pipe die.

ORIFICE SHAPE

The approach used for shaping orifices in the dies is important and requires 3-D evaluation that includes streamlining. When possible, all dies should be groomed to promote streamlined melt flow and avoid the obvious pitfalls associated with areas that could cause stagnation, such as right-angle bends, sharp corners, and sections where flow velocities are diminished and are not conducive to streamlined flow. The aim is to avoid these design faults.

As shown in Figure 17.97, stagnation areas from nonstreamlined flat-plate dies can easily cause accumulation of melt that will degrade and affect the extrudate. Orifices are straight-sided cuts into flat plates that approximate the shape of the desired profile. Adjustments are made in the profile shape to allow for flow characteristics of the particular plastic to be extruded. The plate die is a very inexpensive approach compared to those of other dies. Of the three types reviewed, the plate die is the easiest to make.

Use of these plate dies is limited to low-volume production runs. They can be fabricated to rather tightly controlled profiles. If on initial start-up the profile requirements are not met, perhaps all that is needed is additional cutting. If a new plate is needed, it will be quick to cut another plate. These plants usually have their own machine shops. If processing problems develop due to plastic accumulation interfering with the flow, all that is required is to remove the plastic accumulations.

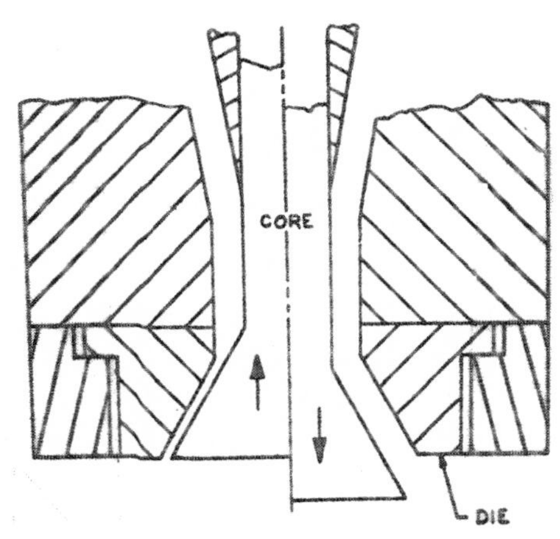

Figure 17.96 Schematic of wall-thickness control for extruding blow molded parisons.

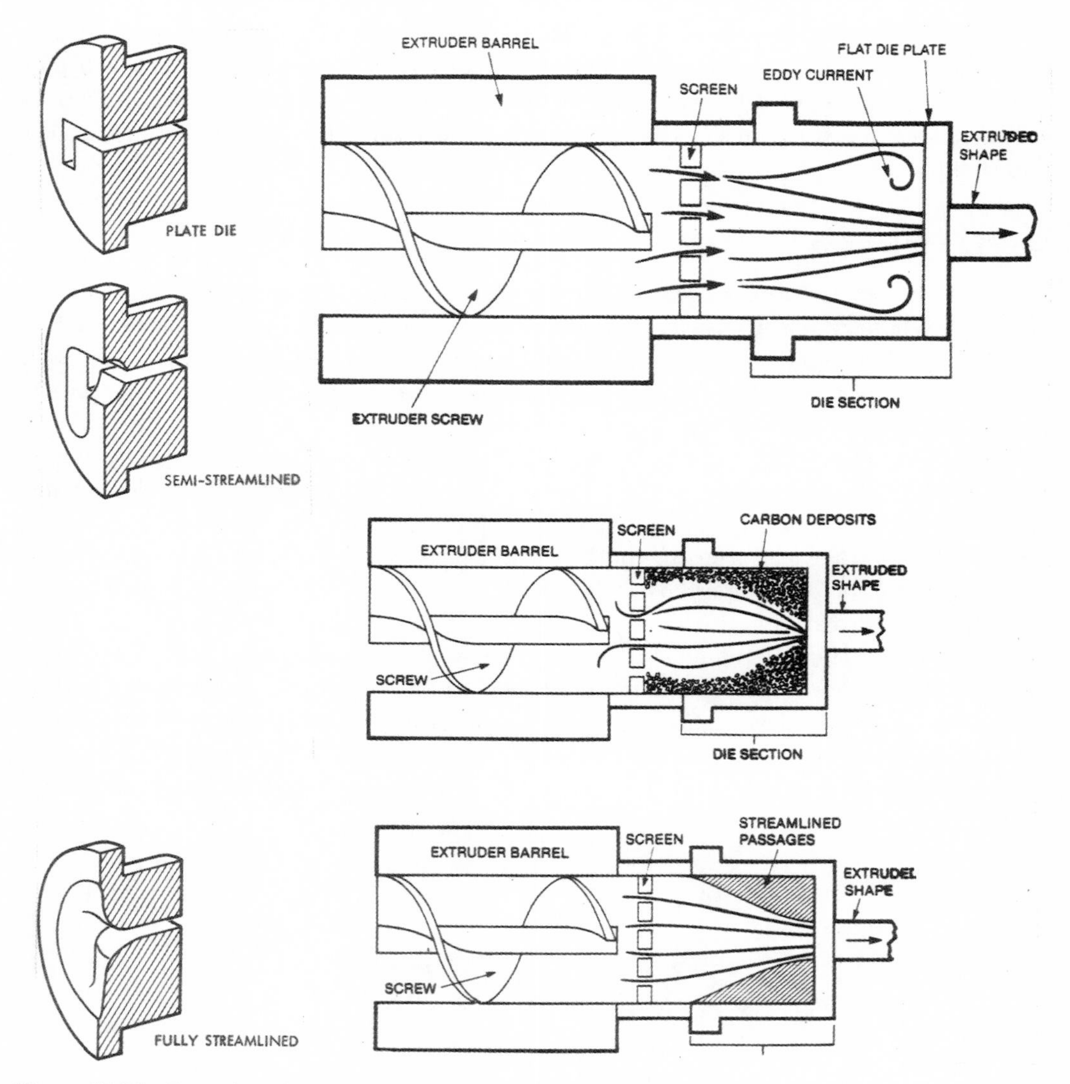

Figure 17.97 Examples of nonstreamlined and streamlined entrances in dies.

There are semistreamlined dies that may be required for limited production. However, the streamlined dies are designed so that melt flow is evenly and uniformly delivered through the die. Fully streamlined dies are used for high-volume, continuous production runs.

An important factor concerning melt flow is the angle or taper of entry and the parallel length of the die land. For most TPs, the entry angle must be as small as possible to ensure good product quality, particularly at high output rates. The abrupt changes (nonstreamlined) in the direction of melt flow tend to cause rough or wavy surfaces and often internal flaw conditions called melt fracture.

There are different approaches to developing the streamlined shapes. They range from totally trial-and-error to finite element analysis (FEA). The trial method usually involves gradually cutting or removing the die orifice metal. Between cuts an examination is made of the extrudate and the metal cavity surface to check on melt hang-ups, melt burning, streaks, and other stagnating problems. With FEA and using appropriate rheological plastic data (chapter 1) one can easily determine an approach to streamline a flow pattern that may be stable even without minor adjustments (chapter 19).

A variety of advantages exist with streamlining, such as the following: (1) dies can operate at higher outputs; (2) pressure drops are lower and more consistent over a range of melt temperatures and pressures; (3) generally the melt uniformity across the extrudate is more uniform and shape control is enhanced; and (4) sometimes crucial for high production output rates where plastics have limited stability and causes hang-ups or degradation going through nonstreamlined dies.

ORIFICE EQUATION

There are many equations that relate to melt flow and in turn to orifice shapes. Off-the-shelf software programs are available (chapter 25) with certain die designers/manufacturers having their own very successful software. Analyzing melt flow in dies is rather complicated and difficult. Using available CAD programs can be extremely helpful but what really helps is experience in the design and the use of different dies with the different plastics. The industry has specialists in-house or specialty die manufacturers that produce efficient operating dies used to extrude all types of products. What makes it difficult is the nature of the plastic melts that are not perfect.

Each melt basically has its own advantages and disadvantages for operating in the die melt channels according to its non-Newtonian behavior (chapter 1). The extruders (and other equipment) have their limitations, such as heat transfer through metal parts and metal parts that are subject to wear. Therefore, what exists is an empirical science that continues to work efficiently. The limitations have always existed but, as material and equipment developments occur, die designs and operating equipment continue to improve (449–455).

As described throughout this book, including this chapter, the melt flow rate is influenced by many variables that start with the plastic's composition (chapters 1 and 2), extruder capability (chapter 5), die geometry (Table 17.58; 454), die performance, and downstream equipment (chapter 18). In studying a complex die system, one must take the following actions:

1. Examine each component where the geometry is constant.
2. Apply the appropriate formula (approach) to compute factors, such as pressure drop, at the desired melt temperature and flow rate.
3. Sum all the pressure drops to determine the pressure that must be provided by the extruder.

A very basic, simple, and useful approach to designing orifices had its start in France in the nineteenth century. In 1868 M. J. Boussinesq first developed formulas for pressure drops in simple flow die channels of simple shapes, such as circular or rectangular channels, of liquids to solids. The more complex channel equations had not yet developed because researchers required extremely complicated mathematics that they could not handle in any reasonable time frame.

The following review uses an equation obtained from a high-speed computer study during the early 1960s by G. P. Lahti (449). He did this work at DuPont and later went to NASA. It provides an excellent foundation using an empirical approach that pertains to extrusion-die channels of several shapes. As shown in Figure 17.98, different equations can be used.

$$Q = (1/\mu)(\mu P/L)(BH^3/12)(F) \text{ or } \Delta P = (12\mu QL/BH^3)(1/F),$$

where

Parameter	*Effect*
Nonuniform exit velocity across the width	Extrudate curling, twisting, rippling Non-uniform drawdown, resulting in residual stresses in the extrudate
Nonuniform exit flow thickness across the width	Nonuniform tranversal extrudate thickness before and after swelling Nonuniform exit velocity across the width (see above)
Flow thickness varying longitudinally (tapers, steps, choker bars)	Significant extensional effects, with possible melt fracture, and contribution to extrudate swell Flow acceleration along the taper, with velocity instabilities at the entrance and exit of the taper Possibility of stagnation points
Flow thickness of length varying transversely	Pressure drop or flow rate varying transversely (thus extrudate curling, twisting and rippling) Nonuniform residence time and relaxation
Set (and melt) temperature varying transversely	Differences in shear heating, promoting transverse flow Transverse components of flow Nonuniform extrudate swell and surface finish Nonuniform shear heating effects Possibility of material degradation Pressure drop or flow rate varying transversely (thus, extrudate curling, twisting and rippling)
Transverse flow	Nonaxial exit velocities Changes in the relative exit velocities and swelling Distortion of the extrudate section Possibility of stagnation points and corresponding material degradation

Table 17.58 Examples of operational effects and geometrical variables on melt flow conditions in a die

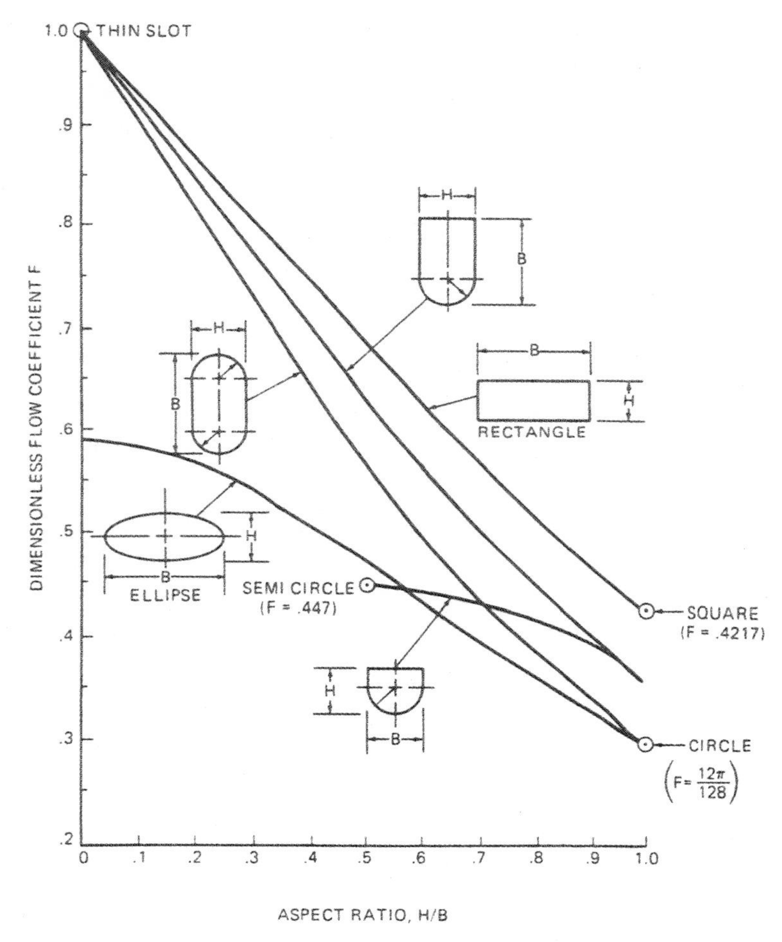

Figure 17.98 Flow coefficients calculated at different aspect ratios for various shapes using the same equation.

Q = volumetric flow rate;
ΔP = pressure drop;
L = length of channel;
B = maximum dimension of cross-section, $\geq H$, in (mm);
H = minimum dimension of cross-section, in (mm); and
F = flow coefficient.

Using this approach and accounting for the entrance effect when a melt is forced from a large reservoir, the channel length (L) must be corrected or the apparent viscosity must be used once it has been obtained from shear rate-shear stress curves for the L/H value of the existing channel. Entrance effect becomes negligible for $L/H > 16$.

By developing the product geometry and determining plastic viscosity and pressure drop, the volumetric flow rate (Q) can be calculated as shown in Figure 17.99. The following calculation is used for the shape shown:

1. Flow in two or three directions that exist in a tapered die
2. Include a detailed discussion that involves the limitations and assumptions for regular and irregular shapes for low-viscosity melts and so on.

To date, various organizations have expanded the capability of CAD software programs for die designs. An example is EDI, where some of their tools to evaluate the flow-channel designs use a computational fluid-dynamics computer code. Using this code, EDI prepares flow models that utilize its powerful capabilities based on a FEA program for calculating the flow of fluids.

EXTRUDATE PERFORMANCE

Extruded melts go through some degree of swelling. To eliminate or significantly reduce the swell to an acceptable amount, stretching or drawing the extrudate to a size equal to or smaller than the die opening occurs. The dimensions are targeted to be reduced proportionally so that the drawdown section is the same as the original section but smaller proportionally in each dimension. However, the effects of melt elasticity mean that the plastic does not drawdown in a simple proportional manner; thus adjustments are made in the orifice opening, melt condition, and/or downstream equipment. These types of variations are significantly reduced in a circular extrudates, such as pipes and wire coatings.

The process of designing of a die can be summarized as follows: (1) minimize head and tooling interior volumes to limit stagnation areas and residence time; (2) streamline flow through the die, with low approach angles in tapered transition sections; and (3) polish and plate interior surfaces for minimum drag and optimum surface finish on the extrudate. The die provides the means to spread the plastic being processed under pressure to the desired width and thickness in a controllable,

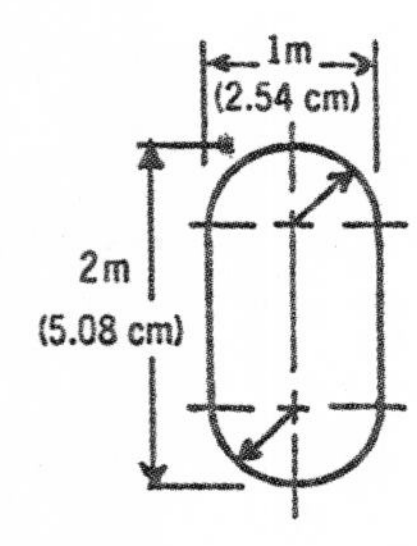

Figure 17.99 Calculation for the volumetric melt flow rate for this specific shape.

uniform manner as the extrudate. In turn, this extrudate is delivered from the die (targeted with uniform velocity and uniform density lengthwise and crosswise) to takeoff equipment in order to produce a shaped product.

DIE EFFICIENCY

Any analysis of die efficiency must include a careful examination of the compatibility of the die with the extruded products. If a die is designed for sheet thicknesses of 3.8 to 9.5 mm (0.150 to 0.375 in), it is extremely difficult to extrude 5 mil (0.005 in or 0.001 mm) film. As there is no die design that could be called a universal die; it is very inefficient to expect an operator to run a die beyond its capabilities. The result would be poor gauge control and various other problems. If the geometry of the flow channel is optimized for a plastic under a particular set of conditions (heat, flow rate, etc.), a simple change in flow rate or heat can make the geometry very inefficient. Except for circular dies, it is essentially impossible to obtain a channel geometry that can be used for a relatively wide range of plastics and a wide range of operating conditions.

LAND

The die land is the parallel section just before the exit of the die head in the direction of the melt flow. It is usually expressed as the ratio between the length of the opening in the flow direction and the die opening; this can be expressed, for example, as 10:1.

It is vital to shaping the extrudate and providing thickness-dimensional control. A very important dimension is the length of the relatively parallel die land. In general, it should be made as long as possible. However, the total resistance of the die should not be increased to the point where excessive power consumption and melt overheating occur (Figs. 17.100 and 17.101).

The required land length depends on not only the type and temperature of the TP melt, but also on the flow rate. The deformation of the melt in the entry section of the die invariably causes

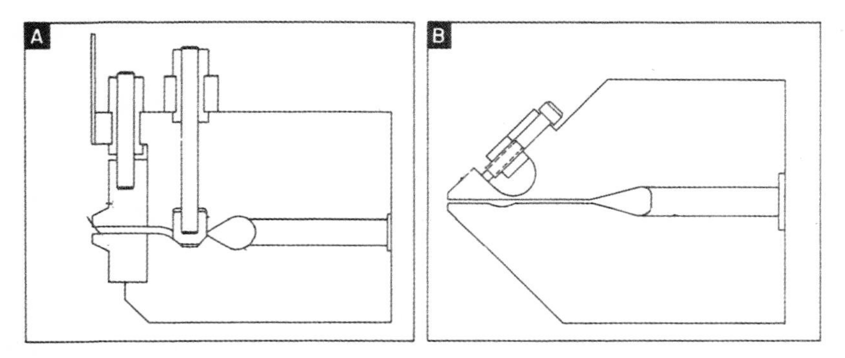

Figure 17.100 Shown are the (more conventional) rigid and die-lip lands.

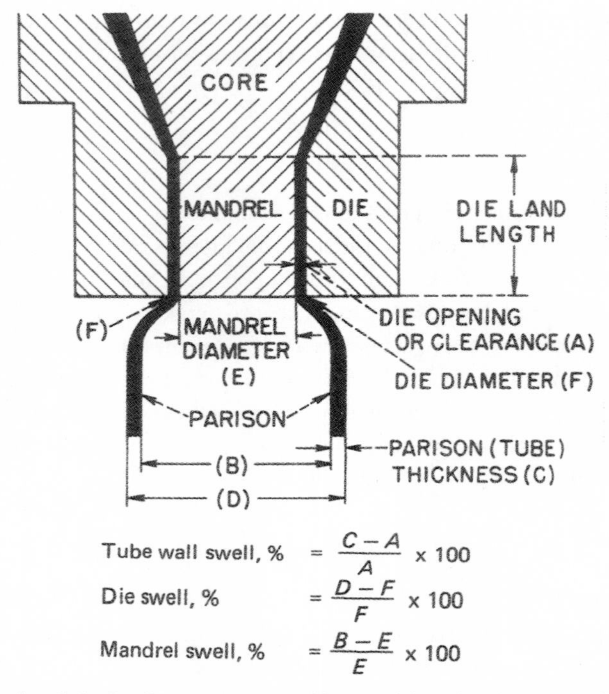

Figure 17.101 Example of the land in an extrusion blow molding die that is usually from 10:1 to 20:1 ratio.

strains that only gradually decrease with time (relaxation). Usually the aim is to allow the melt to relax before leaving the die. Otherwise the product dimensions and the mechanical properties may vary, particularly with rapid cooling.

There are various basic rules for the length. One states that it must be sufficiently long for the residence time of the melt in the die at a mean velocity to be at least equal to the relaxation time. Another relates to long lands as a means to develop sufficient back pressure by using a minimum land of ten times the die opening or (in the case of rod or tubular dies) one diameter of the die opening. A maximum approach angle for tapered sections is 30° and many designs incorporate multiple, decreasing-angle sections or radius sections to minimize drag and plastic hold-up time. There are rigid and flexible die-lip lands.

Figure 17.102 provides examples of die designs to produce different profiles. Their descriptions and their relations to their lands are as follows:

a. The method of balancing flow to produce this shape requires having a short land where the thin leg is extruded. This design provides the same rate of flow for the thin section as for the heavy one.

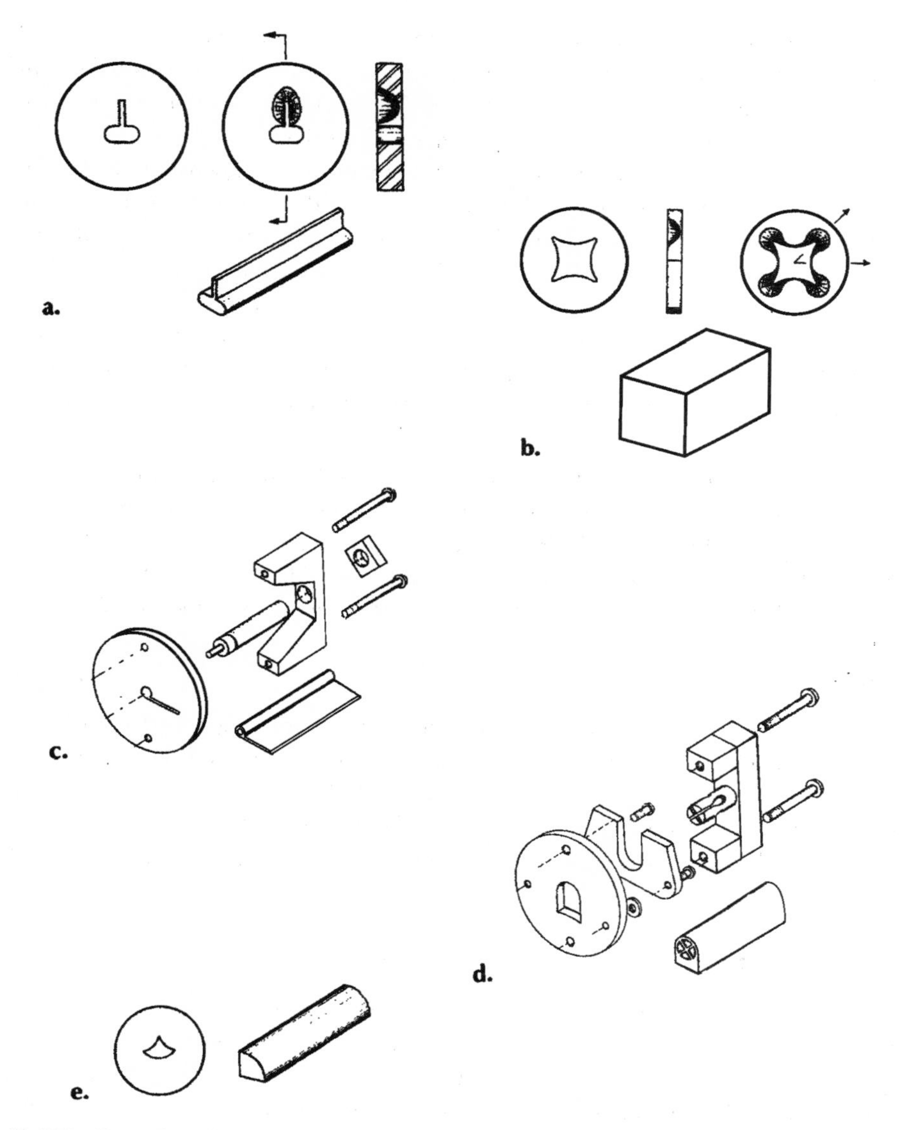

Figure 17.102 Examples of different profiles that include using lands of different configurations.

b. This die for making square extrusions uses convex sides on the die opening so that straight sides are formed upon melt's exit; the corners have a slight radius to help obtain smooth corners. The rear and sectional views show how part of the die has been machined away to provide short lands at the corners to balance the melt flow.

c. In this die for a P shape, a pin mounted on the die bridge forms the hole in the P with its land. The rate of flow in thick and thin sections is balanced by the shoulder dam behind the small-diameter section of the pin. The pin can be positioned along its axis to adjust the rate of flow to meet the melt characteristics.
d. In this die to extrude a rather complicated, nonuniform shape, a dam or baffle plate with its land restricts the flow at the heavy section of the extrudate to obtain uniform flow for all sections. The melt flows between the die plate and the dam to fill the heavy section. The clearance between the dam and the die plate can be adjusted as required for different plastics with different melt behaviors.
e. In this die for extruding a quarter-round profile, the die opening has convex land sides to give straight sides on the right-angled portion, and the corners have a slight radius to aid in obtaining smooth corners on the extrusion.

The machining required to produce all die parts, including the land area, is important. For example, great precision is required in an extrustion coating die. Figure 17.103 shows on top a honing stone and holder for honing the die lands. The angle between holder and stone (A) corresponds to the angle between die land and die face (B), usually, as in this drawing, 90°. The bottom view shows a die land that is tightly clamped in a vise, ready to be honed with the tool shown above it.

MANIFOLD

The manifold (feedblock) directs melt from the extruder to its die (Figs. 17.104 to 17.106). Different features are united in a manifold die. There are the center-fed and side-fed manifolds. In a center-fed manifold, the melt flow is inline from the extruder and if a core exists in the die, melt distribution around the core is uniform. With the side-fed manifold, the melt flow is not as uniform; however, with proper design and enough length to travel through the die, a certain degree of uniformity will exist.

PROCESS CONTROL

Provisions should be made to accurately control melt flow via temperatures, pressures, and rate of flow in all parts of the manifold and die using sensors such as stock thermocouples and pressure transducers. Provisions should be made to accurately control temperatures in all parts of the head and die. Variations in temperature of one degree Celsius or Fahrenheit are typically used in today's temperature-control systems. If there is a cold area in the die, the melt flow in that area will be slow and the result will be thinly gauged. A hot area results in more flow and the potential to burn (degradation) the exiting plastic.

Figure 17.107 shows heaters for flat films and sheets, profile dies, and blown film dies. Heating zones in the flat sheet die are highlighted. The result is that the die-end sections of melt are higher than the center area equalizing the cooling rate of the melt as it passes through the die. The result is the equalization of the flow rate and stock temperature of the sheet across the width of the die.

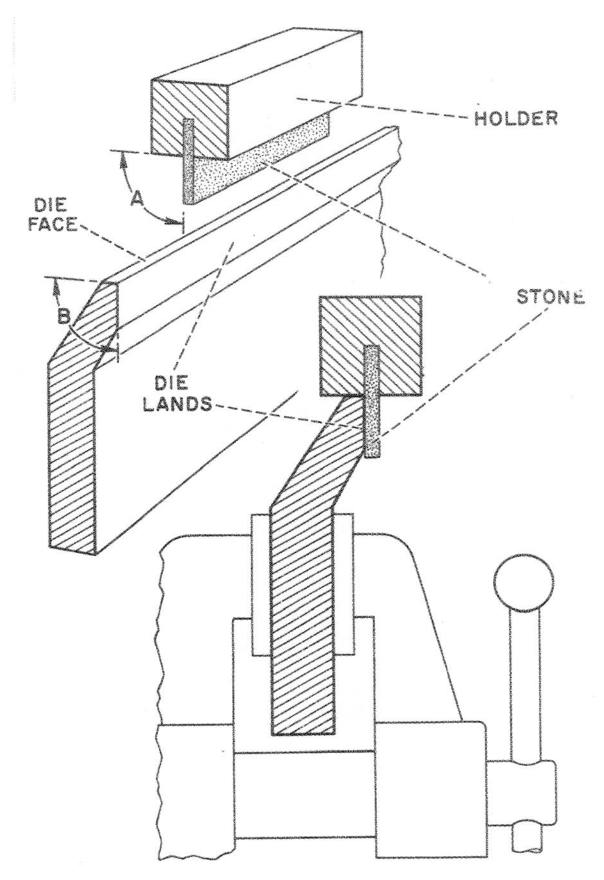

Figure 17.103 Honing extrusion coater die land.

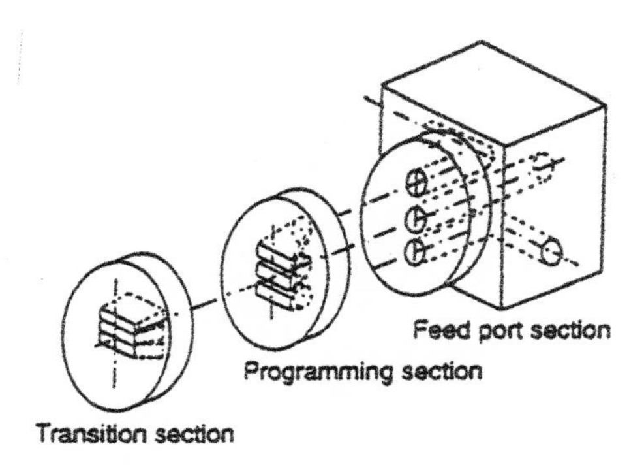

Figure 17.104 Schematic of feedblock sheet die.

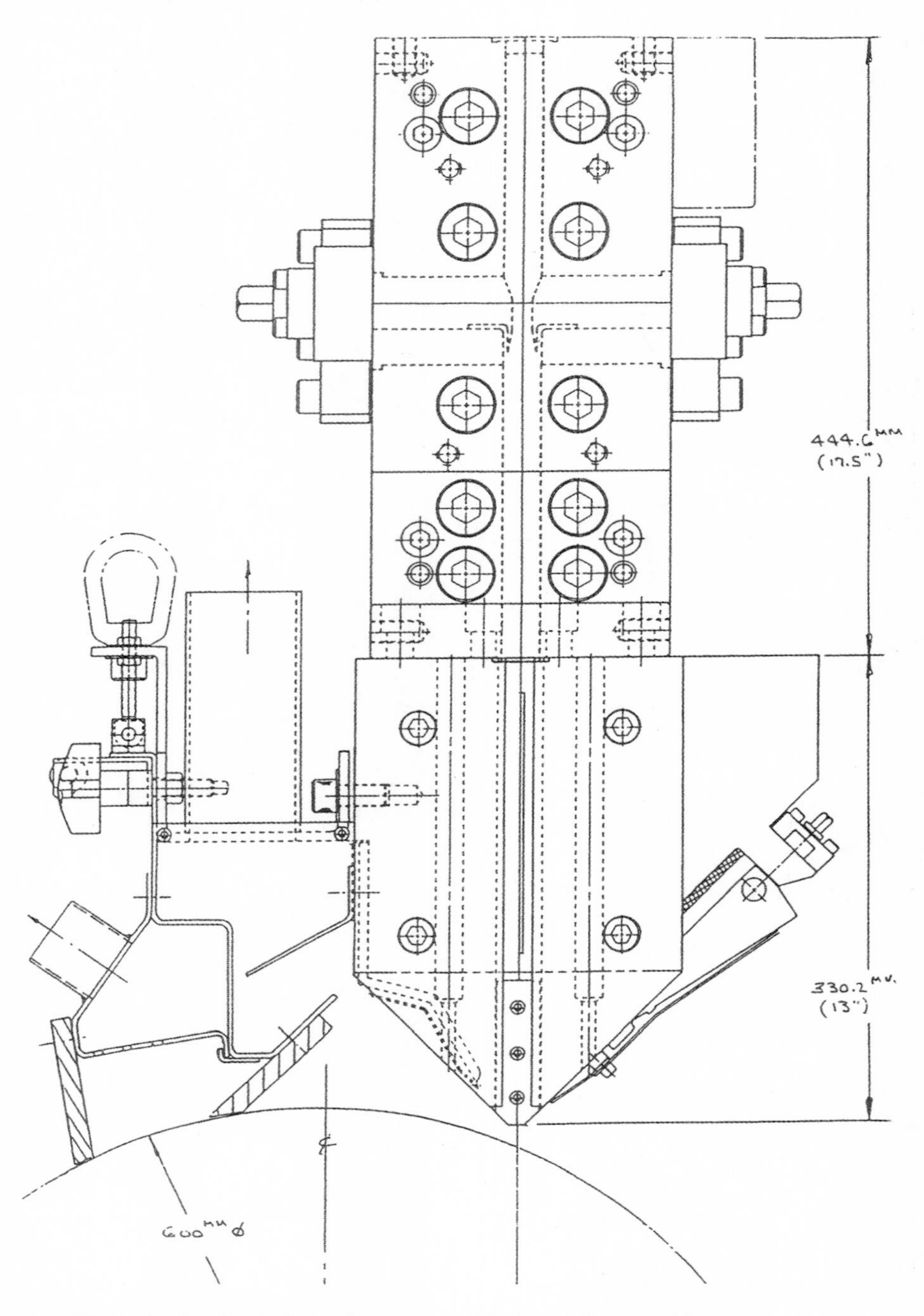

Figure 17.105 Example of a dual chamber of a feedblock and die assembly.

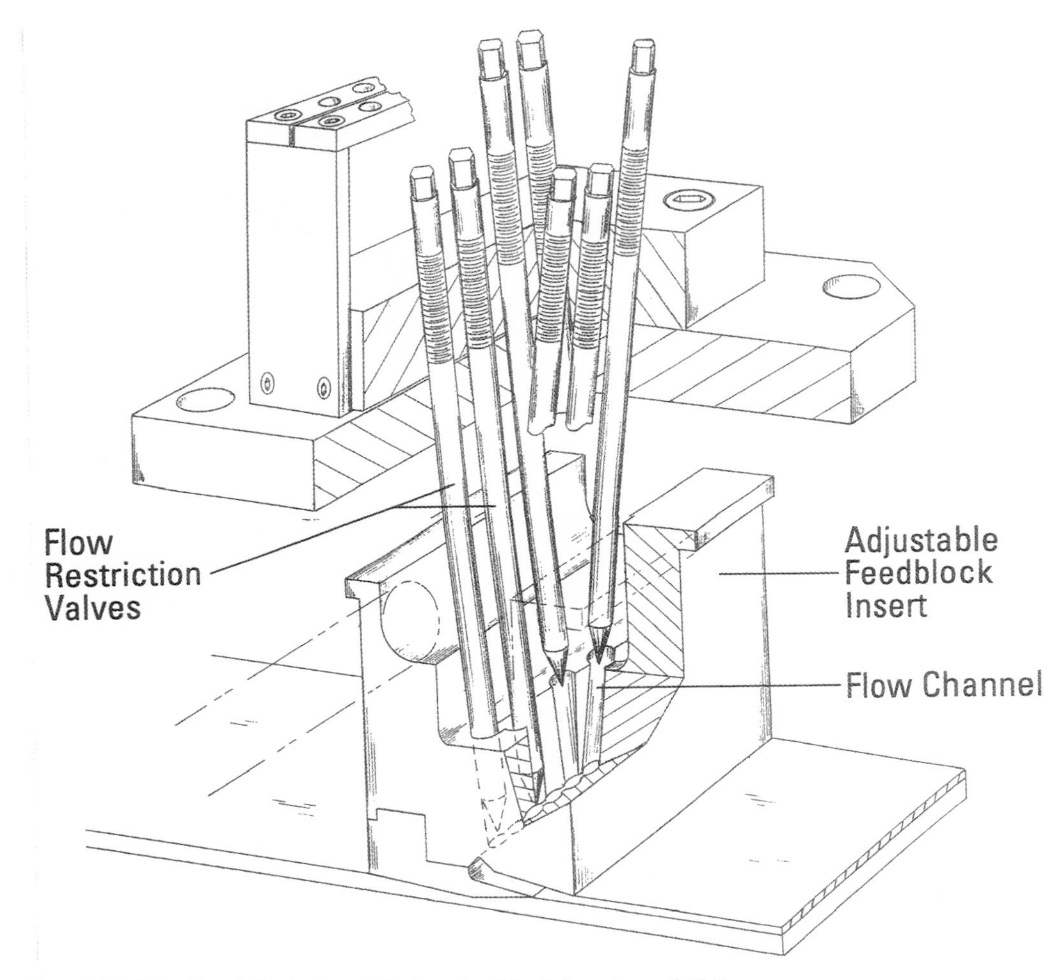

Figure 17.106 Specially designed Proteus feedblock (courtesy of EDI).

A heat pipe (also called thermal pins, heat transfer devices, or heat conductors) transfers the heat. It is a means of heat transfer, either to remove or to add heat in dies (molds, and other kinds of fabricating equipment). Compared to metals, they have an extremely high heat-transfer rate. They are capable of transmitting thermal energy at near-sonic isothermal conditions and at near-sonic velocity. Sizes range from very small to large diameters with limited lengths; however, pins can be put in series, and so on. They are tubular structures closed at both ends and containing a working fluid. For heat to be transferred from one part of the structure to the other, the working liquid is vaporized at a hot spot. The vapor is condensed at a cool spot and then travels via a wick to the hot (depleted) end where the condensate becomes a vapor again to repeat the process.

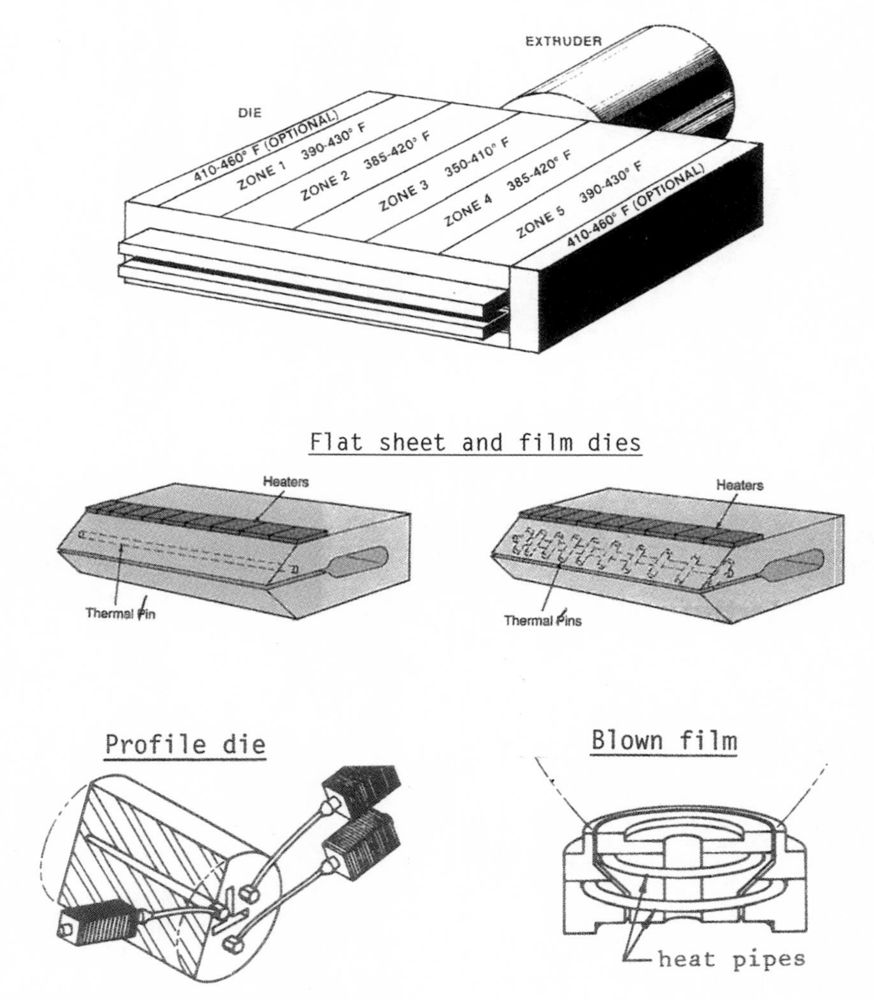

Figure 17.107 Example of heating different dies.

The most common working fluid is water. Since the vessel is closed, the pressure and temperature are not independent. Since a phase change is involved, the heat transfer is highly efficient. This permits one to maintain uniform temperatures along the length of the pin. For this reason, heat pipes are commonly used in extruder dies to equalize the temperature. Figure 17.108 shows the relation of pressure to melt flow rate based on the die's orifice opening, shape (such as a tear shape), and the venturi effect when the melt flow cross section is restricted.

There are many melt flow equations. The example that follows provides a guide that will help to predict the melt action in a die. The land length can be estimated by finding first the shear rate

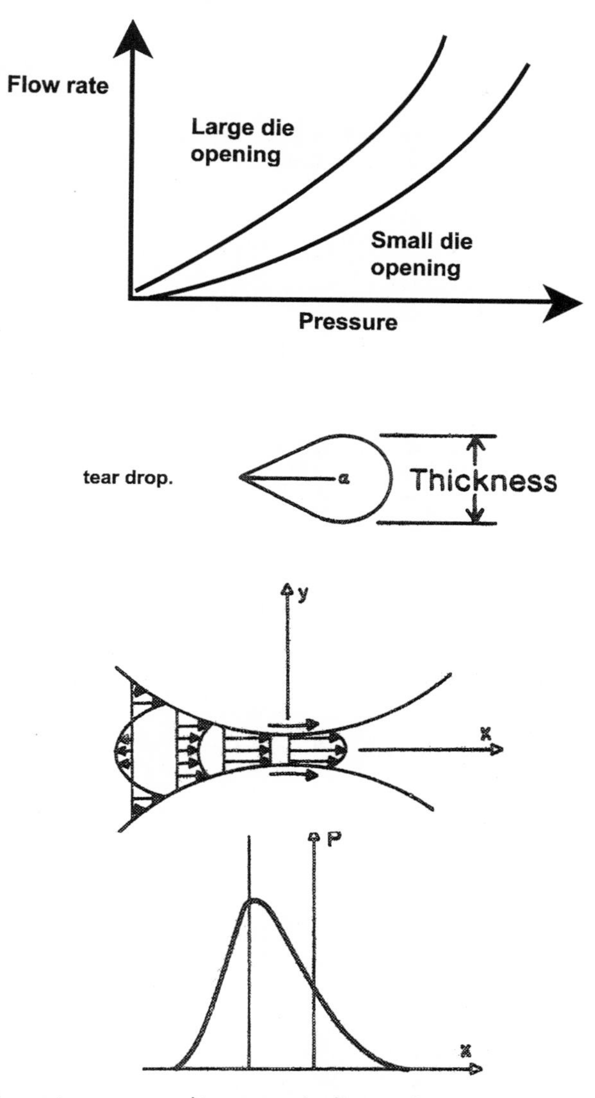

Figure 17.108 Melt flow rates versus melt pressure in die openings.

and then the pressure drop at the die lips for a particular profile. Using raw material data that can be provided by the material supplier (Tables 17.59 and 17.60), the following rheological equations have been developed (chapter 1) for the computation that would be desirable in dies for profile slits, rods, and tear shapes that are use in sheet dies (Fig. 17.108).

Material	Description	Max allowable shear rate 1/sec
ABS	Acrylonitrile Butadiene Styrene	40,000
EVA	Ethylene Vinyl Acetate	30,000
PS	Polystyrene	40,000
PE	Polyethylene	40,000
PA6	Nylon 6	60,000
PA66	Nylon 66	60,000
PBT	Polybutylene Terephthalate	50,000
PC	Polycarbonate	40,000
PMMA	Polymethyl Methacrylate	40,000
POM	Acetal	20,000
PP	Polypropylene	100,000
PVC	Polyvinyl Chloride	20,000
SAN	Styrene Acrylonitrile	40,000
TPO	Thermoplastic Olefin	40,000

Table 17.59 Examples of melt shear rates

	Die swell ratio at 200°C (392°F) for the following shear rates			
Plastic	*10 s⁻¹*	*100 s⁻¹*	*400 s⁻¹*	*700 s⁻¹*
PMMA-HI	1.17	1.27	1.35	–
LDPE	1.45	1.58	1.71	1.90
HDPE	1.49	1.92	2.15	–
PP, copolymer	1.52	1.84	2.1	–
PP, homopolymer	1.61	1.9	2.05	–
HIPS	1.22	1.4	–	–
HIPVC	1.35	1.5	1.52	1.53

Table 17.60 Examples of the effect of shear rate on the die swell of TPs

For a slit die opening the shear rate, $sec^{-1} = 4.61 \times 10^{-2}$ (flow, lb/h)/(density, g/cm^3)(die width, in)(slit thickness, in^2). Pressure, psi $= 2.9 \times 10^{-5}$ (shear rate, sec^{-1})(viscosity, poise)(land length, in)/(slit thickness, in).

For a rod die opening shear rate, $sec^{-1} = 9.8 \times 10^{-3}$ (flow, lb/h)/(density, g/cm^3)(radius, in). Pressure, psi $= 2.9 \times 10^{-5}$ (shear rate, sec^{-1})(viscosity, poise)(land length, in)/(radius, in)

For a tear shape drop shear rate, $sec^{-1} = 0.123 \times$ (flow, lb/h)/(density, g/cm^3)(thickness, in) $(\pi/2 \times 1/\tan\alpha)$. Pressure, psi $= 1.45 \times 10^{-5}$ (shear rate, s^{-1})(viscosity, poise)(land length, in)$(\pi/2 \times 1/\sin\alpha)$/(thickness, in).

Shear rate relates to the dimensions of the die and the profile's opening with the shear increasing as the volumetric flow rate increases and decreases while the profile widens; doubling the width of the profile will halve the shear rate. Pressure drop, though not important by itself, determines the land length and the thickness of the profile, as seen in these equations. By adjusting the measurements of the die lip and the land length, the designer can adjust the amount of the pressure drop.

A large pressure drop is inevitable in extrusion dies since pressure outside the die is zero. If the pressure in the die is too high, such as over 4000 psi (28 MPa), the high pressure may cause the die to leak or force the material to back up over the screw. For example, pressures below 28 MPa (4000 psi) are considered suitable for PVC profile extrusion.

Dies meet different product requirements. Well-built dies can include adjustments to meet the performance requirements for the extrudate. They can include restricter/choker bars, temperature zoning, pressure sensors, adjustment bolt/ device, and/or other devices.

Dies are designed to process plastics that have specific melting/rheological characteristics. For example, a conventional LDPE blown film die with 0.030 in (0.8 mm) die gap will not process linear low-density polyethylene (LLDPE) or metallocene polyethylene (mPE) satisfactorily at a high output rate.

The higher-viscosity LLDPE increases back pressure significantly, thereby decreasing the throughput. With this change, there may be melt fracture (shark skinning), which produces a rough surface finish. Processors of LLDPE can overcome these problems with wide die-gap openings of about 0.090 in (0.3 mm). By increasing the die gap, the head pressure decreases, allowing significant increase in output (Fig. 17.108). Another approach to improve the surface characteristics of the film without impacting the cooling is to increase the lip temperature.

The mPEs generate a lot of shear heat, so overheating will put them in an unfavorable situation, where changes in their viscosities occur. They required a learning period for the operator to properly set their controls. What have made mPEs very attractive is their high-performance physical properties.

TEMPERATURE CONTROL

All dies have some form of temperature control. It can be as simple as insulating to prevent heat loss or, more commonly, it can consist of multiple zones with accurate monitoring and control. Since most dies cannot add or subtract significant amounts of heat from a melt due to short residence time and low thermal conductivity (most plastics being good thermal insulators), the die's temperature control has only two primary functions: to provide heat for start-up and to control temperatures within relatively narrow ranges to minimize the effect on the plastics.

Oil, water, steam, fire-resistant fluids, and induction heating occasionally are used for heating dies, but by far the majority of applications are satisfied by electrical resistance heaters in the form of wires, encased rods, flat elements, blocks, and custom shapes.

Controlling the heat input can be as simple as using variable-voltage transformers that will supply relatively constant heat input or as complicated as using full proportional-integral-derivative (PID) controllers with thermocouple temperature sensing. When temperature sensing is used, zones are established based on logical groupings of heaters and the functional areas to be controlled. Each zone has a set point and through its control device accurately maintains set temperatures.

With microprocessor-based extruders and process lines, a die's temperature can be controlled easily without discrete controllers. Microprocessor controls generally require less operator attention and result in higher levels of reliability, in the case of changing groups of set points. Other advantages are automatically programmed start-up sequences, overtemperature alarms, thermocouple loss alarms, heater failure alarms, and better temperature control.

DIE TYPE

Different types of dies are required to produce the many different shapes produced by extruders worldwide. Extruder dies process more plastics than any other fabricating process (466). Table 17.61 lists a few types of dies designed and manufactured by Extrusion Dies Inc.

Dies can be categorized by their product performance. There are straight-through, crosshead, and offset dies. To be more specific, they can be classified as the following:

1. Axial or straight-through extrusion heads with symmetrical flow channels, particularly tube and pipe heads, circular rod, and monofilament dies
2. Angled dies, particularly crossheads and angular heads for wire and cable covering, crossheads and offset heads for tube and pipe, and film-blowing heads
3. Profile dies that include slot dies for flat films and sheets and multiorifice heads for monofilaments
4. Dies for special products such as netting

The following general classifications may be helpful as a guide to film and sheet thickness selection for a die. Different groups within the different industries may have their own thickness definitions as well as their own terminology. The general classifications are as follows:

1. Film dies, which are generally applicable for thicknesses of 0.010 in (0.003 mm) or less
2. Thin-gauge sheet dies, which are normally designed for thicknesses up to 0.060 in (0.015 mm)
3. Intermediate sheet dies, which may cover a thickness range of 0.040 to 0.250 in (0.01 to 0.06 mm)
4. Heavy-gauge sheet dies, which extrude thicknesses of 0.080 to 0.500 in (0.02-to 0.13 mm)

The coupling between a barrel and a die can be created in various ways using bolts or locking devices. The methods include the following:

EDI Model No.	Flex Range	Restrictor Bar	Film 10 ml & Below (254 µm & Below)	Coating & Laminating	Thin Sheet 10 ml–60 ml (254 µm–1524 µm)	Midrange Sheet 10 ml–90 ml (254 µm–2286 µm)	Heavy Sheet 60 ml & Above (1524 µm & Above)	Lab. Applications	Options
Ultraflex L 40	0.040 in (1.0 mm)		■	■				■	**Lip adjustments**
Ultraflex L 75	0.075 in (1.9 mm)				■			■	Micro push
Ultraflex 40	0.040 in (1.0 mm)		■	■					Micro push/pull
Ultraflex H 40	0.040 in (1.0 mm)		■	■					**Material of construction**
Ultraflex H 75	0.075 in (1.9 mm)				■				Stainless steel
Ultraflex H 100	0.100 in (2.54 mm)				■	■			Other upon request
Ultraflex HM 40	0.075 in (1.9 mm)		■	■					**Platings**
Ultraflex HM 75	0.075 in (1.9 mm)				■				Electroless nickel
Ultraflex HM 100	0.100 in (2.54 mm)				■	■			Polymer impregnated chrome
Ultraflex H 40 EPC	0.040 in (1.0 mm)		■	■					Polymer impregnated nickel
Ultraflex H 40 EPC C/L	0.040 in (1.0 mm)		■	■					**Deckling**
Ultraflex LR 40	0.040 in (1.0 mm)	45°	■					■	**Removable lips**
Ultraflex LR 75	0.075 in (1.9 mm)	45°			■		■	■	**Extended lips for**
Ultraflex R 75	0.075 in (1.9 mm)	45°*			■		■		**close approach**
Ultraflex HR 75	0.075 in (1.9 mm)	45°*			■		■		**Roll guard**
Ultraflex RC 75	0.075 in (1.9 mm)	45°*			■		■		**Wrench guard**
Ultraflex HRC 75	0.075 in (1.9 mm)	45°*			■		■		**Insulation jacket**
Ultraflex HRMC 75	0.075 in (1.9 mm)	45°*			■		■		**Lip heaters**
Ultraflex R 100	0.100 in (2.54 mm)	45°*			■	■	■		**Heat tubes**

* Also available with 90° restrictor bar.

Table 17.61 Examples of extrusion dies from Extrusion Dies Inc.

1. Fitting a flange with a clamp ring on the barrel and fitting a fixed flange on the die
2. Fitting flanges on the barrel and die with tapered links and two bolted half-clamps, or placing a ring clamp hinged at one side and bolted to the other side
3. Using a swing-bolt flange connection between the barrel flange and a die flange

FLAT DIE

The flat dies, or slot dies, are used to produce webs in a variety of processes. They all have an interior manifold for distributing the plastic and lips for adjusting the final profile of the web (extrudate). Some dies have movable restrictor bars for changing the manifold for proper melt distribution (Fig. 17.109). All flat dies have flexible lips that can be adjusted by bolts to remove humps or bumps in the web's profile. Die lips can have their adjustment bolts be push-only, where internal plastic-melt pressures are adequate to keep the lips positioned against the bolts, or they can be push-pull for low-pressure applications. Direct acting or differential-thread designs (for minute adjustments) are available. Profile variations of at least ±3% or less can be achieved with flat dies.

To compensate for variable neck-in or to change web widths, deckles can be fitted to the lips at the ends of the die slot. Deckles cannot be used with degradable materials since there is a stagnant region formed behind the deckle that will eventually decompose the plastic. Deckles can be designed to be adjusted while running or adjusted when off-line (Fig. 17.110).

Computer-controlled automatic profile dies with electrically controlled sensors in closed-loop control systems have developed greater efficiency and accuracy to extrusion coating, cast film, and sheet lines (Fig. 17.111). A scanner measures the web thickness and signals the computer, which then converts the readings to act on thermally actuated die bolts. The individual adjusting bolts expand or contract as ordered by the computer to control the profile. The more sophisticated systems measure the temperatures in the adjusting bolts and provide faster response time with less scrap and quicker start-ups. The scanner is typically an infrared, nuclear, or caliper-type gauge.

CAST FILM DIE

A coat hanger interior-manifold design with a center entry promotes good flow patterns without using restrictor bars. Lip gaps are relatively small since most cast films are thin. Push-only bolts are usually enough to control die lips. Draw-down ratios (DDRs) of 20:1 to 40:1 and rates of 20 lb/in of opening are common. Deckling is usually not required. The lips are usually ground to a 16 RMS finish since the chill roll's downstream equipment determine the final finish of the film.

SHEET COAT HANGER DIE

Many characteristics of cast film dies carry over into sheet dies but because of generally thicker materials, die-lip openings are much larger and do not generate enough back pressure for accurate distribution of melt. Therefore, many sheet dies have a restrictor bar (Fig. 17.112).

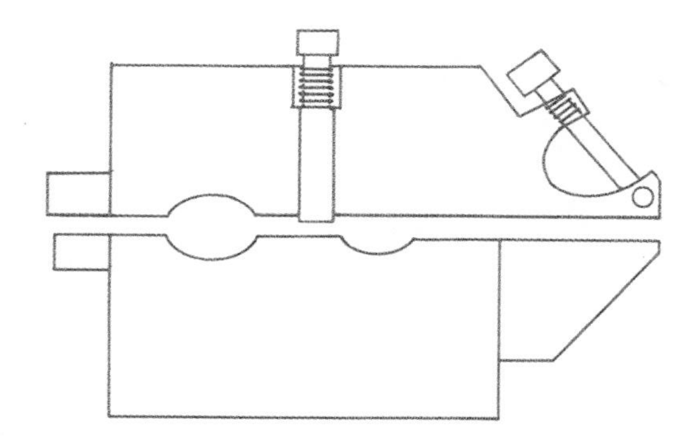

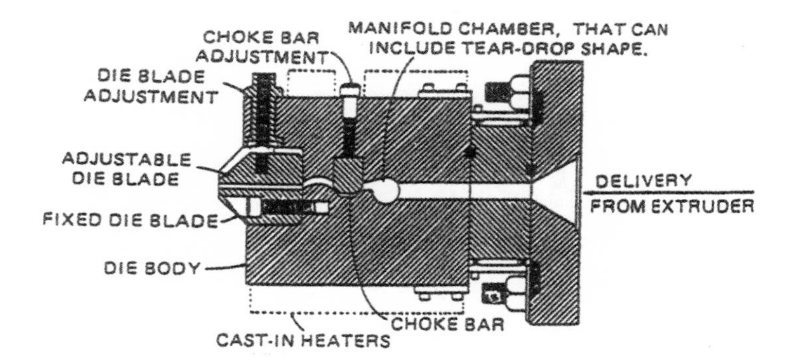

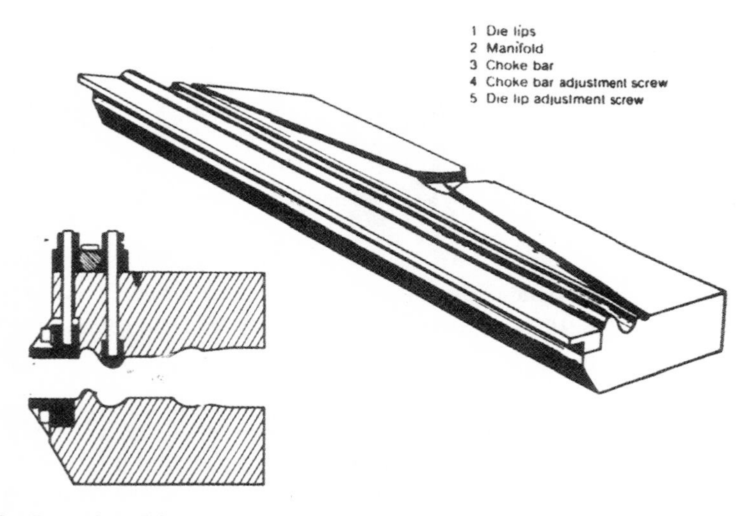

Figure 17.109 Examples of flat dies with its controls.

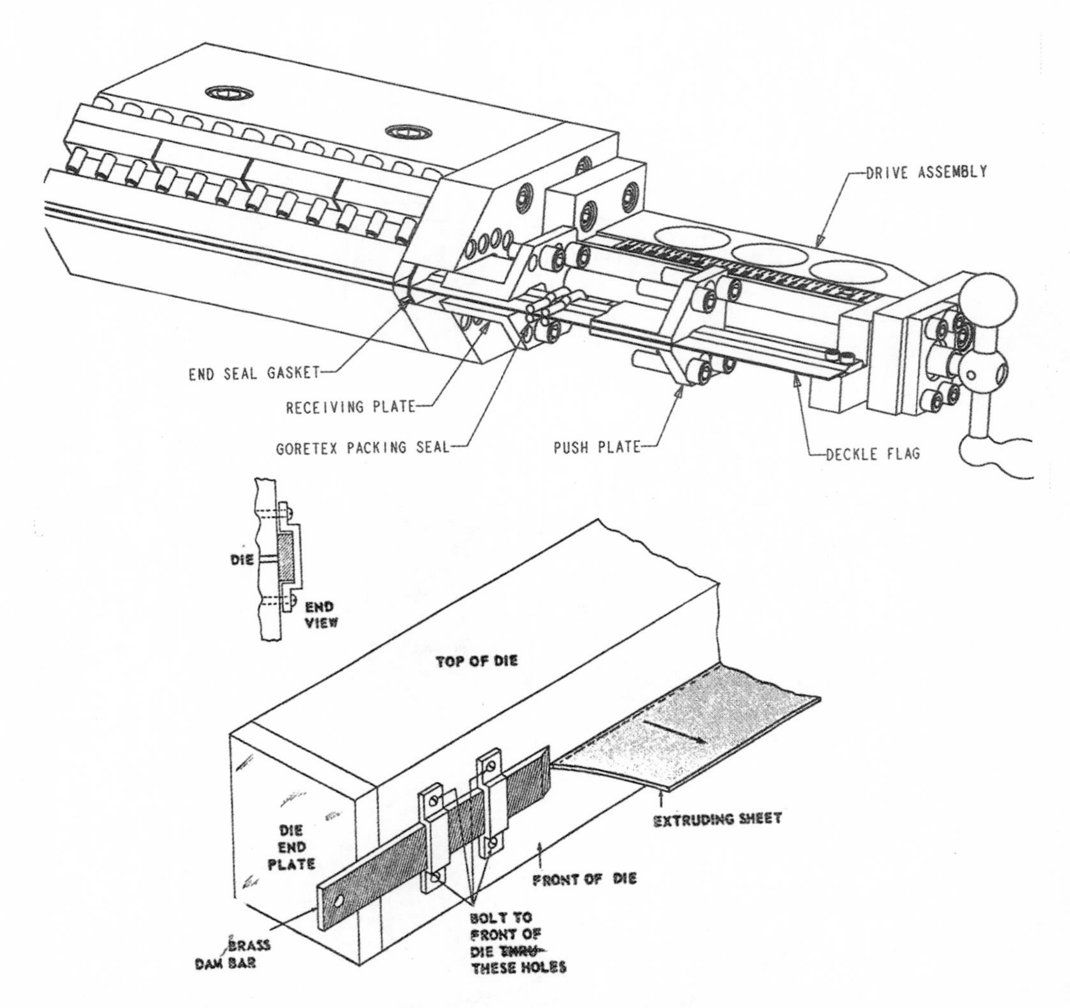

Figure 17.110 Examples of deckles that are adjusted during processing (top) and manually adjusted off-line.

The interior of sheet dies, particularly near the lips, is generally highly polished (2 to 4 RMS). Feedblocks are usually used for sheet coextrusion. Automatic profile control is usually not required in cases of thick sheet (more than 100 mils), where the final profile of the sheet is determined by the cooling and polishing units.

The popular coat hanger die, used for extruded flat sheets and other products, illustrates an important principle in die design. The melt at the edges of the sheet must travel farther through the die than the melt that goes through the center of the sheet. A diagonal melt channel with a triangular dam in the center is an approach to some degree of restricting the direct flow. The principle of

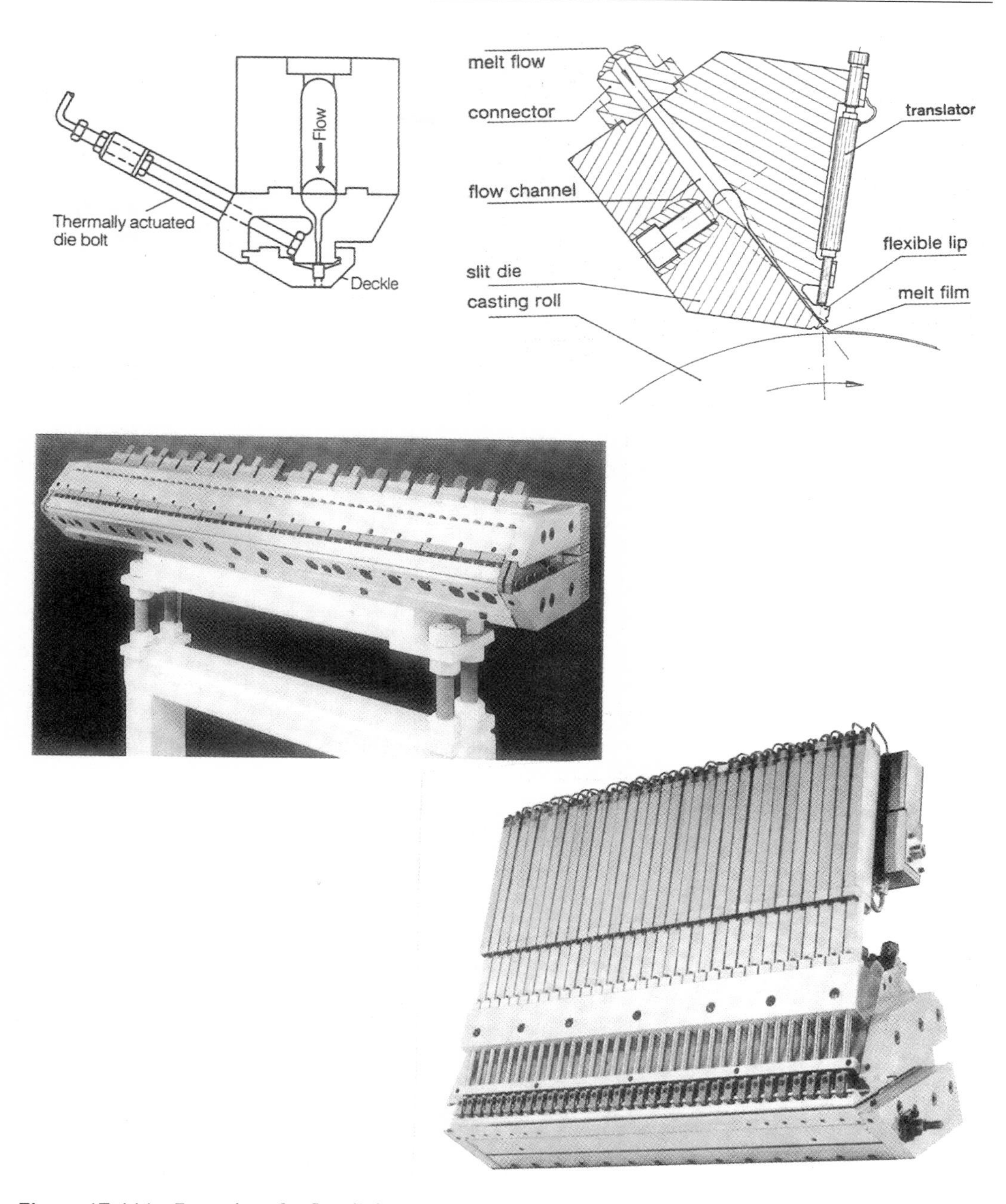

Figure 17.111 Examples of a flat die's automatic control systems.

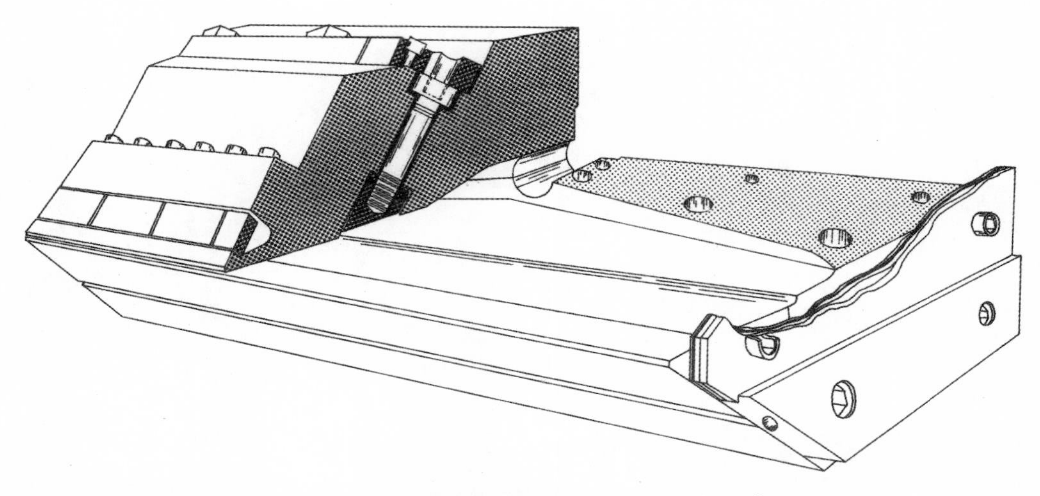

Figure 17.112 Cutaway view of a coat hanger sheet die with a restrictor bar.

built-in restrictions is used to adjust the melt flow in dies. With blow molding and profile dies, the openings require special attention to provide the proper product shape.

The coat hanger manifold design can be tuned for a processing "window" of different flow rate plastics. This manifold is designed by establishing the same resistance to flow across the manifold and land at various points across the die. Its shape is typically a teardrop that reduces in size from the center to the ends of the die. This reduction in volume reduces residence time in the die and is of importance when processing thermally degradable plastics. After the melt leaves land, the melt flows into a secondary manifold, an area of the flow channel that allows the plastic to move laterally again if required. It also is used to control the total pressure drop of the die in conjunction with the final lip land.

There are always advantages and disadvantages to any design. In the coat hanger manifold, the back line of the manifold is farther from the exit of the die at the center than it is on the ends. This could result in a die that will deflect its opening more at the center, causing uneven melt distribution. The cause is a result of the linear flow channels within this type of die.

COATING AND LAMINATING DIE

Extrusion coating dies are simple because most plastics run through them are polyolefins, which are substantially nondegradable. This allows the use of noncontoured, straight manifolds (Figs. 17.113 and 17.114) and deckles, but requires high precision because of the extremely thin coatings desired. In addition, very high operating temperatures can create warpage, corrosion, and control problems.

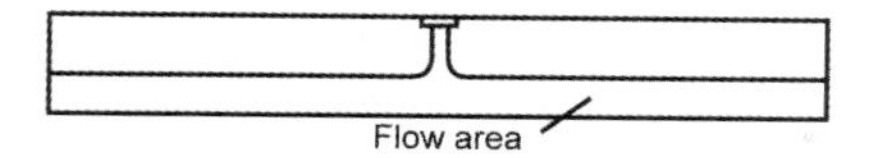

Figure 17.113 Example of a straight coating or laminating manifold die.

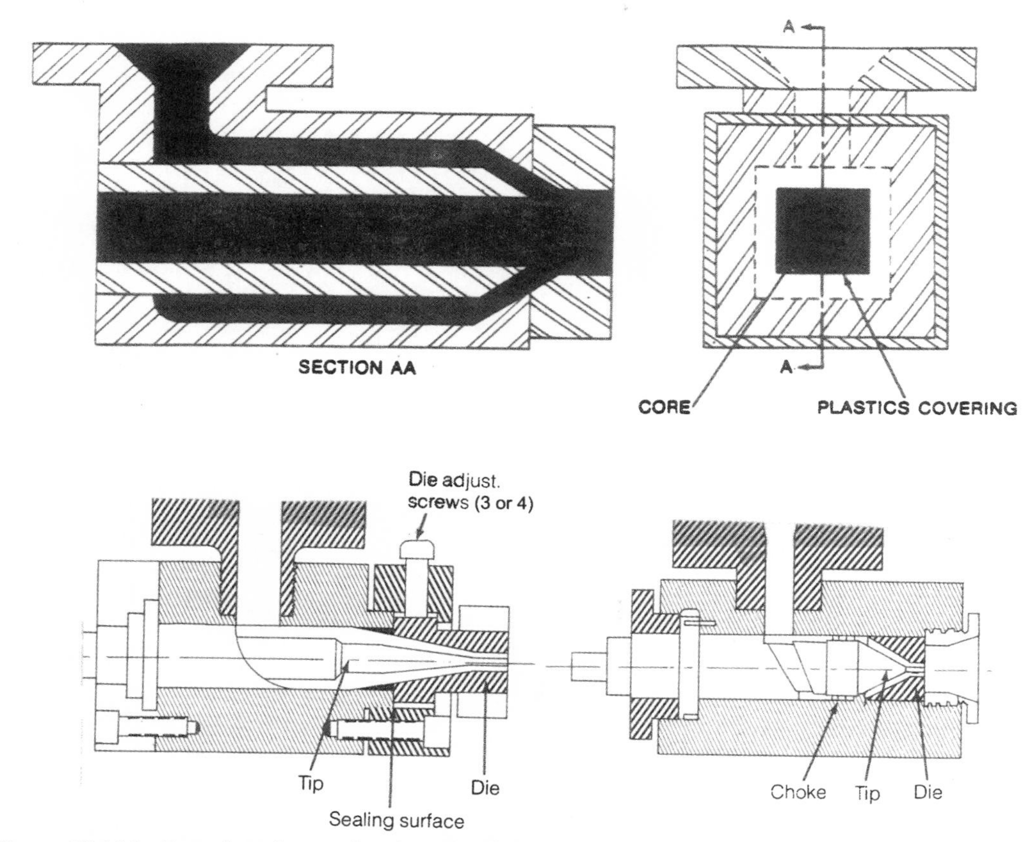

Figure 17.114 Examples of a crosshead coating dies.

Deckles that are adjustable while they run are most common and edge-bead-reduction techniques can be incorporated in the design. Push-pull die bolts usually are necessary since temperatures are high and viscosities and internal pressures are low. Multimanifold dies for coextrusion are more commonly used in extrusion coating and automatic profile control.

TUBULAR DIES

BLOWN-FILM DIES

The spiral mandrel die is used to eliminate spiders in the die and the inherent film weakness (Fig. 17.115). This design usually is calculated by a computer since the flows and pressure drops are complicated.

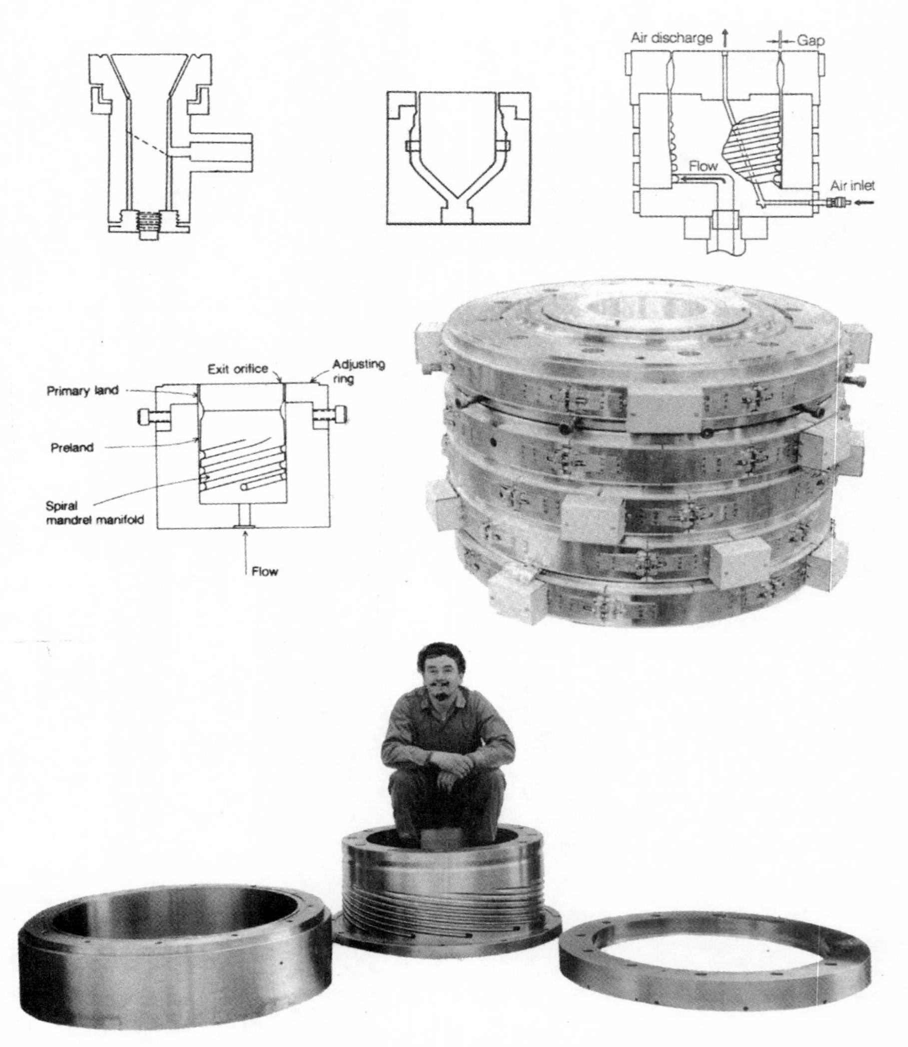

Figure 17.115 Examples of single-layer blown-film dies include side-fed typex (top left), bottom-fed types with spiders (top center), and spiral-fed types.

Shifting the outer ring of the die controls the profile, but since the ring is difficult to deflect locally, only gross adjustments are generally possible. The accuracy of machining can be the primary determination of gauge. Variations of $\pm 5\%$ or higher are typical for this type of die. When compared to other blown-film dies, they each have advantages and disadvantages, as indicated in the following.

SIDE-FED DIE

Some advantages are the following:

1. Low initial cost
2. Adjustable die opening
3. Ability to handle low-flow materials

Some disadvantages are the following:

1. Mandrel deflects with extrusion rate, necessitating die adjustment
2. Die opening changes with pressure
3. Nonuniform melt flow
4. Cannot be rotated
5. One weld line in film

BOTTOM-FED SPIDER DIE

Some advantages are the following:

1. Positive die opening
2. Can be rotated
3. Will handle low-flow plastics

Some disadvantages are the following:

1. High initial cost
2. Very hard to clean
3. Two or more weld lines in film

SPIRAL FEED DIE

Some advantages are the following:

1. No weld line in film
2. Positive die opening

3. Easy to clean
4. Can be rotated
5. Improved film optics

Some disadvantages are the following:

1. High head pressure
2. Will not handle low-flow resins without modification

Automatic profile control for tubular dies can be applied to the blown-film process generally by heating or cooling small segments of the die lips or by selectively cooling the bubble at specific spots via the air ring. The intent is to effect temperature and viscosity changes of the plastic and therefore stretch ratios within closely defined limits, without affecting optics and other properties. Since the residence time in the die lip area is very short, heat gain or loss in the melt is slight. This technique is generally suitable only for very thin films.

Pipe die

Processing can use spider dies, spiral mandrel dies, or basket-type dies that support the inner mandrel with a perforated sleeve through which the melt flows. Figure 17.116 provides examples of different pipe die designs.

Foam die

Spider dies are used to a large extent because of their low cost, and for many applications in thermoforming the spider lines can be aligned with edges and center material, which is trimmed and recycled.

Spiral mandrel dies are used when spider marks are unacceptable. The center mandrel normally is adjusted to control overall gauge by having tapered exit lips and adjusting the mandrel axially to change the gap.

Profile die

Solid profiles can be simple flat-plate dies with finished land geometry and preland dimensions determined by experience and trial in conjunction with sizing plates. If hollow shapes are extruded, supports are necessary, and tubing applications can have inflation air holes. Most of the profile dies, particularly those used in long production runs, require precision dies to meet very close tolerance requirements. Examples are shown in Figure 17.117.

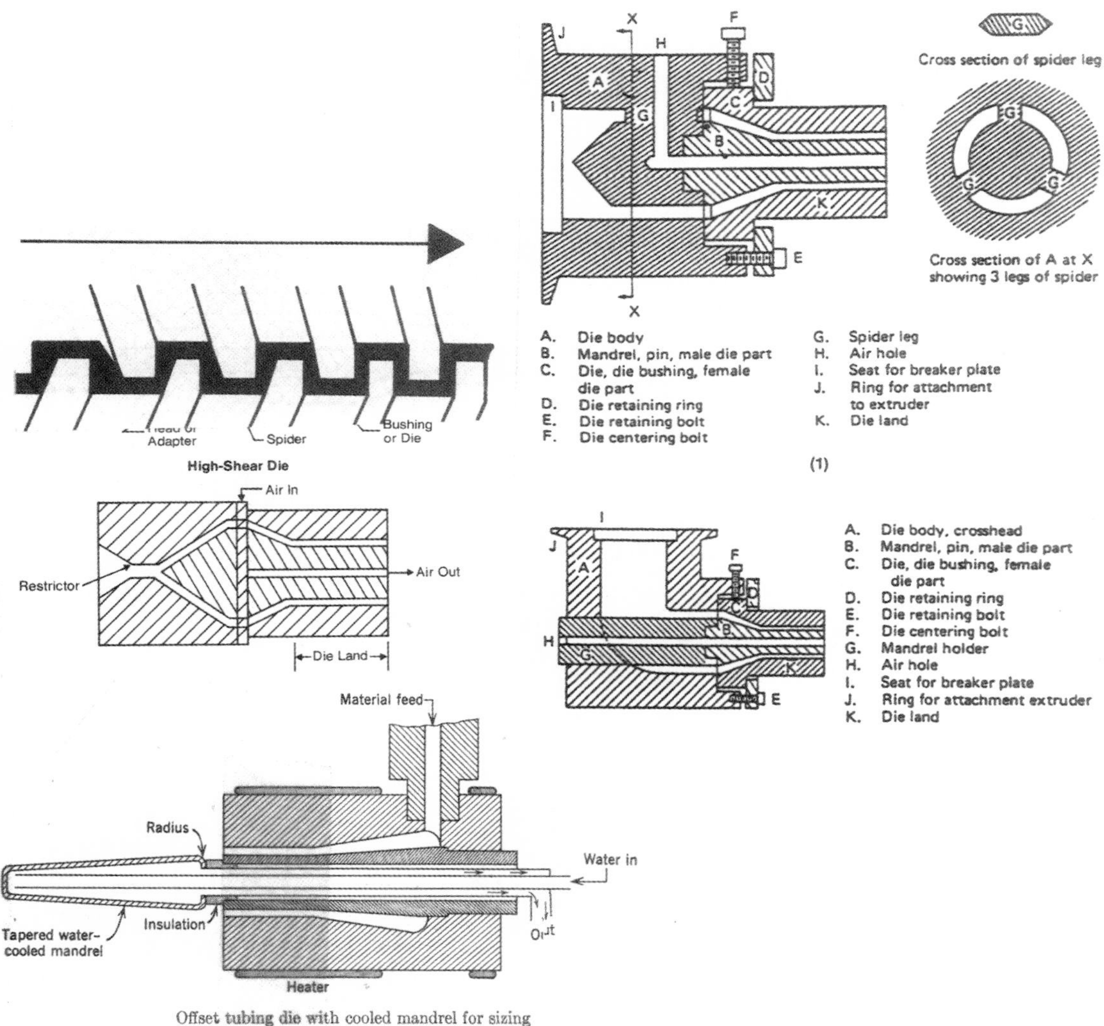

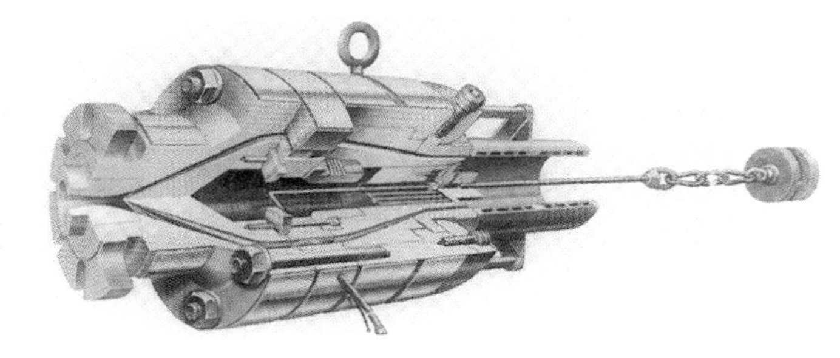

Figure 17.116 Examples of different pipe die designs.

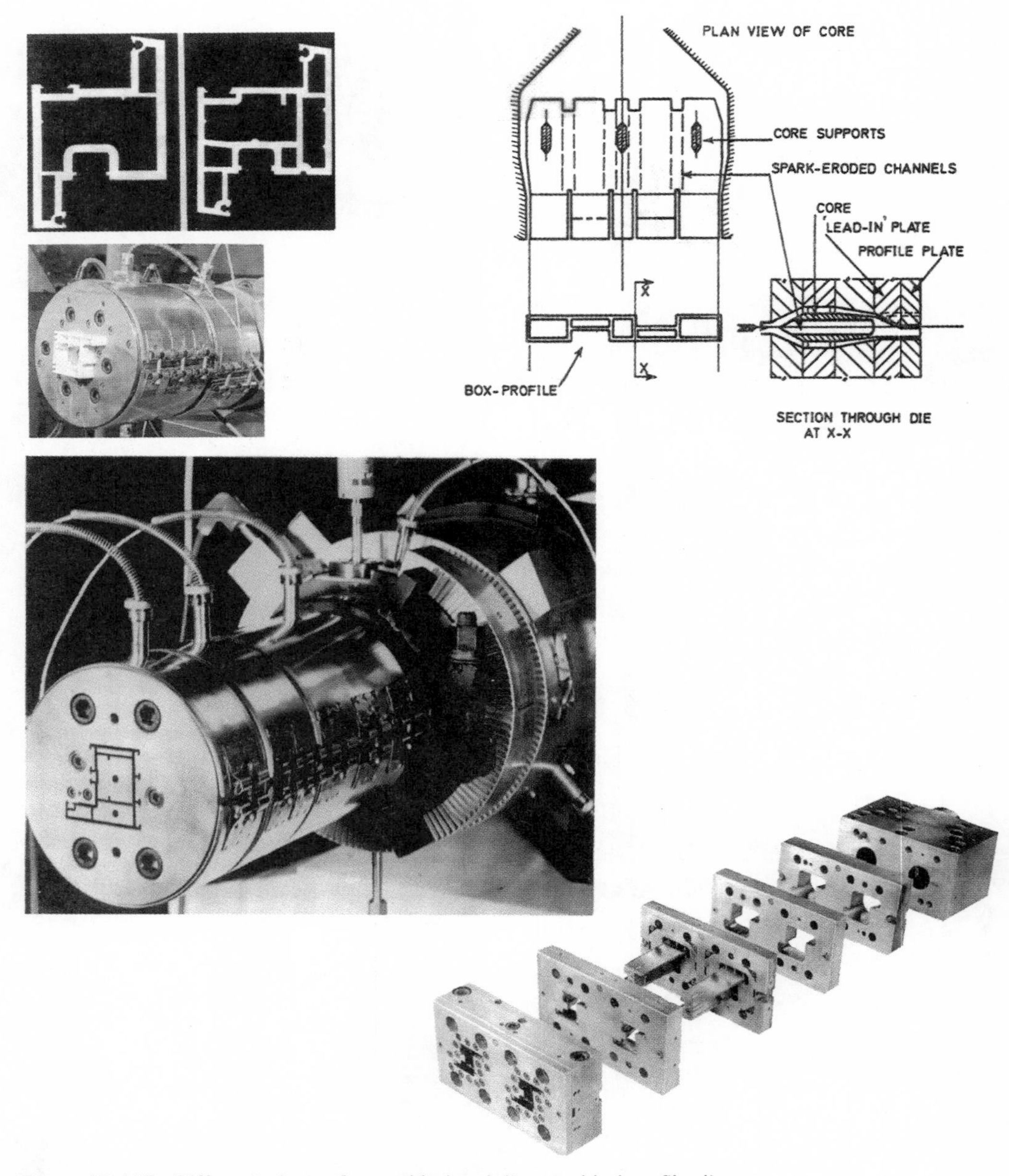

Figure 17.117 Different views of assembled and disassembled profile dies.

WIRE COATING DIE

A specialized case of profile extrusion exists when coating wire. The wire is fed through a hardened insert in the center of the die at high speed and the plastic is extruded around it through a manifold or multiple ports. Most dies are subjected to very high internal pressures since the uncommon pressure of over 5000 psi (35 MPa) is required.

The usual crosshead die has a 90° angle between the wire line and the extruder body axis (Fig. 17.118). Different angles are also used to improve processability. With this setup, the entire length of the extruder projects sideways from the coating lines. To help melt flow from developing dead spots in the melt channels with certain plastics, 30° or 45° crossheads can be used. They provide a more streamlined interior and the extruder location is better adapted to some plant layouts. Regardless of the angle used the process relates to draw ratio balance (DRB) and DDR to ensure

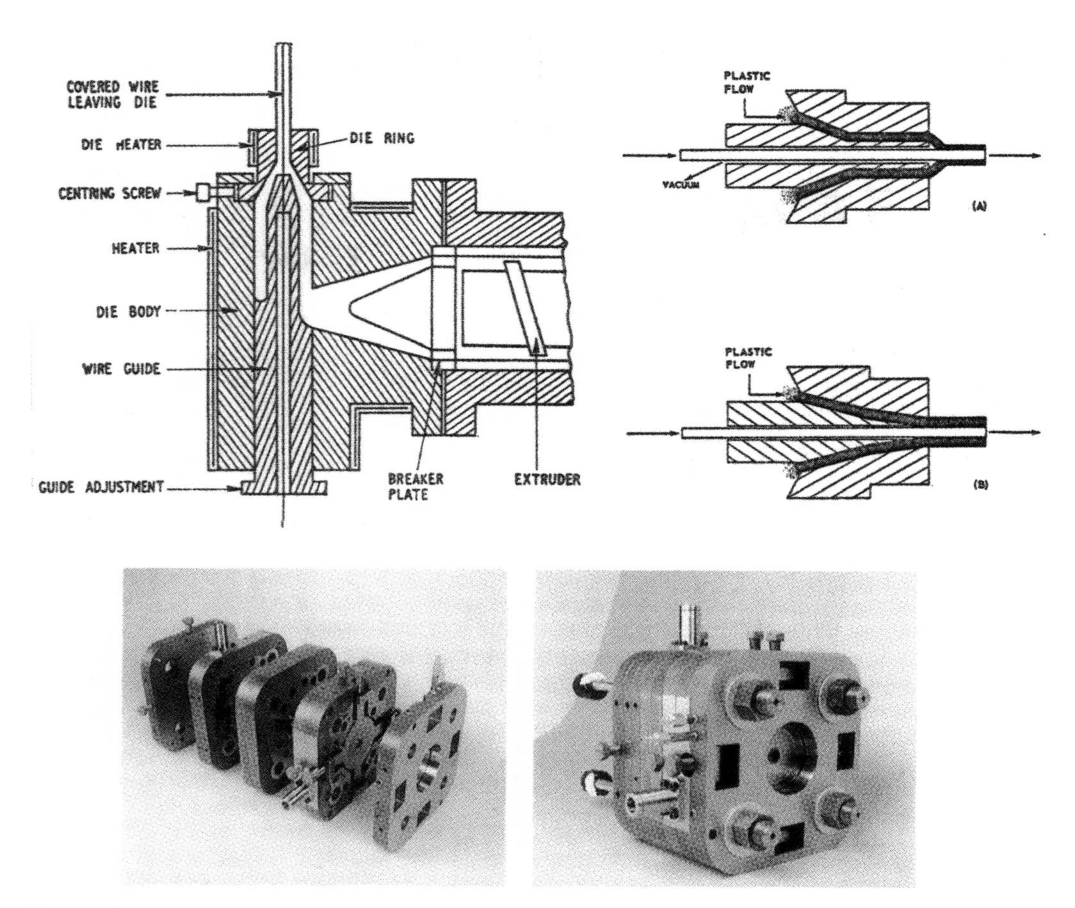

Figure 17.118 Examples of wire coating dies.

proper coating. Plastics have different DRBs and DDRs that can be used as guides to processability and to help establish their various melt characteristics.

DRAW RATIO BALANCE

The aim is to set uniformity and balance in the plastic coating. This DRB aids in determining the minimum and maximum values that can be used for different plastics (Fig. 17.119).

To determine the DRB, use the equation

$$\mathrm{DRB} = (D_D/d_{cw}) \ / \ (D_T/d_{bw}) \geq 1,$$

where

D_D = diameter of die opening,
D_T = diameter of guide tip,
d_{cw} = diameter of coated wire, and
d_{bw} = diameter of bare wire.

The value of the DRB will range around 1, with the ± close to 1. Being outside the set limits can cause at least out of round and plastic degradation.

DRAW-DOWN RATIO

The DDR in a wire die or a circular die, is the ratio of the cross-sectional area of the die orifice/opening to the final extruded shape (Fig. 17.120).

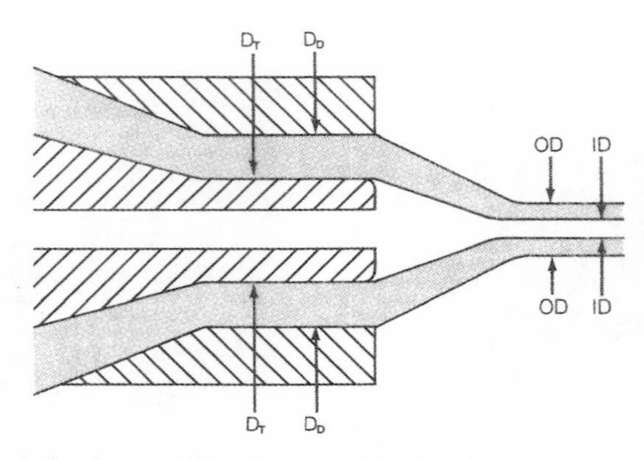

Figure 17.119 Schematic for determining wire coated DRB in dies.

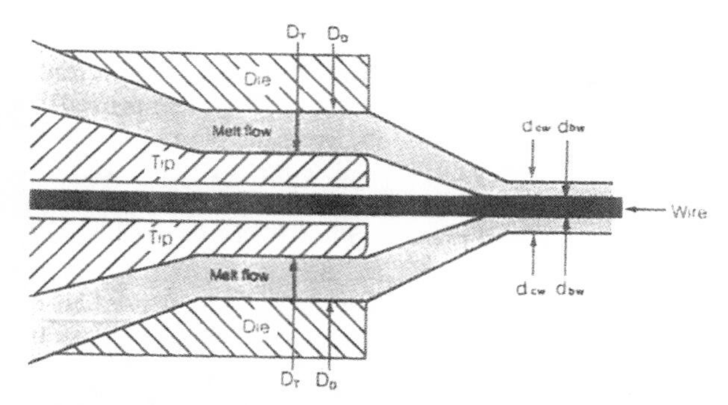

Figure 17.120 Schematic for determining wire coating DDR in dies.

To determine the DDR use the equation

$$\text{DDR} = (D_D^{\ 2} - D_T^{\ 2})/(d_{cw}^{\ 2} - d_{bw}^{\ 2}).$$

With the DDR too high, a rough surface and/or internal stresses in the coating will exist. Typical satisfactory DDR values for LDPE are 1.5. A satisfactory DDR value for HDPE is 1.2; the value for PVC is 1.5, and the value for nylon is 4.0.

FIBER DIE

The spinneret is a type of die principally used in fiber manufacture. It is usually a metal plate with many small holes through which a melt is pulled and/or forced. They enable extrusion of filaments of one denier or less. Conventional spinneret orifices are circular and produce a fiber that is round in cross-section. They can contain from about 50 to 110 very small holes. A special characteristic of their design is that the melt in a discharge section of a relatively small area is distributed to a large circle of spinnerets. Because of the smaller distance in the entry region of the distributor, dead spaces are avoided, and the greater distance between the exit orifices makes for easier threading.

NETTING AND SPECIAL FORMING DIE

The dies are designed to produce different melt flow patterns such as flat, tubular flat netting; corrugated flat tubing; and perforated tubing (Fig. 17.121).

For a circular output, a counterrotating mandrel and orifice can have semicircle-shaped slits through which the melt flow emerges. The slits can be of any shape. If one part of the die is held stationary, then a rhomboid or elongated pattern is formed. If both parts of the die rotate, then a true rhombic mesh is formed. During the time when the melt extrudes through the orifice and the slits overlap, a crossing point is formed, where the emerging threads appear to be welded but a uniform melt flows through the matching/aligned slits. For flat netting, the sliding action is in opposite direction.

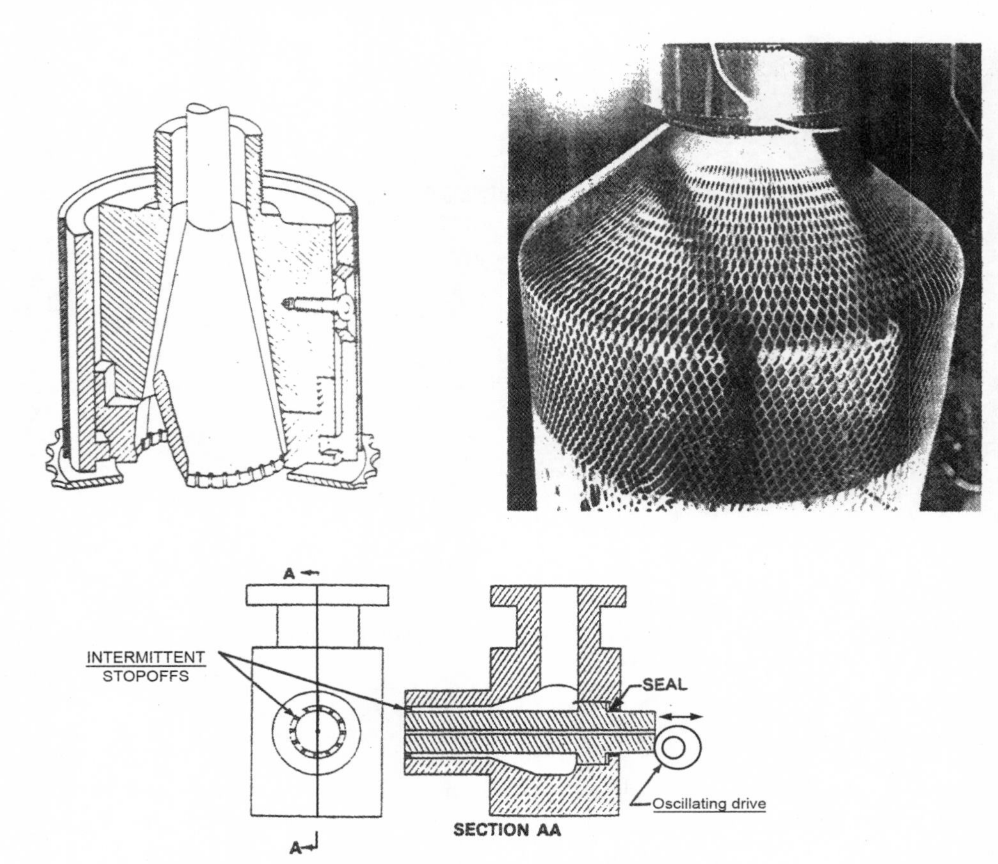

Different perforated tubing pattern; using oscillating mandrel die

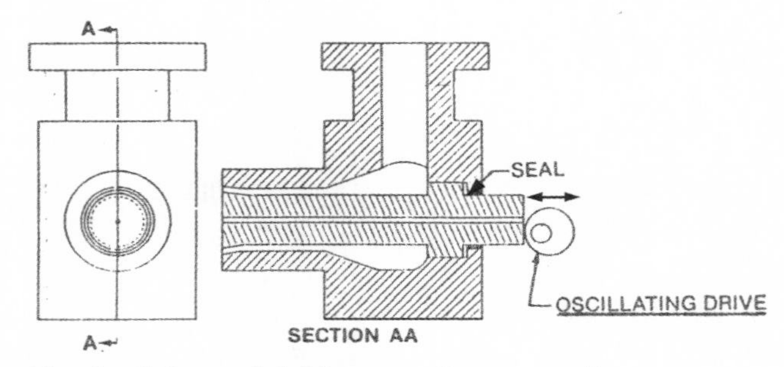

Varying tube wall thickness using an oscillating mandrel in cross head tube due

Figure 17.121 Examples of netting and other special forms.

Mechanical movement action in a die is used to extrude different profiles such as tubing or strapping with varying wall thicknesses or perforated wall. It is usually accomplished by converting a rotary motion to a linear motion that is used to move or oscillate the mandrel. For certain profiles, such as perforated tubing, the orifice exit would include a perforated section usually on the mandrel.

PELLETIZER DIE

With these die-faced pelletizers, the extrudate is cut on or near the die face by high-speed knives. Table 17.62 provide information on the shapes and processes used to fabricate pellets.

There are several different designs:

1. An extruder pump that melts through a straining head into the die. It passes through round holes in its die plate, where a wet atmosphere exists. Upon exiting the plate, a spinning knife blade cuts the extrudate into pellets. The pellet/water slurry is pumped into a dryer where the pellets separate from the water. Water is reclaimed for repeat use.
2. Very popular are the wet-cut underwater pelletizers (Fig. 17.122). The die face in this design is submerged in a water housing and the pellets are quenched with water

SHAPE WD = 1		D, L	D, L					
GEOMETRIC DESCRIPTION	SPHERES	CUBE	CYLINDER	CYLINDER WITH ROUNDED EDGES	CYLINDER BICONVEX	CYLINDER CONCAVE/ CONVEX	CYLINDER BICONCAVE	SHAPE D WITH FLANGE
PROCESSING INFLUENCES	SPECIAL SHAPE OF E —SURFACE STRESSES	NO PARTICULAR INFLUENCES	NO PARTICULAR INFLUENCES	VISCOUS FLOWING	VISCOUS FLOWING AND ELASTIC STRESSES THE COLDER THE ORIFICE THE MORE CONVEX THE SHAPE	DENSITY CHANGE AND CONTRACTIONS WITH COLD FACE	VISCOUS FLOWING AND ELASTIC STRESSES THROUGH HOT ORIFICE	SHAPE AS D. BUT WITH FLANGED EDGE CAUSED BY KNIFE STROKE
PELLETIZING SYSTEM	AIR/WATER PELLETIZER OR UNDERWATER	STRIP DICING	STRAND PELLETIZERS ON AIR PELLETIZER OR FORCED CONVEYANCE PELLETIZER	AIR PELLETIZER	AIR/WATER PELLETIZER OR UNDERWATER OR KNIFE ROTOR PELLETIZER	AIR/WATER PELLETIZER OR UNDERWATER OR KNIFE ROTOR PELLETIZER		AIR/WATER PELLETIZER OR UNDERWATER

Table 17.62 Guide to different pellets that are fabricated from different performing dies

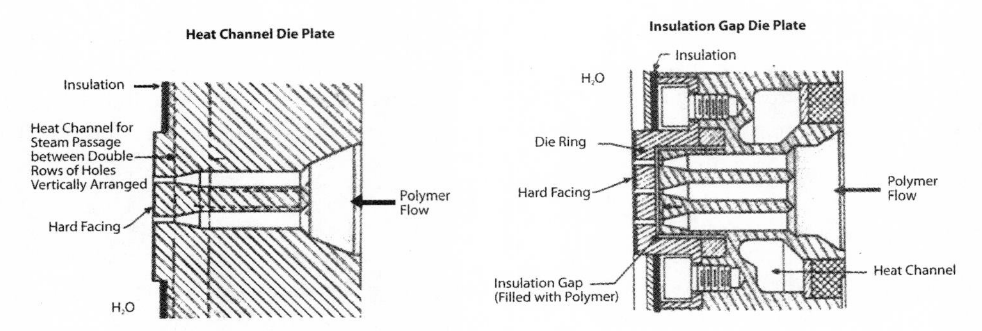

Figure 17.122 Examples of underwater pelletizer dies.

followed by a drying cycle. Throughput rates are at least up to 50,000 lb/h (22700 kg/h). Smaller units are economical and operate as low as 500 lb/h (227 kg/h).

3. The water-spray pelletizer, with a rotating knife, uses a water-jet-spray cooling action as pellets are thrown into a water slurry. Throughput is about 100 to 1300 lb/h (45 to 590 kg/h).
4. The hot-cut pelletizer has melt go through a multihole die plate. A multiblade cutter slices the plastic in a dry atmosphere and hurls the pellets away from the die at a high speed. Usually the cutter is mounted above the die so that each blade passes separately across the die face and only one blade at a time contacts the die. Pellets are then air and/or water quenched (and then dried if water is involved). Throughput is up to at least 15000 lb/h (6810 kg/h).
5. The water-ring unit has melt extruded through a die plate and cut into pellets by a concentric rotating knife assembly. Pellets are thrown into a rotating ring of water inside a large hood. After cooling in the water, they are spirally conveyed to a water-separated and then to a drying operation.
6. With the rotating-die unit, a rotating hollow die and stationary knife is used. The die, which looks like a hollow slice from a cylinder, has holes on its periphery; melt is fed into the die under minimal pressure and centrifugal force generated by the die rotation causes the melt to extrude through the holes. Pellets cut as each strand passes are flung through a cooling water spray into a drying receiver.

COEXTRUSION DIE

Coextrusion can be performed with flat, tubular, and other types of dies (Fig. 17.123). The simplest application is to nest mandrels and support them with spiders or supply the plastic through circular manifolds and/or multiple ports. Spiral mandrel blown-film dies with up to eight layers have

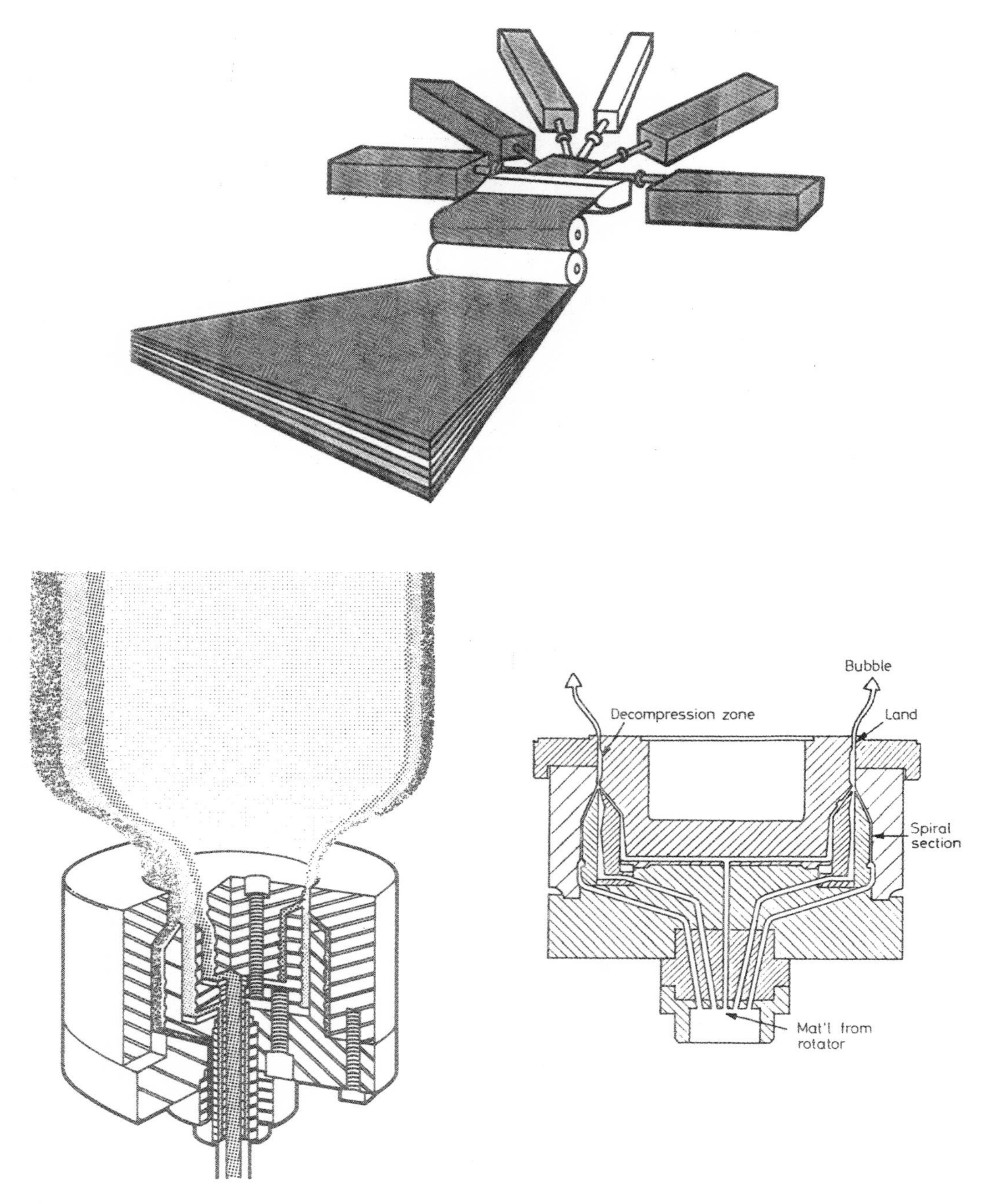

Figure 17.123 Examples of coextruded dies.

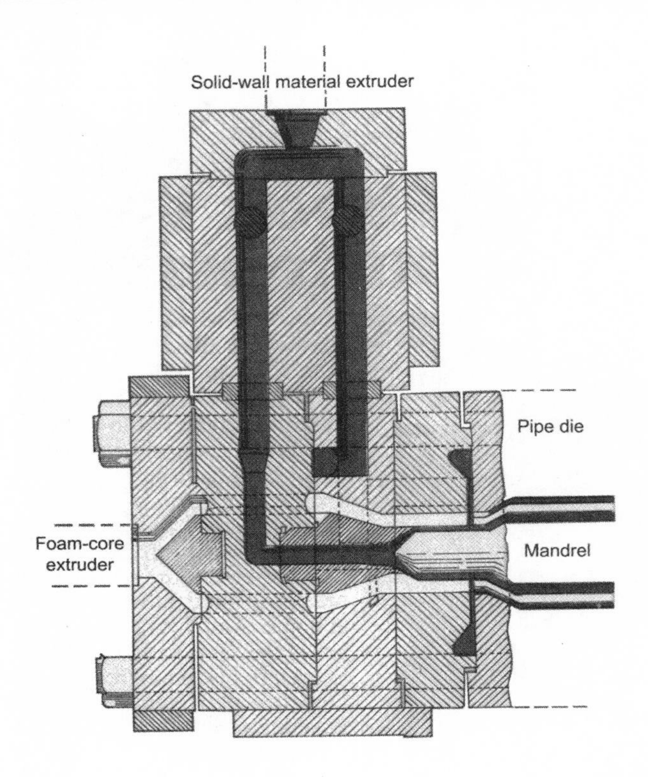

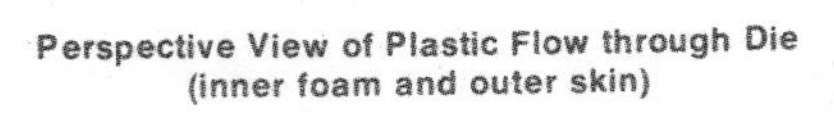

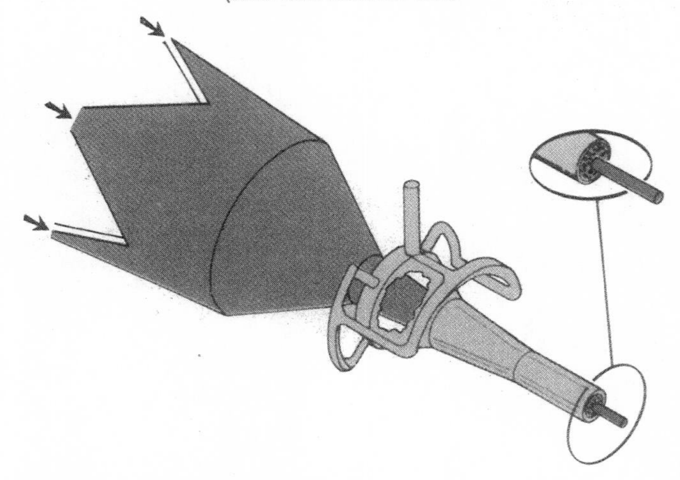

Figure 17.123 Examples of coextruded dies *(continued)*.

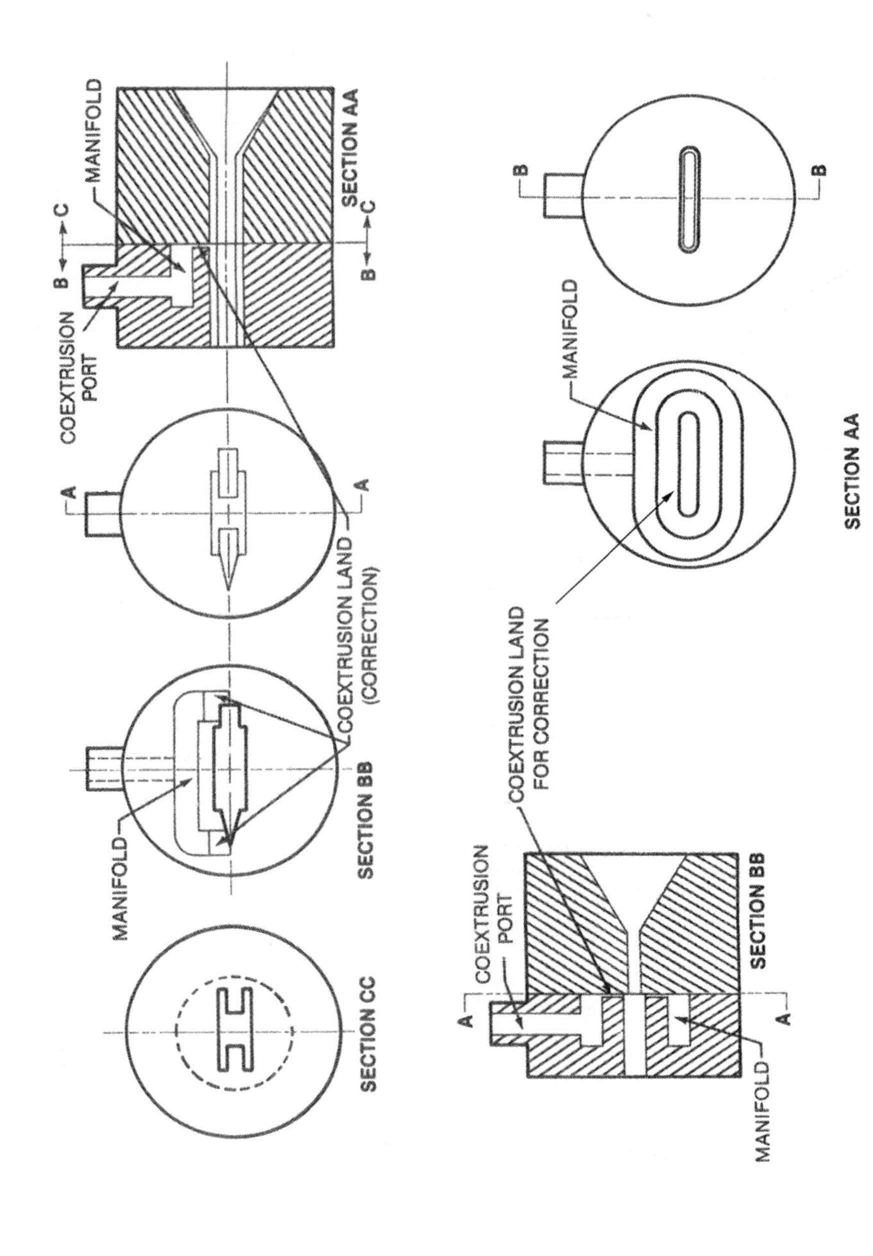

Figure 17.123 Examples of coextruded dies *(continued)*.

been built; they require eight separate spiral flow passages with the attendant problem of structural rigidity, interlayer temperature control, gauge control, and cleaning. Many techniques are available for coextrusion, some of them patented and available under license (Table 17.63).

For flat dies, there are basically the feedblock (single manifold) or the adapter (multimanifold) dies with a third system that combines the two basic systems. This third system provides processing alternatives as the complexities of coextrusion increases.

The feedblock method combines several monolayer manifolds in a common body creating a multimanifold feedblock die. Each manifold processes a distinct layer of product until the flows from all manifolds are merged into a single multilayer flow and extruded from a set of common lips (Figs. 17.124 and 17.125). With the single-manifold die, the plastics meet (combine surface to surface) and spread to a given web width.

Because individual plastics are processed in separate manifolds until just prior to exiting the die, large differences in viscosities, rates, and ratios can be tolerated. Using restrictor bars makes individual layer adjustments. Overall, adjusting the common lips controls the profile. Because of their complexity, size, and cost, however, the number of separate manifolds can be limited to about six layers.

In a multimanifold die the plastics are spread to a given web width and then meet slightly upstream from the die lip exit (Fig. 17.126). Because the plastic layers are kept separate until they are fairly close to the die exit, interface deformation and interfacial instability are generally not a problem. This due to the fact that there is a limited time the plastics are allowed to flow together so that any temperature differential has a short duration. With proper die designs, flow rate can be adjusted for each melt via the usual fixed or adjustable restrictor bars.

		Diameter Range		
Applications	*Layers*	*(m)*	*(mm)*	*Materials*
Form/fill/seal	3, 5–8	6–16	150–405	EVA, LD, LLD, LLD-M, PA, EVOH, ADH
Stretch	1, 3	22–40	560–1000	LLD w/PB
Lamination	1, 3, 5	12–28	305–710	EVA, LD, LLD, LLD-M, PA, EVOH, ADH
Construction and agriculture	1, 3	25–90	635–2300	LD, LLD
In-line bags	1, 3	6–30	150–750	LD, LLD, LLD-M, HD
High barrier	5–8	6–28	150–710	EVA, LD, LLD, LLD-M, PA, EVOH, ADH
Geomembrane	3	70–90	1780–2300	HD, VLD, LLD

Table 17.63 Examples of blown-film applications for coextrusion

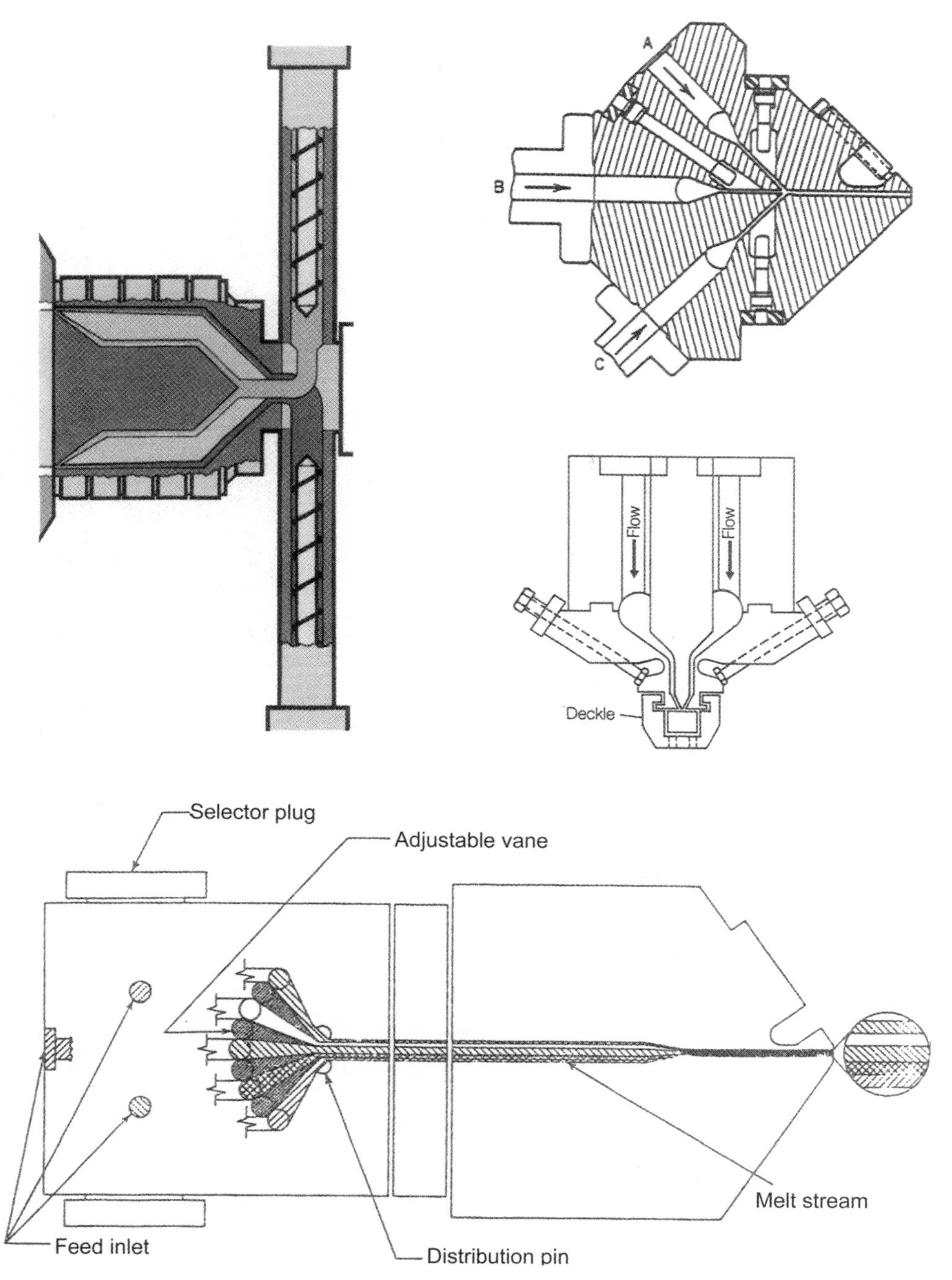

Figure 17.124 Examples of feedblock multimanifold coextrusion dies.

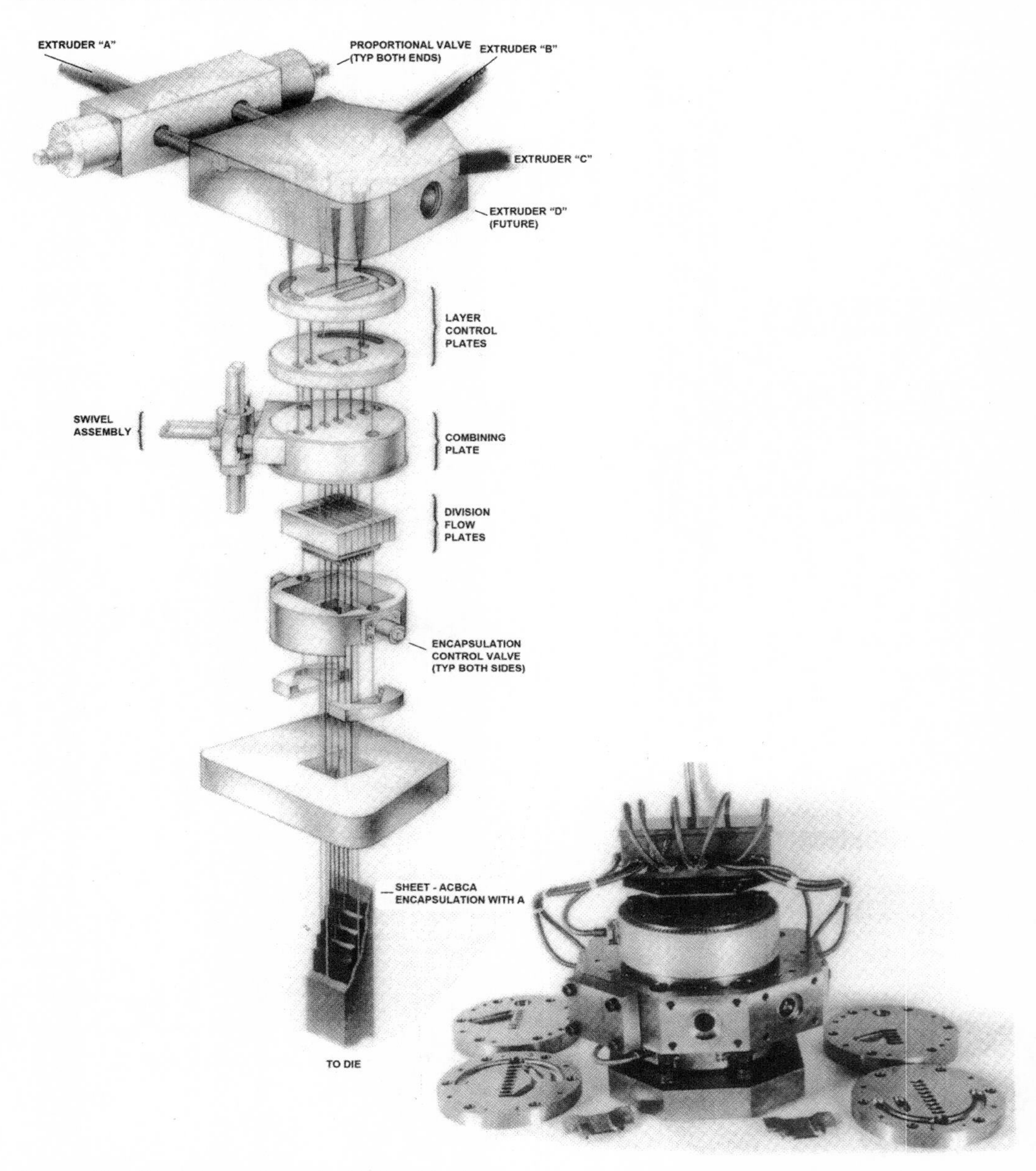

Figure 17.125 Schematic of the RV feedblock showing melt paths and assembled RV feedblock with layer control plates and skin flow inserts in the foreground (courtesy of Davis-Standard).

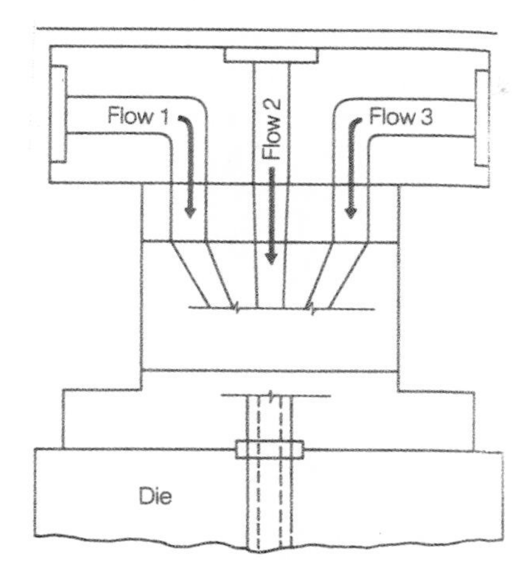

Figure 17.126 Example of a coextrusion combining adapter.

Once combined, the melt flows enter a single manifold die for forming into a flat web. Feed-block technology allows an almost limitless number of layers and combinations. Only a narrow range of viscosities is tolerated by this method, however. Certain aspects of the technology, both in the hardware and in the intellectual property related to this technology are patented.

Assuming both basic types have good manifold designs, the multimanifold can process a broader range of melt flow plastics. No matter which method is used, it is important to maintain the melt heat of each layer above the"freezing" temperature of all layers. If temperatures are not properly set, adhesion between the layers will be poor. The number of layers usually limits the multimanifold die. To date, this restriction is due to the required size of the physical equipment. To compensate for this restriction, one or more feedblocks are used, resulting in more layers of materials. Compatible plastics can flow through a single manifold, reducing any potential problem downstream in the multimanifold. Combinations can be made to provide different laminated designs. However, the final exiting layer thickness distributions can be affected by the amount of die body deflection if the die is not properly designed to take the required loads. Any deflection causes distortion that influences melt flow channels.

COEXTRUDED PLASTIC LAYERS

At least 115 layers of coextruded plastics have been produced. Mechanical movement action converting rotary motion to a linear motion is used to move or oscillate the mandrel in a die. The result is to extrude different profiles such as tubing or strapping with varying wall thicknesses or perforated walls. This Dow Chemical-patented process generates hundreds of layers, each one thinner than the wavelength of light (458). The tubing die generates a large number of layers by rotation of annular die boundaries.

It can be accomplished by novel coextruded blown-film (or flat-film) die. Product produces iridescent effects simply by taking advantage of some basic optical principles. At least 115 alternating layers of two plastics, such as PE and PP, produce an extruded film 0.5 mil (0.013 mm) thick. Individual plastic components are forced through a feedport system into a die in alternating layers extending radially across the annular gap (Figs. 17.127 to 17.129). Simultaneously, the rotation of the die's inner mandrel and outer ring deform the layers into long, thin spirals around the annulus.

The increased interfacial surface area related to rotational speed multiplies the number of layers. Overall the number of layers and layer thickness is determined by the dimensions of the

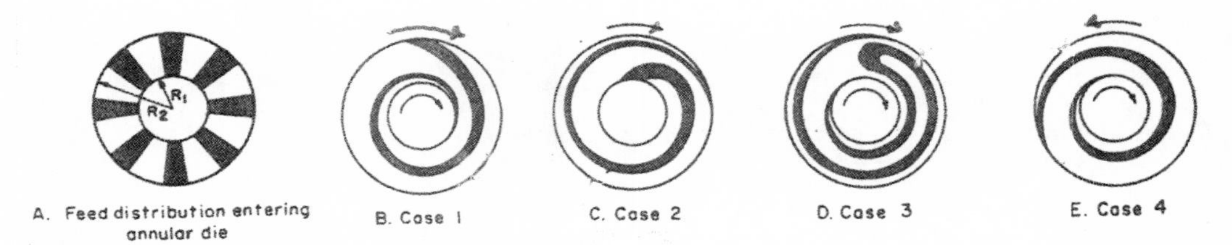

Figure 17.127 Examples of layered plastics based on four modes of die rotation.

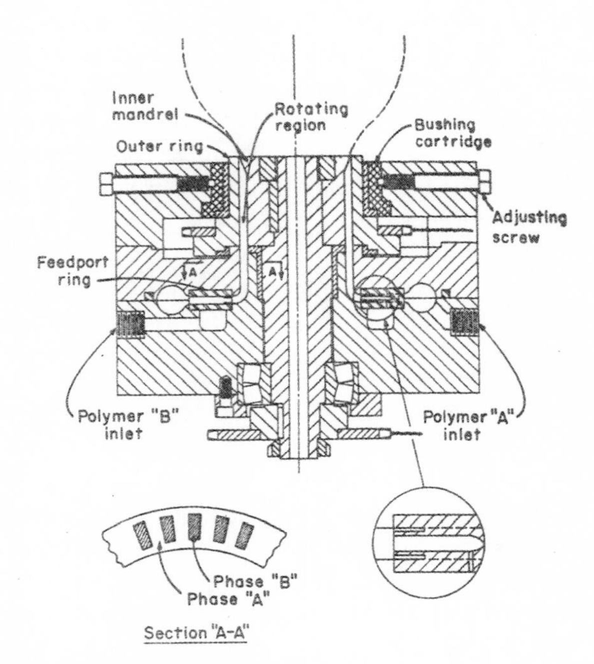

Figure 17.128 Example of the multilayer blown-film die.

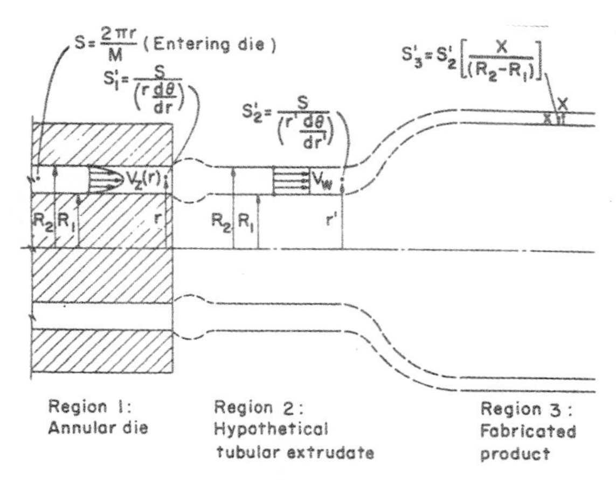

Figure 17.129 Displacement of layers leaving an extruder film die.

annulus, the number of feed ports for each phase, the extrusion rate, and the rotational speed of the die mandrel and ring relative to the feed ports. The resulting four basic layer patterns are generated by four modes of the die rotation (Fig. 17.127). Case 1 has the inner die mandrel rotating while the outer ring is stationary, where layers are thicker near the outer ring. Case 2 has the inner die mandrel stationary while the outer ring rotates with layers thinner near the outer ring. Case 3 has both inner and outer die members rotate at the same speed and direction; the result is that layers of curved, open-ended loops and thicker layers are in the center. Case 4 has inner and outer die members counterrotating at equal speed, generating the maximum number of symmetrical layers with the thickest in the center. All these examples have layers that are concentric. The deformation is usually so large that the spiral characteristic is indistinguishable when examining the extrudate in the cross-section.

NEW DIE DESIGNS

As reviewed in other chapters, there is an endless abundance of new developments in plastic and equipment. Dies are no exception. One of the latest new concepts is from Dual Spiral Systems Inc., Hamilton, Ontario, Canada (http://www.dualspiralsystems.com).

This new film-die design improves on the classic stackable die (Fig. 17.130). Using an innovative dual-spiral design to split single layers into multiple layers, it creates structural improvements in the final product, improves barrier properties, and saves money. Improved streamlining and the lack of sharp bends eliminates dead spots. The die gives the film producer the flexibility to meet future needs without buying another die (456).

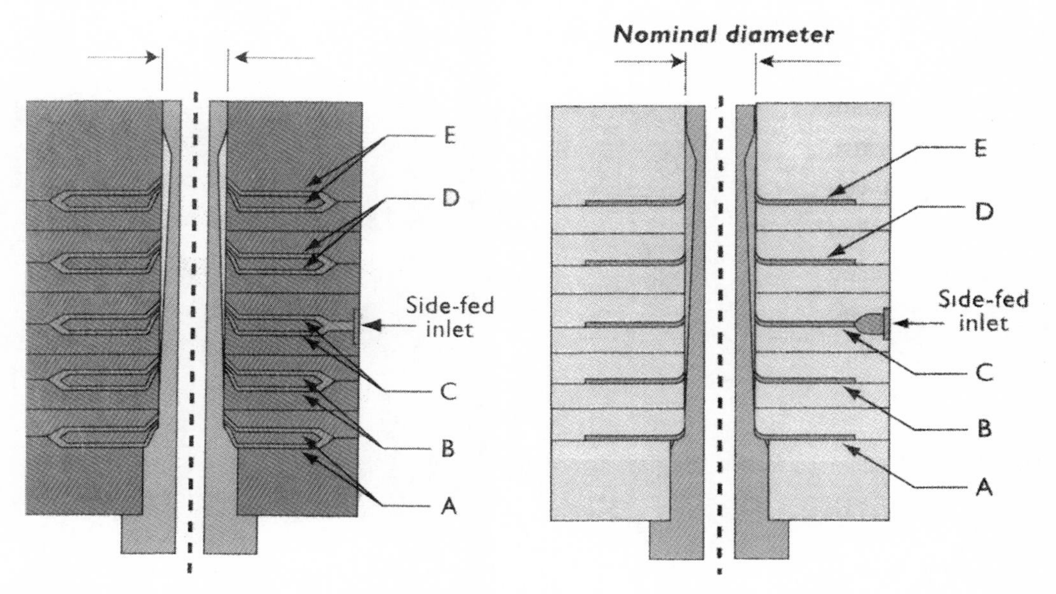

Figure 17.130 New coextrusion die design (left) is compared to the traditional flat-plate die.

COMPUTERS

The use of computers has become part of the lifeline in producing tools and products via its displays and/or developing physical prototypes. Creating physical models can be time consuming and provide limited evaluation; however, they can be less expensive. By employing kinematic (which describes a branch of dynamics that deals with aspects of motion apart from considerations of mass and force) and dynamic analyses on a design within the computer, time is saved and often the result of the analysis is more useful than experimental results from physical prototypes. Physical prototyping often requires a great deal of manual work, not only to create the parts of the model, but also to assemble them and apply the instrumentation needed as well.

CAD prototyping uses kinematic and dynamic analytical methods to perform many of the same tests on a model. The inherent advantage of CAD prototyping is that it allows the engineer to fine-tune the design before a physical prototype is created. When the prototype is eventually fabricated, the designer is likely to have better information with which to actually create and test the prototype model.

Engineers perform kinematic and dynamic analyses on a CAD prototype because a well-designed simulation leads to information that can be used to modify design parameters and characteristics that might not have otherwise been considered (chapter 19). Kinematic and dynamic analysis methods apply the laws of physics to a computerized model in order to analyze the motions within the system and evaluate the overall interaction and performance of the system as a whole. It allows the engineer to overload forces on the model as well as change location of the forces. Because the model can be reconstructed in an instant, the engineer can take advantage of the destructive testing data. Physical prototypes would have to be fabricated and reconstructed every time the test was repeated. There are situations in which physical prototypes must be constructed, but those situations can often be made more efficient and informative by the application of CAD prototyping analyses.

CAD prototyping employs computer-aided testing (CAT) so that progressive design changes can be incorporated quickly and efficiently into the prototype model. Tests can be performed on the system or its parts in a way that might not be possible in a laboratory setting. It can also apply forces to the design that would be impossible to apply in the laboratory.

TOOL ANALYSIS

Computer tool analysis programs allow simulating tool performance prior to cutting steel and aluminum. The programs allow optimizing the tool design without the traditional prototyping trial-and-error methods. Graphics perform analysis in much less time than it would were it to be performed manually. The techniques include plastic-flow analysis from the sprue into the cavity, determining venting locations, determining pressure-load capability on the tool, and tool cooling and/or heating analysis. Graphs are automatically generated to illustrate the effects of varying the operating analyses in a tool.

A finalized tool design generally consists of the computer summation resulting from the preliminary tool design and analysis. It includes an interface the tool with the molding machinery. Dimensions are finalized, section views are produced, and the necessary drawings are created so that enough information is obtained to proceed to the manufacture of the tool.

MODEL CONSTRUCTION

With the tool model constructed, the system will contain all the geometric (mathematical) information required for producing NC input. CAD/CAM/CAE/CIM systems can furnish the NC data to postprocessors and generate the NC tapes or electronic systems to drive individual machine tools. CAD/CAM/CAE/CIM, with the ability to dynamically simulate tool cutting action, results in a significant reduction in NC programming errors. Additional benefits resulting from NC capabilities include reduced lead time, standardization of machining practices, and reduced costs.

Their ability to detect potential problems and correct them at the design stage reduces the usual heavy reliance on costly, time-consuming physical testing. A CAE package integrates information that passed from stage to stage through a computer database, eliminating redundant effort and misinterpretations.

The most cost-effective applications for this technology are those with complex product and tool designs, stringent end-user requirements, close tolerances, extremely short mold cycle times or fast extrusion rates, and other applications with demanding constraints. For example, the process of developing injection molded plastic tools and products or extruded die and product has been streamlined by using isolated computer techniques, notably CAD/CAM automated drafting systems for producing engineering drawings faster. However, CAE software provides a single tool that ties together all the steps in the development process.

SOFTWARE

Many software programs are available to serve the different needs of manufacturing functions. The products that have evolved over this period are tools that serve to replace the hit-and-miss approach of the past with analytical analyses based on sound theoretical principles. This results in cost-effective tools that operate efficiently. These tools are most effectively applied prior to construction of the tool, but they can be applied after the fact to solve process-related problems.

CAE analysis tools will at least provide melt flow analyses, cooling analyses, and economic and plant-operating analyses. These analyses supplement the knowledge of a trained individual, making one more productive and more accurate in predictions. The approach behind the CAE technique is that a design or process is initially proposed. The engineer then constructs a model, or representation, of the specific design. The computer rapidly evaluates the results of both the input conditions and model that the engineer has described. The computer lists the output conditions and the engineer evaluates the consistency of results-based experience. If required the design is modified to meet the best product performancewise and costwise. The skill and experience of the designer are still reflected in the final results.

Software provides advantages such as simplifying the design approach. It solves process problems such as meeting tolerances, warping, dimensional inconsistency, and reduction in output. Manufacturing time and costs—start-up costs, fabricating costs, material costs, rework costs, scrap and regrind costs—are all reduced.

MATERIAL SELECTION SOFTWARE

Different software programs are available from different sources that are usually the metal and plastic material producers. For example, the PLA-Ace software package from Daido Steel Co., Tokyo, Japan, provides the basic information that encompasses selections that include a mold base, cavity, and core pins. It also provides different inputs, including types of plastic being processed, plastic content (glass fiber, flame-retardant, etc.), cavity texture or finish, comparative wear and hardness capabilities, minimum mold thickness, corner radius, existence of ribs, product category (electronic, auto, etc.), product type (panel, TV chassis, etc.), and causes of mold defects due to improper material selection both for its tool steels and those of other suppliers. The software provides exceptional help to processors and tool shops with limited experience in selecting the optimum material

TOOLING AND PROTOTYPING

Rapid tooling (RT; mold and die) and product rapid prototyping provides reducing development cycles. RT and rapid prototyping are characterized by any method or technology that enables one to produce a tool or product quickly. The term rapid tooling refers to RT-driven tooling. A prototype is a 3-D model suitable for use in the preliminary testing and evaluation of a mold, die or product. It provides a means to evaluate the tool's or product's processing performances before going into production. The ideal situation is for the prototype to be the actual tool made in production. However, techniques such as machining stock material to using RT or rapid prototyping methods can make prototypes for preliminary or final evaluation prior to manufacturing the tool or product (Table 17.64).

Conventional machining operations can be used preferably from the same material to be used in the tool (392, 459–461). Other methods, such as different casting techniques that provide low cost even though they are usually labor intensive, can be used. The casting of unfilled or filled/ RP used include TP or TS polyurethane, epoxy, structural foam, room temperature vulcanization (RTV) silicone, ceramics, and metals that include die-cast metals.

Routinely produced, accurate, and durable rapid prototyping parts are used to produce finished, manufactured parts in quantities of hundreds and thousands. This permits manufacturers to deliver the products while waiting for tooling; in some cases, it also eliminates the molds altogether. Companies use rapid prototyping to model early and often when design changes are required; this early use of rapid prototyping results in inexpensive operations.

Wohlers Associates publishes an annual progress report on the rapid prototyping and tooling industry. The 2001 year report indicates worldwide sales of 1320 rapid prototyping machines in 2000, up from 1178 in 1999 and only 34 in 1988. The installed systems produced more than 3 million models and prototypes during the year 2001. It is reported that users have documented many cases where rapid prototyping has saved significant amounts of time and money. Rapid prototyping has also helped prevent critical errors in the tooling phase that would have derailed time-sensitive projects. Others claimed that through the increase in the number of prototypes offered by rapid prototyping, the designs result in higher-quality products. The Wohlers report estimates that at the end of 2000, more than 45% of all rapid prototyping machines installed worldwide were in North America. Some 28.6% were in operation in the Asia-Pacific region, and 24.6% in Europe.

The technology of RT and rapid prototyping provides a quick way between design and creativity ideas and the fabricated product. More precision tooling and prototype materials continue to become available while system speeds keep increasing. Plastics and other industries are actively using these rapid systems. For example, international space agencies are experimenting with rapid prototyping to quickly replace parts in space vehicles (293).

Manufacturer	Process name	Material & structure generation
3D Systems Inc., Valencia, CA, U.S.A.	Stereolitho-graphy Appar-atus (SLA)	Photopolymer system; point-by-point irradiation with a HeCd resp. an argon ion laser
CMET, Japan	Solid Object UV Plotter (SOUP)	Photopolymer system; point-by-point irradiation with an argon ion laser
SONY-Japan Synth-etic Rubber, Tokyo, Japan	Solid Creater	Photopolymer system; point-by-point irradiation with an argon ion laser
SPARX, Molndal, Sweden	Hot Plot	Self-adhesive film; cutting of the films layer by layer with a thermal electrode
Stratasys Inc., Minneapolis, WI, U.S.A.	Fused Deposit ion Modelling (FDM)	Thermoplastic filaments (PA, etc.)as well as wax; melting the plastic in a mini extruder
Light Sculpting Inc., Milwaukee, WI, U.S.A.	LSI	Photopolymer system; irradiation of the entire surface with a UV lamp
Mitsui Engr' & Shipbuilding Ltd., Tokyo. Japan	COLAMM	Photopolymer system; point-by-point irradiation with a HeCd laser

Table 17.64 Rapid prototyping processes

RAPID SYSTEM

As a result of ever-increasing advances in product design and the shortening of this process due to competitive market pressure worldwide, prototyping houses, mold and diemakers, and tool rooms have experienced a continuing mounting urgency to shorten lead times. Various rapid program methods have been successful in offering fast tooling or product prototypes. A major advantage of RT is in the ability of the design to be seen and felt by the designer and less technically adept personnel, especially when esthetic considerations must come into play.

Cubital Ltd., Herzlia, Israel	Solider 5600	Photopolymer system; irradiation of the entire surface with a UV lamp
DTM Corp., Austin, TX, U.S.A.	Sinterstation 2000	Powderized thermoplastics (PA, PC), wax; local melting of the powderized plastic by laser energy
DuPont license to Teijin Seiki, Tokyo, Japan	SOMOS	Photopolymer system; point-by-point irradiation with an argon ion laser
EOS GmbH, Plaegg/ Munich, Germany	STEREOS	Photopolymer system; point-by-point irradiation with an argon ion laser
Helisys Inc., Torrence, CA, U.S.A.	Laminated Object Manufacturing (LOM)	Self-adhesive paper and plastic films; cutting of the films layer by layer with a CO_2 laser beam

Table 17.64 Rapid prototyping processes *(continued)*

The rapid technology used for building physical models and prototypes from 3-D CAD data is popular. Even though these systems are usually more expensive than the past methods, they provide the desirable end result to the industry that is a much quicker way to obtain tooling or prototypes (hours instead of days/weeks). They are used in the preliminary evaluation of form, design, performance, and material processing. It is the forming of 3-D tools and products (to date principally for IM, where the larger market exists) from the design concept to production using computer-controlled laser beams producing the final prototypes of simple or complex shapes.

Models can be made of cast or RPs, steel (including sprayed steel), copper-based alloys, powdered metals, or other materials such as MIT's starch and sugar material. With powder metal molds, they can be used as inserts in a mold ready to produce prototype products. RT processes can quickly generate tooling suitable for production of up to millions of products.

Methods to produce the more costly RT and rapid prototyping machines include those that produce models within a few hours. They include photopolymerization, laser tooling, and their modifications. The laser-sintering process uses powdered TPs rather than chemically reactive liquid photopolymer used in stereolithography. Metals (steel, hard alloys, copper-based alloys, and powdered metals) are also used in these different processes. These systems enable a user to have precise control over the process and allows the construction of products with complex geometries.

With many new RT processes, identifying the ideal application can be relatively simple or incredibly confusing, so it is important to understand what is available and in particular the different capabilities of the RT techniques. There are many factors to consider when determining the appropriate RT process. The marketing information promising cost savings, reduced lead time, and quantity of products often leaves out important characteristics of the processes.

Make sure that you set up all your requirements, include those required by the RT process. Application opportunities are broad and vary between industries, molding processes, manufacturing processes, product geometry, product size, manufacturing materials, and time constraints. All these factors play a part in shaping one's decision-making process. The most fundamental of decisions are cost, delivery time, and quality.

Rapid Tooling

Various RT methods are available. Two prime groups exist that are identified as indirect (or transfer) and direct. The indirect methods involve the use of a master pattern from which the tool is produced. Reduction in time to produce tools and repeatability and meet tight dimensions influence the use of direct methods. Ultimately, companies want to produce the molds directly, although most of the direct tooling methods are not without limitations. Many different companies worldwide are actively pursuing RT approaches and eliminating or decreasing limitations.

Indirect tooling methods are many. Examples include cast aluminum, investment metal cast, cast plastics, sprayed steel, spin-castings, plaster casting, electroforming, and others that are explained in the next sections.

Silicone rubber

One of the popular tooling applications is the production of RTV silicone rubber molds, which cure at room temperature. Silicone is a versatile material that can be molded around a master pattern to produce a cavity. They provide fast, relatively inexpensive molds, excellent cosmetics, and the option of using multiple materials. They are also suitable for small and large parts.

Rubber

The ModelMaker 11 and PatternMaster equipment from Solidscape Inc., Merrimack, New Hampshire, concentrates on producing tooling-grade master patterns for rubber toolmaking and direct investment casting. The delivery system consists of two piezoelectric jets with a capacity of 12000 droplets per second. One jet delivers a styrene-based build material, while the other delivers a sacrificial wax for support of undercuts and overhanging features. Selection of configuration software provides a selectable layer thickness from 0.0005 to 0.0030 in for the patterns generated in a 12 × 6 × 8.5 in building area with an accuracy of ±0.0010 in/in across the *x*, *y*, and *z* dimensions. Pattern features can be as small as 0.0010 in.

VACUUM CASTING

Products can be vacuum cast by placing a silicone tool in a vacuum chamber with a plastic such as TS polyurethane. The two-part plastic is mixed and degassed before being poured into the silicone cavity. After pouring, the vacuum is released and the tool is removed to a postcuring oven (chapter 11).

REACTION INJECTION

Unlike vacuum casting, the RIM process does not rely on expensive vacuum chambers and mixing units (chapter 12).

CAST KIRKSITE

Cast Kirksite cavities from models provide excellent simple to complex tools for prototype and bridge to production. One of the advantages of Kirksite (aluminum/zinc alloy) is the ease of making geometry modifications by either welding or adding inserts. Their cavities are normally standard, preengineered mold bases with standard injector and runner systems, or they can be used on a stand-alone basis. Heating or cooling systems can be either cast in place or added later by drilling.

SPRAY METAL

Spraying metal is used for the production of soft tooling. It involves spraying a thin shell of about 2 mm (0.080 in) in thickness over a pattern and backing this with epoxy, RPs, and other reinforcements to give it rigidity. A major advantage is that very large molds can be produced. Spray metal tools can produce more than 1000 products depending on the process and material being used.

RAPID SOLIDIFICATION

The development of a steel spray process at Idaho National Environmental and Engineering Lab (INEEL) is called the rapid solidification process (RSP). It differs from other sprayed metal processes because it can deposit hundreds of pounds of material per hour, while the conventional wire-feed systems deposit approximately 7 kg (15½ lb) per hour. This means that, potentially, the RSP process could be used to build the entire tool as opposed to a thin shell that requires back filling, thus providing advantages such as lower mold costs.

There are also many new direct methods. The next sections explain a few that are already in use.

STEREOLITHOGRAPHY

This process is the most widely used technology for rapid prototyping. It builds plastic parts or objects one layer at a time by tracing a laser beam on the surface of a vat of liquid photopolymer. A moveable table (Z adjustable elevator platform) is initially placed at a position just below the surface of a vat filled with liquid photo polymer. When light of the correct wavelength strikes the surface of the photopolymer, localized polymerization is initiated. The most common photopolymer materials used require an ultraviolet light, but resins that work with visible light are also utilized. A laser beam is guided in the X-Y direction over the surface of the liquid photo polymer by a scanner system, which traces the geometry of the cross-section of the object, causing the liquid to polymerize in areas where the laser strikes. The exact pattern that the laser traces is a combination of the information contained in the CAD system that describes the geometry of the object and information from the rapid prototyping application software that optimizes the faithfulness of the fabricated object. As each layer is completely traced and polymerized by the laser beam, the table is lowered into the vat a distance equal to the thickness of a layer. This process is continued until the object is completed. It is then removed from the vat for postcuring and surface finishing as necessary.

RapidSteel

In the same way that a cavity can be generated directly by stereolithography, it also is possible to build cavities directly using the laser-sintering process. With DTM's (Austin, Texas) RapidSteel (also referred to as RapidTool), digital models of the core and cavity geometries are created and sent to a sinter-station machine for fabrication in RapidSteel powder. This material consists of particles of mild stainless steel that are coated with a thin layer of a plastic binder material. The sinter station produces green parts that are then fired in a furnace. The furnace removes the plastic binder and infiltrates bronze into the mold inserts through capillary action.

Direct Metal Laser Sintering (DMLS) involves the processing of metal powders in a laser-sintering machine. Typically, the machine is used for the production of tool inserts, but it also produces metal components. Two materials are available for the process: bronze-based materials that are used for IM of up to 1000 products in a variety of plastics, and steel-based material that is useful for up to 100,000 plastic injection molded products.

POM Group Inc., Auburn Hills, Michigan, has focused on providing RT services and equipment. Its next-generation machine, the DMD 5000, incorporates a 5000-watt CO_2 laser for higher deposition rates and the next-generation closed-loop feedback control system. This laser-based DMD technology for metal synthesis and metal component fabrication currently is utilized to build high-precision, advanced, thermal management tooling systems for IM, die casting, hydroforming, forging, and stamping processes with capabilities for tooling reconstruction and reconfiguration.

Laminate

Laminated tooling is an alternative to building cavities directly on an RT machine. Using the similar principles to the laminating, layers of sheet metal are cut to replicate slices through a CAD model. Laser cutting or water-jet technologies generally produce the profiles. Laminated tools have been used successfully for a variety of techniques for over a half century that included the laminates in aircraft primary structures. Now they include tools. Tool life is a function of the initial type of sheet material, which can be hardened after cutting and lamination. However, part complexity is bound by layer thickness.

One significant advantage of laminated tooling is the ability to change the design of products quickly by the replacement of laminates (if unbonded). Cooling channels are easily incorporated within the tool design and are good for large tools. The need for machine finishing to remove the stair steps is its main disadvantage.

Cubic Technologies Inc., Carson, California, has a laminated-paper system called laminated object manufacturing (LOM) for building parts with thin adhesive-coated paper. Complex 3-D parts are made by sequential lamination of thin layers of adhesive-coated sheet material and cutting them with a focused laser beam. The resulting 3-D parts have the appearance and properties of wood. High-quality paper (thicker tan 0.006 in) can provide stronger, more dimensionally stable parts and speed up the cycle. The process is especially beneficial for making large parts that are generally used as patterns or molds for vacuum or thermoforming, and blow or rotational molding.

LASER ENGINEERED NET SHAPING (LENS)

The LENS system from Optomec (Albuquerque, New Mexico) originally developed at Sandia National Laboratories. It is a metal-powder-feed laser system. It injects metal powder into a pool of molten metal created by a focused laser beam. The fabrication process occurs in a low-pressure argon chamber for oxygen-free operation. A motion system moves a platform horizontally and laterally as the laser beam traces the cross-section of the part being produced. After forming a layer of the part, the machine's powder-delivery nozzle moves upward prior to building the next layer. Similar to other rapid prototyping techniques, LENS is an additive fabrication method, although it produces fully dense metal parts.

Mold Insert D-M-E Co., Madison Heights, Michigan (248-398-6000), and the ProMetat Division of Extrude Hone Corp., Irwin, Pennsylvania (724-863-5900), have the MoldFusion 3-D metal printing that builds near-net shape mold inserts layer on layer system enables difficult-to-achieve configurations. The type of inserts include internal cooling channels that can yield cycle time reductions of 20% to 35% and improve part quality.

MoldFusion utilizes layer-by-layer build techniques to print a polymer binder onto powder tool steel. Invented at MIT, the process creates near-net-shape mold inserts from 3-D CAD models. After printing, the "green" part is loaded into a sintering furnace that fuses a tool steel skeleton of 60% density and burns off the binder. In a second furnace cycle, the part is filled with molten bronze via capillary action to full density. The near-net-shape part is then ready for final machining by the customer (443). The process can create mold inserts up to 12 in in length and width and 10 in high. Truss or honeycomb structures can be built into mold inserts to improve thermal isolation and reduce tool weight.

SELECTING RAPID TOOLING

As with the variety of processes, there also are a variety of ways to take advantage of being involved in RT. For a hands-off approach, there are a number of talented consultants, rapid service bureaus, and injection molders that can take the project from start to finish. These are businesses that have already invested the resources to go through the learning curves of certain processes and can make decisions based on experience instead of what should be possible.

The next level would basically be to act as a general contractor and go to outside sources for the services that are not available in-house. This could mean having a master built at an RT company, sending it to another company to be cast in a hard or soft material, and putting all the pieces together in-house. This requires a very strong understanding of the individual process capabilities that develops usually through trial and error.

The last option is to actually buy into the available processes and provide a complete package. This approach can vary greatly in cost since it depends on the process to be used. It involves a minimum investment in castable materials to hundreds of thousands of dollars in rapid prototyping equipment and licensing rights.

RT options can fall into one of the following categories: (1) soft tooling, (2) bridge tooling such as substrate inserts with plated or sprayed metal coatings, and (3) production 3-D tooling. In selecting an RT, different considerations have to be considered. For example, the required plastics production material is an important consideration because it usually can immediately eliminate some of the tooling options. Most of the soft tools cannot withstand being filled with a high heat or corrosive plastic.

The issue of the tool life expectancy goes along with the production material, since they both define fabricating the required number of products. Cavity and core wear and damage are influenced by the behavior of the plastic being used and how many times it is recycled. When possible and practical to help this situation, the aim is to use multiple cavities in a mold.

As with conventional tooling, certain features such as product geometry tend to be difficult to handle for certain tools, including areas that require frail insert structures, nondraw features that require slides, minimum draft areas, critical dimension tolerances, and so on. There is usually a process to overcome these distractions with RT. A popular technique is the use of handpicked inserts and hybrid tools that use a combination of insert materials that are suitable for the individual applications.

RAPID PROTOTYPING

The technology of rapid prototyping provides a quick way timewise between design creativity ideas and the fabricated product. More precision prototype materials continue to become available with system speeds that keep increasing.

A novel system of fused deposition modeling is from Stratasys Inc., Eden Prairie, Minnesota (952-937-3000). FDM Maxum rapid prototyping system has a TP material fed through an extrusion plasticator where it is extruded to form a model layer by layer (chapter 5). Each layer rapidly solidifies before the next one is extruded. This approach significantly improves build-up speed, surface finish resolution, and material characteristics. The surface finish resolution has improved by more than 300% since the early systems of the 1990s. The FDM used with ABS plastic provides an approximately 24 in^3 building volume and a build-up rate 50% faster than its predecessor. The system allows a resolution of 0.005 in.

FDM materials now available for building functional prototypes have broadened from the wax-based models (notably for investment casting applications) to include ABS, polycarbonate, elastomers, and durable polyphenylsulfone. The PPSF is sterilizable for medical applications.

There is also the Stratasys's WaterWorks process that creates model supports from a water-soluble material that eliminates the need to manually remove the support material, thus facilitating the production of parts with fine-featured detail without having to clean crevices and risk part damage.

There is the 3D Systems Corp., Valencia, California, stereolithographic solid imaging product line the Viper si2 system that offers standard and high-resolution part-building modes in one package. Capable of building parts from miniature-sized jewelry and electronic parts up to prototypes of 10 in^3 and patterns for IM and investment casting, the system runs on a Windows NT-based operating system, with the company's Buildstation control and 3D Lightyear part-preparation software. Its 100 milliwatt solid-state laser beam allows selection between a standard part resolution mode, for the best balance of build speed and part quality, and a high-resolution mode for ultradetailed small parts and features. The laser beam's width dictates the system's level of part quality; the laser can be focused within a tolerance range of 0.008 cm to 0.0025 cm (0.003 to 0.010 in). The wider beam width is typically used for faster production of larger parts and the narrower width for higher resolution of smaller parts.

SOFTWARE TREND

An important trend in software for design and manufacturing has been a significant reduction in price even as functionality has increased considerably. CAD/CAM software, such as that offered by VX Corp. Palm Bay, Florida (321-676-3222), sells for about one-tenth the price of traditional high-end software for design through manufacturing. The software includes all the necessary tools for designers and engineers to construct a sophisticated computer model of a product and then use the data to prototype it or takes it directly to machining of the tooling for mold-making for just $4000 to $10000.

Until recently, such capabilities required investments ranging from $50000 to $100,000. In fact, the entire CAD/CAM industry has witnessed significant growth in design and virtual prototyping software selling for less than $10000. Software suppliers at the high end are focused on total-enterprise data management that ties together design, manufacturing, supply-chain management, accounting, and other business data (293).

PROTOTYPES VIA SOFTWARE ONLY

Researchers at Germany's Fraunhofer Institute developed a 3-D software system that allows making product projection and foregoes complex, time-consuming model-making for Industrial Engineering IAO, Stuttgart. It makes applications simpler, more efficient, and less expensive. This software system, called Lightning, generates virtual-reality models (457).

Lightning is a freely configurable system that can work with various 3-D input/output systems. They can be 3-D stereo projection screens, configurable application view storage systems (CAVES), tracking systems, or data gloves. The advantage that Lightning has is that it helps with the generation of all kinds of interactive 3-D designs, which can be easily incorporated into a user environment. It is easy to develop an application that users with little experience can produce.

It is reported that CENIT Systemhaus in Stuttgart has been using Lightning software as the basis for its Virtual Reality Business Unit, which develops applications for companies including Daimler-Chrysler, BMW, Audi, and Porsche. BMW carried out a comparison of variants used to examine and approve tooling. BMW concluded that applications based on Lightning were capable of increasing productivity at its operations tenfold. Their engineers appreciated the system because of its ease of use.

REPAIR VERSUS BUYING

There are pros and cons when this subject is discussed. The outcome depends on various factors such as extent of damage, professional feasibility to rebuild, time to be back in production, and availability of money. Even though the initial capital expenditure is much lower than for a new tool or machine, the long-term economic value can be questionable. For example, retrofits can be tailored to meet the customer's performance requirements at 40% to 70% of a new tool. In order to provide a good basis for a decision, a technical evaluation matrix system using weighted criteria and a time-related method for judging the economic value of an investment are required.

After a certain period of service, most tools become scratched, carburized, and/or discolored due to the plastic temperature and pressure melting action. Tools are difficult to clean and tend to lose their original operating characteristics. If they have been plated (usually chrome), the chrome may be gone in some places or peeling in others. It is best to refurbish in this condition by stripping the old chrome, polishing, buffing, plating, and buffing again. The part will look much better and will also perform better for little cost and a short delivery time. (See Mold repair checklist in the Glossary.)

WELDING

Unfortunately there are times when tools require welding. This subject tends to be complicated if one is not really familiar with welding. A book about how to weld tools was written by Steve Thompson (355). It is not a metallurgist's definitive tool steel chemical composition handbook or a welding engineer's definitive guide. It is a book designed for the shop floor, written from the shop floor, using the authors' 25 years of experience to convey the most practical ways that have been found for tackling the repair of tools.

Mold, tool, and die repair welding is a very complicated process, shrouded in mysteries, myths and closely guarded secrets. This is because two very different trades have overlapped: welding and toolmaking. For many years, toolmakers and tool users have had to rely on the small number of specialist welders who do understand exactly what repair welding involves and who have the hand skills to do it, especially when it comes to welding tools.

Understanding the technical side of tool steels is a big problem for welders and understanding the practical side of welding is a big problem for machinists. The book has been written this book so that either will understand it and learn something from it.

Tool steels should be preheated or even annealed before welding and then cooled carefully or even postheat treated. One has never been allowed to anneal a tool before welding because the customer would never accept the increased downtime and cost. The book, therefore, tells the repair welder how to compromise between what the customer wants and how best to provide what is within the constraints of good welding practice.

Storage

During both short- and long-term storage from hours to months or longer, dies and molds must be protected from water and humidity. Unprotected steel can almost immediately begin to corrode, resulting in damaged tools that will require repolishing, regrinding, and/or repair of at least the surface of the tool. The net result is a cost in both labor and machine downtime. It is most cost-effective to protect the tools. There are excellent rust protectants on the market that operated for different time periods. However, most of these anticorrosion treatments must be completely removed before using the molds. Some may require special cleaners including toxic solvents. Some operations dry off the tool and enclose it in an air-evacuated container. For special protection, vacuum containers are used after the mold is properly dried.

TOOL BUILDERS

There will be changes for mold/die builders. They will begin producing more servo-electric systems to actuate core pulls and other functions. Mold builders will not want to be seen as the sole cause for bringing hydraulic oil contamination to a clean production floor.

The environment of the molding plant will change considerably in a relatively few years, as will the standard of quality that we take for granted in the process. In just over a couple decades, the worldwide industry has gone from the first electric machines to having dozens of electric machine manufacturers in the market. This, alone, is a leading indicator of the market's appetite for cleaner, more precise, more energy efficient molding. (See Moldmaker Directory in the following glossary section.)

GLOSSARY

Please also review the terminology that are located in the Die section.

Adapter plate

The plate holding the mold to the molding machine press or platen.

Air lock

Surface depressions on a molded part caused by trapped air between the mold surface and the plastic material during processing.

Annealing

This is also called hardening, tempering, physical aging, and heat treatment. The annealing of metals or plastics can be define as a heat-treatment process directed at improving performance by removal of stresses or strains set up in the material during its fabrication.

Automatic

An automatic mold is a mold that goes repeatedly through its entire cycle, including ejection, without human assistance.

Back draft

A term used to describe a detail of a cavity that is smaller than the normal opening of the mold cavity. The opposite is a mold undercut.

Backing plate

A plate used to support cavity blocks, guide pins, bushings, and similar mold parts.

Balanced

A mold is laid out/designed with runners and cavities spaced and sized for uniform melt flow, fill, and packing pressure throughout the system.

Base

Assembly of usually steel plates that holds or retains the cavities.

Bluing

Bluing is a mild blemish in the form of a blue oxide film that occurs on the polished surface of a mold as a result of the use of abnormally high mold temperatures.

Bluing off

Checking the accuracy of mold cutoff surfaces by putting a thin coating of Prussian Blue on one-half and checking the blue transfer to the other half. Other techniques used include carbon paper, shims, sprayed-on soap, and so on.

Bottom plate

This is the part of the mold that contains the heel radius and the push-ups (ejection mechanism). It is used to join the lower section of the mold to the platen.

Brass tool

It is used to clear or remove melted plastic that may be trapped or stuck to a tool or screw, in the hopper throat when melt bridges, and so on. The soft brass, copper, and aluminum tools do not damage the metal, whereas steel or other metals would damage the steel mold.

Breathing

Breathing is also called mold bumping, dwell pause, dwell, gassing, and degassing. It is a pause in the application of mold pressure using material that gives off gases during the heating process; it also functions to remove any entrapped air. This on-off-on pressure action occurs just prior to the mold closing completely so as to allow the escape of gas and/or air. It is applied with many TS plastics and the vulcanization of TS elastomers/rubbers.

Cam bar

The stationary angled bar or rod used to mechanically operate the slides on a mold for side-action core pulls. It is also known as an angle pin and a horn pin.

Carbonizing

A low-carbon steel-surface hardening process that resists wear and abrasion. It is used in molds, dies, and other machine parts. The steel is heat-treated in a box packed with carbonizing material, such as wood charcoal, and heated to 1093°C (2000°F) for several hours, then allowed to cool slowly.

Cavity

The space between matched molds that encloses the molded part. It is the depression in the mold that forms the outer surface of the molded part. There can be single or multiple cavities in one mold.

Cavity chase

A cavity chase is an enclosure of any shape, used to shrink-fit parts of a mold cavity in place, to prevent spreading or distortion in hobbing, and to enclose an assembly of two or more parts of a split-cavity block.

Cavity debossed

Depressed or indented lettering or designs in the cavity that produce bossed impressions on the molded part.

Cavity deposit

Material builds up on a cavity's surface due to plate-out of the plastic melt; this is usually attributed to the use of certain additives or alloys.

Cavity draft

This is a draft in the direction of the mold. On most molded parts, there are features that must be cut into the surface of the mold perpendicular to the molding parting line. To properly release the part from the tool, parts almost always include a taper. The amount of mold draft required will depend on factors such as the type of plastic being processed, processing conditions, surface finish, and so on. For example, a highly polished surface will require less than an unpolished mold. Any surface texture will increase the draft at least 1° per side for every 0.003 cm (0.001 in) depth of texture. Special mold cavity surface action can be used. With elastomeric materials it is possible, with their rubbery condition, to perform ejection that does not require any draft.

Cavity duplicate plate

Removable plate that retains cavities that is used where two-plate operation is necessary for loading inserts.

Cavity ejector

Different mechanical means are used to eject or remove the molded part from the cavity.

CAVITY, FEMALE

The indented half of a mold designed to receive the male half.

CAVITY LAND

Refers to the length in the different gate configurations that influence melt flow.

CAVITY, MALE

Also called a plunger or a core. The extended half of a mold designed to match the female half.

CAVITY, MULTI-

A multicavity has two or more cavities to mold two or more parts.

CAVITY REGISTER

Angle faces on the molds that match when the mold halves are closed to ensure their correct alignment.

CAVITY RETAINER PLATE

They hold the inserted cavities in a mold. These plates are at the mold parting line and usually contain the guide pins and bushings that line up the two halves of the mold.

CAVITY, SPLIT-RING

A mold in which a split-cavity block is assembled in a chase to permit the forming of undercuts in a molded part. The part along with the molded part(s) is ejected from the mold and then separated.

CAVITY, UNBALANCED

A nonuniform layout of cavity and runner sizes and/or locations that will potentially cause nonuniform melt-fill rate, packing pressure, and part quality unless a proper design approach is used to incorporate the proper runner, gate, and cooling system.

CAVITY UNIT

Cavity insert(s) designed for quick interchangeability with other cavity insert(s).

Cavity venting

Basically shallow channel(s) or minute hole(s) in the cavity and/or in the mold parting line to allow air and other gases (that may form during mold filling) escape.

Change, quick

Quick mold change (QMC) systems have the ability to remove, for example, the mold in different fabricating machines and replace it automatically with another one in a matter of minutes; they also transfer the mold to storage. Devices such as slotted or shouldered bars assembled in a molding machine are designed to allow the movement of removing and loading molds. The machine's microprocessor controls can be programmed to change the feeding of materials and process controls with the quick change.

Chase

An enclosure of any shape used to (1) shrink fit of a mold cavity in place, (2) prevent spreading or distortion in hobbing, and (3) enclose an assembly of two or more parts of a split block.

Chase floating

A mold member, free to move, that fits over a cavity or a lower plug, and into which an upper plug telescopes.

Chunk

An open face mold.

Cleaning

Mechanical cleaning, such as separating the residual cured plastic, is usually confined to simple, uncomplicated molds. Thermal, chemical, and solvent-bath methods are generally used requiring the available proper equipment to entrap pollutants, and so on. Proper precautions are taken not to damage molds. For example, mold (or die, screw, etc.) damage may occur when flammable components of the plastics burn up in an oxygen-containing enclosure causing uncontrolled exothermic combustion.

Coating in-mold

A melt-processable paint coating forms a skin layer within the mold. It first enters the mold that is a takeoff of coinjection or sandwich IM.

COLD SLUG

A cold slug is the first plastic melt to enter an IMM's cold runner mold; it is called a cold slug because in passing through the sprue orifice it is cooled below the effective molding temperature. Usually a well in the runner system is used to unload this cold slug.

COLD SLUG WELL

The space or cut-out in the runner system (such as opposite the sprue travel of the melt in the mold) to trap the cold slug so that it does enter the cavity.

COLLAPSIBLE CORE

The collapsible or expandable core provides a means to mold internal threads, undercuts (as in tamper-proof bottle caps, etc.), protrusions, cut-outs, and so on. Different patent-approved design systems are used.

COMPLEX SHAPE

These are molded parts with undercuts preventing the part being released in the direction of the mold opening requiring molds with side actions/cores, rotating cores, loose cores or inserts, more than one parting line, wedges, and/or other devices. The choice of method, or a combination of these methods, is dependent by the shape and the properties of the plastic (flexible to rigid, shrinkage, etc.) and also the standard of quality required, such as not having a parting line at a certain location.

CONTACT MOLDING

Contact molding is also called open molding or contact pressure molding. molding or laminate under no or very little pressure (usually less than 70 kPa [10 psi]) unreinforced or usually reinforced TS plastic to obtain the desired shape followed with curing outside the mold. Cure can be either at room temperature using a catalyst-promoter system or by heating in an oven with or without additional pressure.

COOLING

Controlled cooling of the mold for certain materials is an essential feature and requires special attention in the design stage. The usual water cooling medium requires that it be in turbulent flow, rather than laminar flow, in order to transfer heat out of the molded part at a much faster rate.

COOLING BAFFLE

Ribs, plugs, and other devices are used to provide more uniform cooling action within the mold.

COOLING CHANNEL BUBBLER

A device inserted into a mold core that allows water to flow deep inside the core into which it is inserted and also discharged. Uniform core cooling can be developed. For smaller cores where there is insufficient diameter a baffle, a heat pipe, or a copper alloy plug may be installed.

COOLING, CONFORMAL

Basically conventional machining drills and bores in straight lines that produce cooling channels are limited to 2-D thermal cooling action. Conformal cooling allows the cooling medium to flow in all three dimensions and follow the internal contours of the mold-surface geometry to maximize the heat-transfer efficiency of the mold. An example for its construction uses thin metal slices to conform to the contour. These sliced layers are fitted together and can be bonded (brazing, etc.). In addition to using standard steels, high heat-transfer alloys (copper, etc.) can be used. This approach reduces cooling time from 20% to 60% that in turn significantly reduces cycle time and also produces products of higher quality.

COOLING, FLOOD

This system, used particularly in blow molding molds, involves an internal flooding action in a confined open chest that surrounds the mold cavity rather than using drilled holes. However, drilled-hole systems are also used.

COOLING FLOW METER

A device that can be put in-line with supply and return water to measure temperature, pressure, and flow rate of water through the mold.

COOLING/HEATING

Channels or passageways are located within the body of the mold. A cooling medium (of water with or without ethylene glycol when processing TPs) can be circulated through these channels to control temperature on the cavity surface, providing required cooling action to solidify the TP melt into a molded part. With TS plastics, heating the mold can be accomplished by circulating steam, hot oil, or other heated fluid; however, the usual method involves inserting electrical heating elements or probes in the mold rather than using the channels. A few TPs may require the higher heat to complete their cure.

COOLING, PULSE

Those using pulse cooling explain that conventional continuous coolant flow mold-temperature control generates the popular temperature gradients called isotherms (contour/wavy lines that move around the cooling channels). Although heat travels from the plastic to the metal to the coolant quickly, the mass of steel between the channels goes supposedly unused. Pulse cooling, because it is not continuous, eliminates the isotherms that segregate a conventional cooled mold. Heat from the part is absorbed not only by the cooling channels, but also by the large mass of steel on the shop-side of the mold. When the fill stage is complete, coolant circulates quickly, removing excess heat and quickly bring the mold and part back to minimum temperature. It is a cyclic process that aims to reduce cycle time 6% to 10%, reduce molded-in stress, reduce scrap, and reduce energy use.

COOLING RATE

The plastic melt's cooling rate is usually the final control in the variable associated with the final plastic part's performances. This variable influences factors such as melt flow rate, residual stress, and degree of orientation. Heating and cooling rates for amorphous and crystalline plastics differ; if not properly controlled, part performances are either not meeting the maximum or they are defective.

COOLING, SPIRAL

Spiral cooling is a method of mold cooling in which the cooling medium flows through a spiral cooling cavity in the body of the mold located opposite a flat or relatively flat cavity.

COOLING TEMPERATURE

Lowering coolant temperature below the required level is supposed to speed up heat removal; actually, the reverse is often true. Lowering temperature reduces water chiller capacity. If possible, avoid temperatures below 4°C (40°F) since ethylene glycol is added to the water to prevent the water from freezing. Going lower requires more EG. With EG and particularly more EG, the more viscous the solution becomes, resulting in greater pumping power and increased operating costs.

COOLING TIME

In addition to the mold cost of materials requiring cooling during molding, and machine costs, the final cost to mold a part interrelates with the molding cycle. A large part of this cycle with plastics, up to 80%, is due to the time required to cool the part in the mold. Thus, the target is to reduce cooling time recognizing that heat transfer through plastics has its limitations.

CORE

A core is a channel in a mold for circulation of a heat-transfer media; it is also called core pin, and is part of a complex mold that molds undercut parts. Cores are usually withdrawn to one side before the main sections of the mold open. They have passages for heat transfer to the melt in the cavity.

CORED

mold incorporating ducts permits the passage of heating and/or cooling services.

CORE PIN

Pin used to mold a hole in the molded part.

CORING, SIDE

Also called side-draw pin action or cam-pin action. Projections that are used to core a hole (or other shape) in a direction other than the line of closing a mold. It is withdrawn before the part is ejected and/or prior to the mold opening.

CORNER

Sharp corners (radiuses and fillets) should be avoided except at parting line. If possible when required, they should be as generous as possible. They can interfere with smooth melt flow and create possibilities for melt turbulence. Examples of problems that can develop include surface defects and stress concentrations.

COST

Up to about 12% to 20% of cost is for the (raw) material used to manufacturer the mold, mold design about 5% to 10%, mold building hours about 40% to 60%, and profit at about 5% to 10%.

CUT-OFF

Cutoff is also called shutoff, kiss-off, or flash land. It is that part of the mold land that isolates the molding.

DEEP-DRAW

Mold has a core cavity much longer than the mold wall thickness.

DEFORMATION

After the molding pressure stroke and during any after-fill, pressure is built up in the mold cavity. During IM this pressure is generally one-third to one-half of the pressure in the IMM plasticator. The consequence of such pressure must be recognized. It can cause elastic deformation such as a bending of the cavity retainer plates and cores and ejector and guide pins. To reduce this action, sturdy construction of the mold is required. However, the aim is to minimize the amount of mold material because light construction is usually desired for efficient cooling.

DEHUMIDIFICATION

If you use chilled water as a heat-transfer medium during periods of high humidity, you may have to deal with condensation forming on the surface of the mold that usually causes imperfections on the molded part. Chilled water can reduce cycle time so alternate methods such as enclosing the mold with dehumidified dry air and high-velocity dry air stream, can be used to eliminate condensation.

DISHED

A term used to describe a depression in a molded surface.

DOUBLE-SHOT

This is also called coinjection. It is a method for producing two-color plastic or plastics containing two different plastics in a part using an IMM with two plasticators. The part molded first becomes an insert for the second shot. Other processes, such as injection blow molding and compression molding, can be used.

DOWEL

A dowel is also called a mold pin or a retaining pin. It is a metal pin located in one half of a mold that enters a corresponding hole in the other half so that upon closure of the mold, the two halves become correctly aligned.

DOWEL BUSHING

A hardened steel bushing lining a dowel hole.

DRY AS MOLDED

DAM (dry as molded) describes a part immediately after it is removed from a mold and allowed to cool.

DWELL

A dwell is a pause in pressure just prior to the mold completely closing; it can be referred to as mold bumping

EJECTOR

An ejector is an attached mold assembly that is for operating ejection pin or pads to eject the molded part from the cavity. It may operate mechanically, hydraulically, pneumatically, or electrically.

EJECTOR MARK

An ejector mark is a surface mark on the part that is caused by the ejector pin when it pushes the part out of the mold cavity. It may be an unwanted mark or can be located where it is OK.

EJECTOR PIN

An ejecter pin is also called knockout pin. The pin pushes the molded part out of the mold after the mold starts opening.

EJECTOR RAM

A small ram that is mechanically, hydraulically, or electrically operated and fitted to the molding press for the purpose of operating the ejection mechanism.

EJECTOR RETURN PIN

The ejector return pin is also called a surface pin, a safety pin, or position pushback. They are projections that push the ejector assembly back as the mold closes.

ELASTOMERIC MOLD

This is an elastic or stretchable mold made of elastomer (rubber) rather than the usual steel, so that parts of complex shape can be removed without mold-side actions Elastomeric molds are usually used for casting plastics. They can be stretched to remove cured parts having undercuts.

EYEBOLT AND HOLE

These holes, properly located in a mold, are used to easily lift and move molds (and dies and other tools). Balanced lifting occurs with proper location and use of eyebolts; safety devices are incorporated to ensure no accidental dropping occurs so that personnel and the molds are protected. The different bolts have different safety load-carrying capacities.

Family mold

A family mold is a multicavity mold in which each cavity forms a part that often has a direct relationship in usage to the other parts in the mold. The term also applies to molds in which parts for different customers are grouped together in one mold for economy of production.

Feed pushing

hardened steel bushing in an injection mold that forms a seal between the mold and injection unit.

Ferris wheel

Mold halves are attached to a rotating platen. When the platens close, each mold cavity receives melt in a sequence pattern; this occurs in the production of parts such as an acrylic automobile taillight made with three molds. The first mold receives the amber color, the second receives the yellow color, and the third receives the transparent plastic that not only fills its respective cavity opening but also can cover the complete light plate to ensure no moisture or rain leakage occurs when the part is in service.

Film insert molding

Film insert molding (FIM) starts with a cut film that is decorated and/or labeled, thermoformed to shape, and then inserted in the mold.

Filling hesitation

To understand the "hesitation effect," consider the flow patterns throughout IM filling. The melt first enters from the gate and the flow front reaches the first thin-wall section. There is insufficient pressure to fill this thin section as the melt has an alternate route along the thick section. Melt that just entered the thin section sits there losing heat until the rest of the mold is filled. When the mold is almost completely filled, the full injection pressure is available to try to fill the thin section. However, the melt in the thin section has frozen and the thin section is not filled. This problem is caused by the fast/slow/fast (hesitation) filling speed used. Basically if the melt continues to flow at a relatively steady/uniform rate, there is no difficulty in filling the thin section.

Filling monitoring

Flow-front speed during filling is commonly inferred either from screw position or cavity-pressure sensors (387). The quality of the final molded part, however, is determined by the actual flows of molten plastic into the cavity to pack the melt. The ultrasonic technique is an example of monitoring the filling action. This technology includes the use of ultrasonic transducers and software to verify mold-filling patterns and measure flow-front speeds. It permits identifying exactly when mold cavities are filled and switching from injection pressure to packing pressure, saving energy. Ultrasonic beams are emitted from transducers installed at the external surfaces of a steel mold cavity. The beams propagate in the steel mold cavity interface. Before the melt arrives at the transducer's position, ultrasonic energy is totally reflected at this interface. After the melt's arrival, part of the beam energy is transmitted into the melt, indicating that the melt front arrived. The sensor can monitor the gap caused by the shrinkage of the part away from the mold wall, as well as measure the speed of the gap's development. Ultrasonic waveforms show echoes in the solidifying parts, which can be used to obtain temperature profiles across the melt and to study cooling efficiency.

FLASH MOLD

A flash mold is a type of mold with a land surface that permits the escape of excess molding material and has no mold trimming action. Such a molding material relies on back pressure to seal the mold and put the part under pressure. This also identifies a portion of material that protrudes beyond the edge of the finished molding. It is attached to a molding along the mold's parting line, holes, openings, and so on. This occurs principally with TS plastics. With most parts it is objectionable and must be removed before they are acceptable either manually or deflashed by barreling/tumbling, buffing, grinding, and blasting.

FLASH LINE

A flash line is a raised line evident on the surface of a molding and formed at the junction of the mold faces, such as at the parting line after the removal of the excess flash. It is usually removed by high-speed buffing or grinding.

FLASH RING

A flash ring is the clearance between the force male plug and the vertical or horizontal wall of the female cavity in a positive or semipositive mold. It is the ring of excess melt that escapes from the cavity into this clearance space.

FLASH TRAP

A molded-in lip or blind recess on a part that is used for trapping excess melt (flash).

FLOW MOLDING

Leather-like (plastic) materials are made by placing a die-cut plastic blank (usually solid or expanded vinyl with or without a coated substrate) in a mold cavity (usually silicone rubber mold) and applying heat with a high-frequency radio generator. Melted plastic fills up the mold cavity and reproduces the texture in the mold cavity. Modifications to step up production such as a rotating arm with the mold cavity pattern facing outward and vertical to the arm.

FRENCH MOLD

A two-piece mold for irregular shapes, sizes, proportions, and level of detail. such as being tall, top heavy, leaning to one side, and/or with extremely fine detail.

GATE

The gate is the orifice through which the melt flows and enters a mold cavity. Single-gate or multigate arrangements can be used for a single cavity. The gate can have a variety of configurations to meet different melt flow requirements and cavity configurations.

GATE BLUSH

Gate blush is associated with melt fracture around the gate from stresses caused by process conditions and mold geometry. It is a blemish or disturbance in the gate area. To eliminate or reduce this problem, raise the melt temperature, reduce injection speed, check the gate for sharp edges, enlarge the gate, and check that the runner system has a cold-slug well.

GATE DEGATING

The removal of the gate from the molding by automatic or manual means within the mold during molding or after part(s) are removed from the mold.

GATE, DIAPHRAGM

Also called a disc or web gate. This designed gate is used in molding annular or tubular parts. The gate forms a solid web across the opening of the part.

GATE, DIAPHRAGM-AND-DING

This gate is used mainly for cylindrical and round parts in which concentricity is an important dimensional requirement and a weld-line presence is objectionable.

GATE, DIRECT

This gate has the same cross-section as that of the runner. It is commonly used for direct-sprue gating.

GATE, EDGE

There are different edge-gate designs such as those that use a heater coil or rod and single- or multiple-edge gate openings. Edge gating is carried out at the side or by overlapping the part. It is commonly employed for products that are manually operated. Normally, it is possible to remove the complete shot with one hand and in a rapid manner. The products are separated from the runner system by hand with the aid of side cutters or, if an appearance requirement demands it, by such auxiliary means as sanders, millers, grinders, and so on. When degating is performed with the aid of auxiliary equipment, it becomes necessary to construct holding devices.

GATE, FAN

This fan-shaped gate is an opening between the runner and the mold cavity. This fan shape helps to reduce stress concentrations in the gate area by spreading the opening over a wider range.

GATE FLASH

Gate flash is usually a long, shallow, and rectangular gate extending from a runner that runs parallel to an edge of a molded part along the flash or mold parting line. It is especially suitable for flat parts of considerable area (over 7.6 × 7.6 cm [3 × 3 in]). This gate can also be used when the danger of part warpage and dimensional change exists.

GATE, FOUR-POINT CROSS

This gate is also known as a spoke gate. This is used for tube-shaped parts and offers easy degating. Disadvantages are possible weld lines and the fact that perfect roundness is unlikely.

GATE, HOT-PROBE

This may also be called an insulated runner gate and is used in runnerless molding. In this type of molding, the molten plastic material is delivered to the mold through heated runners, thus minimizing finishing and scrap costs.

GATE, INTERNAL WING

This gate is suitable for tube-shaped articles in single-cavity molds.

GATE MARK

A surface discontinuity on a molded part caused by the gate through which material enters the cavity.

GATE NOZZLE

It is a nozzle whose tip is part of the mold cavity, thus feeding the molding melt directly into the cavity, eliminating the need for sprue and runner systems. The nozzle becomes the mold gate.

GATE, PINPOINT

This gate is also called a restricted gate. This small gate minimizes the size of the mark left on a molded part. The gate breaks clean when the part is ejected. Generally used in three-plate and hot runner mold construction, it provides rapid freeze-off and easy separation of the runner from the. The size of such gates may be as great as ⅛ in, provided that the product will not be distorted during gate breaking and separation. A further advantage of pinpoint gating is that it can easily provide multiple gating to a cavity (for thin-walled products), should such a move be desired for product symmetry or balancing the flow. It also lends itself to automatic press operation if the runner system and products are arranged for easy drop-off. For a smooth and close break-off, it is best to have the press opening at its highest speed at the moment when the plates causing the gate to snap are separating.

GATE, RING

This type of gate is used on cylindrical shapes. This gate encircles the core to permit the melt to first move around the core symmetrically before filling the cavity, preventing a weld. There are external and internal ring gates in respect to the cavity.

GATE SCAR

Most mold designs start out using a small gate opening. If the gate is too large, scars in the gate area can occur. However, larger sizes permit faster fill and cycle time.

GATE, SPIDER

It is the multigating through a system of radial runners from the sprue.

GATE SPLAY

Gate splay is also called silver streaking. It is a fanlike surface defect near the gate on a part and is usually due to turbulent melt flow.

GATE STRAIN

At the part location of the gate, strain develops in the plastic. If required to relieve this strain, cooling rate of the melt is reduced but cycle time increases.

GATE, SUBMARINE

This gate is also called a tunnel gate. This type of edge gating is where the opening from the runner into the mold is located below the parting line or mold surface. The conventional edge and other gates is where the opening is machined into the surface of the mold on the mold parting line. With submarine gates, the molded part is cut by the mold from the runner system on ejection of the part. It is often used in multicavity two-plate molds because it degates automatically, so it is particularly suitable for automatic operation. For multiple cavities, an angular gate entrance requires special care in machining during moldmaking, in order to ensure uniformity of the gate opening and consistency in the angular approach for a balanced runner system. The angle of approach is determined by the rigidity of material during ejection and the strength of the cavity at the parting line affected by the gate. A flexible material will tolerate a greater angle of entrance than a rigid one. The rigid material may tend to shear off and leave the gate in place, thus defeating its intended purpose. However, the larger angle will give greater strength to the cavity, whereas a smaller angle may yield a cleaner shearing surface than a larger angle.

GATE, TAB

This is a small, removable tab of approximately the same thickness as the molded part, usually located perpendicular to the part. It is used as a site for edge gating location on parts with large, flat sections, as a site for gating so that if any unacceptable blemishes appear they will be on the tab that is cut off, and where it is desirable to transfer the stress generated in the gate to an auxiliary tab, which is removed in a postmolding operation. Flat and thin parts usually require this type of gate.

GATE TIP, REVERSE

With reverse gating, the gate mark is hidden. The mold is designed so that the gate location is on the core side or opposite the cosmetic surface.

GATE, VALVE

Valve gates are used in injection molds to provide a wider processing window of operation and better product quality. They eliminate gate freezing and are cost effective. The valve usually has a pin to mechanically open and close the gate orifice. An actuating mechanism (spring, adjustable air cushion, mechanical cam, pneumatic or hydraulic pistons, etc.) coordinates the movement of the pin with the molding cycle. In demanding cavity melt packing requirements where precise part weight and/or dimensions are required, this technique is considered.

GATE, VALVE FEED CONTROL

In hot runner molds, valves can be controlled in a closed loop. They can be adjusted to partially open or close aiming to eliminate unbalanced cavity fill. The conventional system forces to tweak and adjust transfer position, injection pressures, and pack with hold times and pressures until the defects in the part are minimized or eliminated. With dynamic feed, the valve(s) can be individually adjusted (set) to deliver less melt at individual controlled lower pressure to eliminate flash, increased hold pressure, eliminate sink, position welds, and so on. A transducer between each valve and gate is used to measure melt pressure as it enters the cavity. The user programs the time and pressure profile desired to fill the part properly. In doing so, variations in viscosity are negated.

Grid

The construction of channel-shaped supporting members within a mold.

Halve

Each part of a mold is called a mold half; however, this usually does not mean that the mold is divided dimensionally into two equal halves.

Hand mold

This is also called a portable or loose mold. A small mold is removed by hand from the press for the purpose of stripping molded parts and/or reloading (material and/or inserts). The operator manually removes a mold and removes part(s).

Heat pipe

Also called thermal pins, heat transfer devices, or heat conductors, heat pipes are a means of heat transfer; either to remove or to add heat in all kinds of fabricating equipment. Compared to metals, they have an extremely high heat-transfer rate. They are capable of transmitting thermal energy at near-sonic isothermal conditions and at near-sonic velocity. Sizes range from very small to large diameters with limited lengths; however, pins can be put in series. They are tubular structures closed at both ends and contain a working fluid. For heat to be transferred from one part of the structure to the other, the working liquid is vaporized at a hot spot. At a cool spot, the vapor is condensed and then travels via a wick to the hot (depleted) end, where the condensate becomes a vapor again to repeat the process.

Heat-transfer device

Transfers localized heat to a heat sink in order to improve mold cooling or transfer heat from the source to a localized area such as hot sprue bushings.

HEATING

Controlled heating of the mold is an essential mold feature and requires special attention in mold design.

HEIGHT

This refers to the vertical distance of a closed mold located on a table so that the parting line is in a horizontal position. Most molds are put in horizontally operating machines (IM, blow molding, etc.) so the height actually is the horizontal distance from platen to platen with the mold closed.

HOLD-DOWN GROOVE

A small groove cut into the side wall of the mold cavity surface to assist in holding the molded part in the cavity while the mold opens.

HOLLOW

This is a type of mold that permits melted plastic to be applied to its inside surface to form hollow-shaped parts.

IMPRESSION

That part of a mold that imparts shape to the molding.

INCHING

This is a reduction in the rate of mold closing travel just before the mating mold surfaces touch each other.

INTELLIGENT TOOLING

Controls are used to perform different functions such as temperature of the tool and melt. When required all heaters can be brought up to set temperatures together to reduce thermal stresses. They also can control thicknesses and operating sequences via different tool movements, melt packing behavior, and so on. Other benefits include controlling sequences, process monitoring, and setup parameters within the tool. Overall they can communicate with the primary and auxiliary equipment in the production line.

INTERLOCK

Guide pins and bushings are only an average method for alignment of the mold halves. For precision tooling, additional interlocks in the form of cones and or pins are used to give high-precision alignment particularly for multicavity, high-precision products meeting close tolerances

JETTING

melt entering the cavity, rather than being more in laminar flow, is in turbulent flow. Causes of this include undersized gate and thin to thick cavity section resulting on poor control of the molded part.

KNIFE EDGE

This term describes a projection from the mold surface that has a narrow included angle. They are undesirable because they are susceptible to breakage under molding pressures.

LAMINATED

A molded product produced by bonding layers of plastic, impregnated material (prepreg) using heat and pressure. The layers are often cut to particular shapes prior to being placed in the mold (chapter 15).

LAND

describes the area of those faces of a closed mold that come into contact with one another.

LATCH

This is a device that holds together two members of a mold that are usually held together mechanically.

LATCH PLATE

This retains a removable core to hold an insert carrying pins on the upper part of the mold.

LEADER PIN AND BUSHING

These are also called guide pins. Pins, usually four, maintain the proper alignment of the male plug and female cavity as the mold closes. One of the pins is not symmetrically placed so that the mold halves can only be aligned one way, eliminating misalignment. Hardened steel pins fit closely into hardened steel bushings.

LETTERING AND SURFACE DECORATION

Lettering and other raised or depressed surface decorations and textures are easily incorporated on plastic products. The mold cavities can be machined, hubbed, or cast.

LIFE

Amold's life refers to the number of acceptable parts that can be produced in a particular mold. There are molds that run a few hundreds to many millions of press openings. Design and construction that relates to cost of a mold depends on the lifetime required.

LOADING TRAY

This is also called a charging tray. A tray automatically moves over the cavity and feeds a mold by "dropping" material and/or inserts into a single or multicavity by the withdrawal over the cavity of a sliding bottom to the trays

LOADING WELL

The top volume of a cavity is usually for compression molding bulky compounds. Its size is dependent on the material's bulk factor.

LOCATING RING

This is also called register ring. It serves to align the nozzle of an injection cylinder with the entrance of the mold's sprue bushing.

LOCKING RING

A slotted plate that locks the parts of a mold together while the material is being injected or placed.

LOOSE PUNCH

The male part of the mold functions in such a way that it remains attached to the molding when the press opens and the molding is removed. It is commonly used for moldings possessing threads or undercuts, when the punch cannot be removed from the molding merely by opening the press.

MALE PLUG/FEMALE CAVITY

The typical two-part mold (for compression molding, coining, etc.) has a female cavity with a matching male plug core that literally fits into the female cavity. In the closed position, the space located between the mating cavity and core provides for the molding material to be compressed. With certain materials, such as TS plastics, flash occurs.

MANIFOLD, DIE

This is also called feedblock. It directs melt from the extruder to its die.

MANIFOLD, MOLD

It is a runner system in a mold that has its own heating and/or cooling insulated section to control the melt and be ready for injection into the cavity.

MOLD REPAIR

See Table 17 65.

MOLDMAKER DIRECTORY

SPI prepares a directory for moldmakers; Table 17.66 provides examples.

MULTI-IMPRESSION

A multi-impression mold has two or more cavities.

OPEN MOLD

The practice of pouring or placing a plastic into a mold cavity and with or without a lid permitted to cure or polymerize into the solid shape of the cavity without pressure.

PARALLEL TO DRAW

Axis of the cored position (hole) or insert parallel to the up-and-down movement of the mold as it opens and closes.

PART CORING

removal of excess plastic from the cross-section of a molded part obtains a more uniform wall thickness.

CORES AND CAVITIES.
Checked for water leaks;
Checked for damage on the molding faces;
Surface finish of cores and cavities;
Have the molding faces been protected against corrosion?

MOLD 'SHUT OUT' FACES.
Damage to the parting line on mold;
Removal of any surface deposit;
Check for flatness of mold plates or inserts;
Check for damage caused by stringing of thermoplastics material from the sprue.

SIDE CORE ASSEMBLY.
Check for straightness of angle pins;
Check for wear on angle pins so that the clearance between the location hole in the side core and pin is not excessive causing scuffing or marks to appear on the molding;
Check for a positive retention of the side cores when in their open position;

EJECTION SYSTEM.
Check for the smoothness of the ejection movement;
Check that the ejector blades are suitably supported during the forward and return strokes;
Check that the correct ejection spigot has been fitted (to suit the molding machine it is to fit on);
Check that the ejection return mechanism is operating correctly.

COOLING SYSTEM.
Check for any water leaks in the cooling arrangement;
Check that the cooling circuit has been correctly connected

ELECTRICAL HEATING SYSTEM.
Check the cartridge heaters for any failure;
Check all plugs and connector blocks for positive location and tightness of fit;
Check the thermocouples for each nozzle and manifold assembly of the runnerless system.

UNSCREWING MECHANISM.
Check the position of the cores in relation to the operation of the mold;
Check that the helix spindle has been set correctly - in relation to the mold open and/or close position;
Check that the unscrewing unit has been adequately lubricated and set correctly;
Check that the shear pins have been replaced;
Check that the hydraulic cylinder is aligned with the guide plates so as to prevent premature seizure of the rack during the unscrewing operation.

ELECTRICAL CIRCUITS ON MOLD.
Check that the microswitches are functioning correctly so as to ensure that the side cores, or additional moving parts of the mold, cannot be operated out of sequence in relation to the operation of the machine.

Table 17.65 Checklist procedure for mold repair (courtesy of Synventive Molding Solutions)

NORTHEAST

COMPANY NAME AND CONTACTS	CONTACT NUMBERS	ADDRESS	PRIMARY PROTUCTS
ABA-PGT, Inc. Samuel Pierson Voting Rep. Roger Anderson Alternate Rep.	860/649-4591 860/643-7619 Fax info@abapgt.com abapgt.com	1395 Tolland Turnpike P.O. Box 8270 Manchester, CT 06040	**Mold Types:** Injection; Compression & Transfer; Die Cast; Prototype **Custom Services:** Product Design; Mold Design; CAD/CAM; IGES/Data Transfer; Mold Try-Out; Mold Bases & Components
Athena Plastics Mold Richard Beres Voting Rep. Harold Zuber Alternate Rep.	732/367-5700 732/363-5085 Fax	1785 Swarthmore Avenue Lakewood, NJ 08701	**Mold Types:** Injection; Prototype **Custom Services:** Mold Design; Mold Try-Out
Aztec Tool Company, Inc. Stewart Swiss Voting Rep. James Evarts Alternate Rep.	516/243-1144 516/243-1149 Fax aztec.thomasregister.com	180 Rodeo Drive Edgewood, NY 11717	**Mold Types:** Injection; Prototype **Custom Services:** Product Design; Mold Design; CAD/CAM; IGES/Data Transfer; Mold Try-Out
Cacoosing Industries, Inc. Peter Schlegel Voting Rep. Robert Schlegel Alternate Rep.	610/678-3441 610/670-5154 Fax	333 South Hall Street P.O. Box 2178 Sinking Spring, PA 19608	**Mold Types:** Injection **Custom Services:** Mold Design; CAD/CAM; IGES/Data Transfer; Mold Try-Out

Table 17.66 Example of SPI's moldmakers directory for services

Lion USA Enterprises Corporation Thomas Hsiao Voting Rep. Hope Chen Alternate Rep.	410/730-4280 410/997-4770 Fax lionusa@aol.com	5013 Castle Moor Drive Columbia, MD 21044	**Mold Types:** **Custom Services:**
Marpac Industries, Inc. Gary Gendron Voting Rep. Brian Williams Alternate Rep.	914/336-8100 914/336-5006 Fax	77 Kukuk Lane Kingston, NY 12401	**Mold Types:** **Custom Services:**
Mold Base Industries, Inc. Samuel Shiffler Voting Rep. Mary Forbes Alternate Rep.	800/241-6656 717/564-7705 Fax sales@moldbase.com moldbase.com	7450 Derry Street Harrisburg, PA 17111	**Mold Types:** N/A **Custom Services:** CAD/CAM; IGES/Data Transfer; Mold Bases & Components
MXL Industries, Inc. Steven Cliff Voting Rep. Matthew O'Connell Alternate Rep.	717/569-8711 717/569-8716 Fax mxl-industries.com	1764 Rohrerstown Road Lancaster, PA 17601	**Mold Types:** **Custom Services:**
New Jersey Tool & Die Company Jack Kemps Voting Rep.	908/245-0020 908/245-8822 Fax jkemps@njtoolcom	800 Colfax Avenue Kenilworth, NJ 07033	**Mold Types:** Injection; Compression & Transfer; Prototype **Custom Services:** Mold Design; CAD/CAM; IGES/Data Transfer; Mold Try-Out; Mold Bases & Components

Table 17.66 Example of SPI's moldmakers directory for services *(continued)*

SOUTHERN

COMPANY NAME AND CONTACTS	CONTACT NUMBERS	ADDRESS	PRIMARY PRODUCTS
Altira, Inc. Ramon Poo Voting Rep. Faustino Poo Alternate Rep.	305/687-8074 305/688-8029 Fax	3225 NW 112th Street Miami, FL 33167	**Mold Types:** Blow, Prototype **Custom Services:** Product Design; Mold Design; CAD/CAM; IGES/Data Transfer; Mold Try-Out
Cavaform, Inc. David S. Massie Voting Rep. Robert Massie Alternate Rep.	727/384-3676 727/384-0523 Fax cavaform@aol.com cavaform.com	2700 72nd Street North St. Petersburg, FL 33710	**Mold Types:** Injection; Prototype **Custom Services:** Mold Design; CAD/CAM; IGES/Data Transfer; Mold Bases & Components
Delta Mold, Inc. Eric Mozer Voting Rep. Donald Wisch Alternate Rep.	704/588-6600 704/588-5237 Fax emozer@deltamold.com deltamold.com	9415 Stockport Place Charlotte, NC 28278	**Mold Types:** Injection; Thermoforming; Blow; Compression & Transfer; Prototype **Custom Services:** Mold Design; CAD/CAM; IGES/Data Transfer; Mold Try-Out
E&M Precision Mold and Die Company George Arnold Voting Rep.	804/353-7160 804/353-7161 Fax	1705 Dabney Road Richmond, VA 23230	**Mold Types:** Injection; Blow; Die Cast; Prototype **Custom Services:** Product Design; Mold Design; CAD/CAM; IGES/Data Transfer; Mold Try-Out; Mold Bases & Components
Greeneville Tool & Die Co. Inc. Jurgen Rademacher Voting Rep.	423/639-1271 423/639-6130 Fax greene.xtn.net/com/gtd	965 E. Andrew Johnson Highway Suite 2 Greeneville, TN 37745	**Mold Types:** Injection; Die Cast; Prototype **Custom Services:** Product Design; Mold Design; CAD/CAM; IGES/Data Transfer; Mold Try-Out

Table 17.66 Example of SPI's moldmakers directory for services *(continued)*

MIDWEST

COMPANY NAME AND CONTACTS	CONTACT NUMBERS	ADDRESS	PRIMARY PRODUCTS
Accura Tool & Mold Brian Beringer Voting Rep. Dody Beringer Alternate Rep.	815/459-5520 815/459-4434 Fax molds@accuratool.com accuratool.com	101 West Terra Cotta Avenue Crystal Lake, IL 60014	**Mold Types:** Injection; Die Cast **Custom Services:** Mold Design; CAD/CAM; Mold Try-Out; Mold Bases & Components
Advanced Quality Molds (AQM) Mike Castek Voting Rep. Jim Snell Alternate Rep.	515/424-0370 x281 515/424-1945 Fax mikecastek@aqmmolds.com aqmmolds.com	1511 S. Benjamin Avenue Mason City, IA 50401	**Mold Types:** Injection; Prototype **Custom Services:** Mold Design; CAD/CAM; IGES/Data Transfer; Mold Try-Out; Mold Bases & Components
Allied Tool, Inc. Gregory Hogle Voting Rep. Terry Morris Alternate Rep.	517/764-5554 517/764-0973 Fax alliedtool.com	4941 Page Avenue Michigan Center, MI 49254	**Mold Types:** **Custom Services:**
Alpha Mold LLC Randy Korn Voting Rep. Peter Winner Alternate Rep.	937/233-5670 937/233-0478 Fax alphamold@alphamold.com alphamold.com	7611 Center Point 70 Blvd. Huber Heights, OH 45424	**Mold Types:** Injection; Blow, Prototype **Custom Services:** Mold Design; CAD/CAM; IGES/Data Transfer; Mold Try-Out

Table 17.66 Example of SPI's moldmakers directory for services *(continued)*

E.L. Stone Company, Inc. Mark Micire Voting Rep. Elma Micire Alternate Rep.	330/825-4565 330/825-8699 Fax elstone@elstone.com	P.O. Box 1012 Barberton, OH 44203	**Mold Types:** Thermoforming; Blow; Compression & Transfer; Rotational; Die Casting; Prototype **Custom Services:** N/A
FOBA North America, Inc. Steve Bone Voting Rep. Scott Ludwig Alternate Rep.	816/525-6030 816/525-6151 Fax FOBA@att.net	1212 S.E. Broadway Lee's Summit, MO 64081	**Mold Types:** **Custom Services:** **Other:** Laser Systems
Glenn Beall Plastics, Inc. Glenn Beall Voting Rep.	847/549-9970 847/549-9935 Fax	32981 N. River Road Libertyville, IL 60048	**Mold Types:** N/A **Custom Services:** Product Design; Mold Design
Hobson Mould Works, Inc. Gerald Hobson Voting Rep.	319/885-6521 319/885-6653 Fax hobsong@netins.net hobson.com	511 North Cherry Street Shell Rock, IA 50670	**Mold Types:** Injection; Thermoforming; Blow; Rotational; Die Cast; Prototype **Custom Services:** Product Design; Mold Design; CAD/CAM; IGES/Data Transfer; Mold Try-Out; Mold Bases & Components
Holland Plastics Corp. Glenn Anderson Voting Rep.	616/844-2505 616/844-2267 Fax glenn@lakeshoreconn.com	14000 172nd Avenue Grand Haven, MI 49417	**Mold Types:** Injection; Die Cast; Prototype **Custom Services:**
Illinois Precision Corp. Peter Clarke Voting Rep. Richard Davis Alternate Rep.	630/665-8840 630/665-6032 Fax	303 Delles Road Wheaton, IL 60187	**Mold Types:** **Custom Services:**

Table 17.66 Example of SPI's moldmakers directory for services *(continued)*

WESTERN

COMPANY NAME AND CONTACTS	CONTACT NUMBERS	ADDRESS	PRIMARY PRODUCTS
AAA Tool & Die Sam Gulizia Voting Rep. Donald McAlister Alternate Rep.	909/657-7440 909/657-8413 Fax	3111 Indian Avenue Perris, CA 92571	**Mold Types:** Injection **Custom Services:** Product Design; Mold Design; CAD/CAM; Mold Try-Out
ALBA Enterprises, Inc. Tony Brusca Voting Rep.	909/941-0600 909/940-0190 Fax albaplas@ix.netcom.com sunsup.com/alba	9624 Hermosa Avenue Rancho Cucamonga, CA 91730	**Mold Types:** N/A **Custom Services:** Mold Design; Mold Bases & Components
Allen Mold, Inc. Clayton Allen Voting Rep.	714/538-6517 714/538-6325 Fax	1100 W. Katella Avenue Orange, CA 92867	**Mold Types:** **Custom Services:**
Alpha Mold West Dave Whittington Voting Rep. David Staub Alternate Rep.	303/465-1701 303/466-7719 Fax dwhitt@alphamoldwest.com alphamoldwest.com	7005 W. 116th Avenue Broomfield, CO 80020	**Mold Types:** Injection; Die Cast; Prototype **Custom Services:** Mold Design; CAD/CAM; IGES/Data Transfer; Mold Try-Out
Berry Plastics Corporation Frank DeVore Voting Rep.	714/633-1494 714/538-6325 Fax sales@irwinint.com	1312 North 16th Avenue Yakima, WA 98902	**Mold Types:** **Custom Services:**

Table 17.66 Example of SPI's moldmakers directory for services *(continued)*

Hexcel Industrial Structures Rick Warner Voting Rep. Fred Isley Alternate Rep.	253/395-4898 253/395-6650 Fax rick.warner@hexcel.com	7819 S. 196th Street Kent, WA 98032	**Mold Types:** **Custom Services:**
JK Molds, Inc. Terry Corbert Voting Rep. Gregg Eastwood Alternate Rep.	909/981-0993 909/985-7180 Fax terryc@jkmolds.com jkmolds.com	2048 West 11th Street Upland, CA 91786	**Mold Types:** Injection **Custom Services:** Mold Design; CAD/CAM; IGES/Data Transfer; Mold Bases & Components
L&H Mold and Engineering, Inc. Stan Hillary Voting Rep. Steve Hillary Alternate Rep.	909/930-1550 909/930-1552 Fax stanlhmolds@earlink.net	2031 Del Rio Ontario, CA 91761	**Mold Types:** Injection; Die Cast **Custom Services:** Product Design; Mold Design; CAD/CAM; IGES/Data Transfer; Mold Try-Out
Mastercraft Companies Arle Rawlings Voting Rep. Dave Larson Alternate Rep.	602/484-4520 602/484-4525 Fax webmaster@mastercraft-companies.com mastercraft-companies.com	3301 West Vernon Avenue Phoenix, AZ 85009	**Mold Types:** Injection; Die Cast; Prototype **Custom Services:** Mold Design; CAD/CAM; IGES/Data Transfer; Mold Try-Out
Oak Products Robert Cook Voting Rep.	714/841-0100 714/841-2793 Fax rdc@oakproducts.com oakproducts.com	17782 Georgetown Lane Huntington Beach, CA 92647	**Mold Types:** Injection; Prototype **Custom Services:** Mold Design; CAD/CAM; IGES/Data Transfer

Table 17.66 Example of SPI's moldmakers directory for services *(continued)*

CANADIAN

COMPANY NAME AND CONTACTS	CONTACT NUMBERS	ADDRESS	PRIMARY PRODUCTS
F.G.L. Precision Works Ltd. Frank Meisels Voting Rep. Tom Meisels Alternate Rep.	905/738-2424 905/738-2423 Fax fgl@istar.ca fglmolds.com	215 Drumlin Circle Concord, Ontario Canada L4K 3E4	**Mold Types:** Injection; Blow; Compression & Transfer; Rotational **Custom Services:** Product Design; Mold Design; CAD/CAM; IGES/Data Transfer; Mold Try-Out; Mold Bases & Components
Husky Injection Molding Systems Glenn Atkinson Voting Rep. Michael Urquhart Alternate Rep.	905/951-5000 905/951-5360 Fax gatkinso@husky.on.ca	500 Queen Street South Bolton, Ontario Canada L7E 5S5	**Mold Types:** Injection; Prototype **Custom Services:** Mold Design; CAD/CAM; IGES/Data Transfer; Mold Try-Out; Mold Bases & Components
RYKA Blow Molds Ltd. Michael Ryan Voting Rep. Alfred Stein Alternate Rep.	905/670-1450 905/670-2621 Fax rykabm@aol.com rykabm.com	1608 Bonhill Road Mississauga, Ontario Canada L5T 1C7	**Mold Types:** Blow; Prototype **Custom Services:** Product Design; Mold Design; CAD/CAM; IGES/Data Transfer; Mold Try-Out

Table 17.66 Example of SPI's moldmakers directory for services *(continued)*

PART COSMETIC

The Molders Division of SPI publishes and updates a bulletin entitled *Cosmetic Specifications of Injection Molded Parts.* Its purpose is to provide quantitative definitions and recommended methods of inspection and measurement of the cosmetic quality attributes in the absence of customer-provided specifications. Guidelines include black specks, flow lines, and others.

PART EJECTION

The Usual ejection takes place through with a mechanical knockout device. To aid the removal of certain shaped parts, particularly if the parts are large, air assist is provided. Air is forced between the molded part and cavity wall.

PART SEPARATOR

machine or system used to separate parts from the solidified runner system.

PART SETTING UP

This is also called setup. It is a term used to describe the hardening of material in the mold prior to removal of the molding.

PARTING LINE

This is also called cutoff or spew. It is a line established on a 3-D model from which a mold is to be prepared, to indicate where the mold is to be split into two halves (sections) representing where they meet on closing.

PARTING LINE FLAT (SEAL-OFF AREA)

The two halves of the mold have a flat parting line. When the two halves meet, each half is literally making contact by one flat surface against another flat surface. Pressure on the injected melt in the cavity is through the plasticator's pressure-ram action on the melt. The clamping force acts on the seal-off area to keep the mold closed. The developed compressive stress that is the clamping force divided by the seal-off area must be less than the compressive stress of the steel or the mold will be crushed.

PARTING LINE SENSOR

Instrument to monitor relative movements of the mold halves in response to plastic melt pressure

PILLAR SUPPORT

The general construction of a mold base usually incorporates an ejection housing. If the span in the housing is long, the forces during molding can cause a sizable deflection in the plates that are supported by the ejector housing causing flashing, and so on. To overcome this problem, pillar supports are included so that deflection does not occur.

PIN

Different pins are used, including dowel pins, ejector pins, leader pins, return pins, side-draw pins, and sprue-draw pins.

PLATE-OUT

an objectionable coating gradually formed on metal surfaces of molds, dies, calendering rolls, embossing rolls, and so on. It is caused by the extraction and deposition of certain components in the plastic, such as pigments, lubricants, plasticizers, and/or stabilizers.

POROUS

A porous material is one made up of bonded or fused aggregate (powdered metal, coarse pellets, etc.) such that the resulting mass contains numerous open interstices of regular or irregular size allowing air or liquid to pass through the mass of the mold. It is used in molds for different processes (IM, thermoforming, etc.).

POSITIVE

A compression mold designed to have total applied pressure on the part being molded, the thickness of the part being determined by the amount of charge, is described as positive. The mold is designed to trap all the molding material when it closes.

POT

chamber to hold and heat plastic material for transfer to fabricating equipment such as IM, extrusion, transfer molding, compression molding, and rotational molding.

PREENGINEERED

Standardized mold components have been available at least since 1943. They provide for exceptional quality control on materials used, quick delivery, interchangeability, and lower cost. These preengineered molds and mold parts provide high-quality manufacturing techniques that result in consistent quality and reduced mold cost. The different manufacturers of these preengineered mold bases and components provide similar but also different products. The variations can provide unique and different approaches to meeting complex product designs. Advantages to the molder include time and cost savings because of fast delivery.

PRESSURE

Pressure is developed by a screw or ram to push melt into a mold cavity, or pressure is maintained on the melt after the cavity is filled until the gate freeze-off, allowing the complete transformation to a solid state. Pressure is also applied on the material after it fills the closed injection or compression mold cavity.

PRESSURE, CONTACT

also called open molding or contact pressure molding. molding or laminate under no or very little pressure (usually less than 70 kPa [10 psi]) unreinforced or usually reinforced TS plastic to obtain the desired shape followed with curing outside the mold. Cure can be either at room temperature using a catalyst-promoter system or by heating in an oven with or without additional pressure (chapter 15).

PRESSURE, HIGH

a molding or laminating process in which the pressure used is greater than 7 to 14 MPa (1000 to 2000 psi).

PRESSURE, LOW

In general, parts molded in the range of pressures from 2.8 MPa (400 psi).

PRESSURE PAD

A metallic reinforcing device designed to absorb pressure on the land areas of the mold when the mold is closed; reinforcement of hardened steel distributed around the dead/open area in the faces of a mold help the mold land absorb the final pressure of closing without collapsing.

PRESSURE REQUIRED

This is the unit pressure applied to the molding material in the mold cavity. The area is calculated from the projected area taken at right angles under pressure during complete closing of the mold, including areas of runners that solidify. The unit pressure is calculated by dividing the total force applied by this projected area. It is expressed in MPa (psi). To determine the pressure required for a specific material, the melt pressure is based on past experience and/or from the material supplier.

The pressure is multiplied by the projected area. The result is the total clamping pressure required. To ensure proper pressure is applied, consider using a safety factor (SF) of having available another 10% more pressure. This SF can be reduced or even eliminated by an operator with experience.

Pressure transducer

A pressure transducer is an instrument mounted in different parts of a mold (cavity, knockout pin, etc.) to measure melt pressure.

Quenched molding

It is a method of rapidly cooling TP molded parts as soon as they are removed from the mold. Submerging the parts in water is generally the approach to this method of cooling.

Release agent

Tool release agents are a substance placed on an interior mold cavity or die opening surface and/or added to a fabricating compound, such as silicone. A release agent's purpose is to facilitate removal or sliding of the fabricated product from the tool. Using silicone and certain other agents can cause adhesion problems if products are to be decorated or bonded in a secondary operation or may interfere with electrical circuits.

Rotary

A rotary is also called a rotary press. It refers to a type of IM, blow molding, compression molding, and so on, utilizing a plurality of mold cavities mounted on a rotating platen or table. This process is not to be confused with rotational molding.

Roto molding

Roto molding is powder molding in a hollow, closed mold that rotates on two axes.

Runner, balanced

A balanced runner exists in a multicavity mold when the runner's linear distances of the melt flow from the sprue to the cavity gates is the same.

Runner, cold (for TP)

In TPs, a cold runner is a mold in which the melt within the mold (sprue to gate) solidifies by the cooling action of the mold requiring their removal and runners are usually recycled.

RUNNER, COLD (FOR TS PLASTIC)

mold in which the melt within the mold (sprue to gate[s]) is cooled in the mold maintaining its free melt flowing characteristic so that the next shot starts from the gate(s) rather than the nozzle. The cavity and core plates are heated to solidify the plastic but the runner system is kept insulated from the cooler manifold section. This action eliminates TS scrap that is similar to a hot runner system for TPs.

RUNNER, HOT (FOR TP)

In TPs, a hot runner is a mold where the melt within the mold (sprue to gate) is insulated from the chilled cavity and core. They remain hot, producing no scrap, and the next shot starts from the gate rather than the nozzle.

RUNNER, HOT (FOR TS PLASTIC)

In TS plastics, a hot runner is a mold in which the melt within the mold (sprue to gate) is hot as it is in the cavity and core; all solidify by the heating action. The solid that travels from sprue to gate can be recycled at least as filler.

SCREW THREAD

It is easy and feasible to design a mold, though basically complicated. Three basic methods are used. One method is to use a core that is rotated after the molding cycle has been completed, thus unscrewing the part. The second method is to put the axis of the screw (only for external threads) at the mold's parting line, where potential flash can occur and a parting line could exist. The third method is to make the threads few, shallow, and of rounded form so that the part can be stripped from the mold without unscrewing.

SEMIPOSITIVE

A combination of the positive and flash (vertical or horizontal flash) compression molds. It operates as a flash mold until within a short distance of the final closure, when the force plug telescopes within the chase to exert a positive pressure on the charge during the final closing of the mold.

SHEAR EDGE

The shear edge is the cutoff edge of the mold.

SHEET MOLDING COMPOUND, VACUUM PRESS

Vacuum press molding (VPM) uses a vacuum assist in order to process at low molding pressures. With a molding compound that incorporates a physical thickening agent rather than a more conventional chemical thickening agent, low pressures of only 0.69 to 1.38 MPa (100 to 200 psi) are used. This type of sheet molding compound (SMC) is called low-pressure sheet molding compound (LPSMC).

SHRINKAGE

Shrinkage is the difference in dimensions between a plastic molding and the mold cavity in which it was molded, both being at room temperature when measured, expressed in cm/cm (in/in). Shrinkage usually occurs in the mold while it is solidifying or curing; however, certain plastics may take up to 24 hours at room temperature before they have completed their shrinkage. In designing a product and its mold, it is extremely important to make allowances for shrinkage. Each plastic material has its own shrinkage factor and plastic materials cannot usually be changed once the mold is built since the shrinkage will probably differ.

SIDE ACTION

A mold operates at an angle to the normal open-closed action permitting the removal of a part that would not clear a cavity or core; the mold may have a pin to core a hole that has to be withdrawn prior to opening the mold.

SINGLE IMPRESSION

This is a term used to describe a mold that has only one cavity.

SINK MARK

This is also called a shrink mark or an inverted blister. It is a shallow depression or dimple on the surface of a part due to the collapse of the surface following local internal shrinkage after the melt solidifies. This frequently occurs on the part face opposite to a face in which the section thickness increases, as in a rib.

SOLVENT MOLDING

This is a process for forming TPs by dipping a male mold into a solution or by dispersing the plastic and drawing off the solvent, leaving a layer of plastic film adhering to the mold that is later removed from the mold.

SPLIT RING

This is also called a split mold. A split mold has a split-cavity block that is assembled in a chase to permit the formation of undercuts in a molded part. These parts are ejected from the mold and then separated from the part.

SPEW GROOVE

This is the groove in a mold that permits the escape of excess or surplus plastic.

SPRAYED-METAL MOLDS

These are molds made by spraying molten metal onto a master until a shell of predetermined thickness is obtained. The shell is then removed and backed with plaster, cement, casting plastic, RP, or other types of reinforcements.

SPRUE

A sprue is also called a stalk. Feed opening in a mold that is directing melt into the mold from a molding nozzle.

SPRUE BUSHING

A part of the mold that provides an interface between the IMM nozzle and runner system in the mold.

SPRUE PICKER

This is also called a sprue puller. It is a device to remove solidified plastic sprue from an open mold.

STACK

A stack is a three-plate mold. Rather than the usual two-plate mold to handle a single mold, a stack has a third or intermediate movable plate. It makes center or offset gating of each cavity on two levels possible. Thus it is a two-level mold or two sets of cavities stacked one on top of the other for molding more parts per cycle (385). These molds generally use a hot runner manifold located in the center plate (platen). There are also four-stack molds in use.

STANDARD AND PRACTICE

The SPI continually updates *Standards and Practices of Plastics Molders*, which covers the issues surrounding the design of plastic molded parts. It is useful to designers, purchasing agents, custom molders, processors, and so on. Details presented include engineering and technical guidelines commonly used by molders for injection, compression, and transfer molding processes; lists tolerance specifications for plastic materials in metric and English units; and a glossary of terms. It reviews important commercial and administrative practices for purchasers to consider when specifying and purchasing molded parts. These customs of the trade include mold type, safety considerations, maintenance requirements, contract obligations, charges and costs, inspection limitations, storage, disposals, proper packing and shipping, and claims for defects.

Stop

This refers to devices such as steel blocks. They separate the mold so that they limit the amount of closure. It governs the thickness of the part. Since the plastic could receive less pressure, the part could contain voids.

Stripper plate

This plate strips molded parts from a cavity with or without air support.

Sweating

Molders can be faced with sweat and moisture condensation on their chilled mold surfaces, particularly during the summer months. This can lead to corrosion and rust, which results in poor finishes and parts of inferior quality. In addition, rust on guide pins can cause damage to the mold. By keeping the air in the plant and aroun the mold dry, you can not only improve part quality but also increase your production rate.

Taper, back

The back taper is the reverse draft or undercut used in a mold to prevent molded parts from being removed from the mold freely.

Threaded fastener

A fastener or insert, usually metal, that is molded into the part during the molding process.

Thermal expansion molding

Process in which an elastomeric mold is constrained within a rigid frame to generate consolidation pressure by thermal expansion while curing in an autoclave and other processes. During curing, the plastic expands.

THREE-PLATE MOLD

A three-plate mold is a cold runner mold with two parting lines: one for the runner and one for the products. The products can be gated at the center, and the mold opening action gives automatic part degrading with products and runners falling in different places.

TIME

This refers to curing time.

TRAY

A tray under the mold on to which moldings fall during automatic cycling.

TWO-SHELL

This is a technique to produce hollow parts by molding (injection, compression, blow molding, rotational molding, etc.) two halves with matting flanges—or the equivalent—which are in turn assembled by different techniques.

UNDERCUT

An undercut is a reverse or negative draft, such as a protuberance or indentation in a mold. They may require sliding cores or split molds. External undercuts can be placed at the parting line to obviate the need for core pins. Shallow undercuts often may be stripped from the mold without the need for core pulls. If the undercut is strippable, the other half of the mold must be removed first. Then the mold ejector pins can strip the part.

UNIT

Mold designee for quick changing interchangeable cavities.

VENTING, WATER TRANSFER

This technique is based on negative pressure coolant technology. Mold coolant is being pulled by vacuum; not the more conventional way of pushing under water pressure but having a negative pressure. This system permits venting into the water via the knockout pins or porous metal inserts, difficult locations in a cavity (such as long, thin cores) that entraps air during molding, and so on. They require that the pin or cavity (through a porous metal media) run through the water line. Coolant does not leak into the cavity because it is under atmospheric pressure. In an emergency, it could eliminate water leaks in a cracked mold that extends into the water line.

Water channel

Water channels, through which water circulates to cool the melt in the cavity, are designed to extract heat.

Weld line overflow tab

This is a small, localized extension of a part at a weld-line junction to allow a longer melt flow path for the purpose of obtaining a better fusion bond of the meeting melt fronts.

Well

In processes such as transfer or compression molding, a well is the space provided to take care of the difference in volume between loose molding material and the same weight of material after molding.

Witness line

A witness line is a line on a molded part that appears due to poor alignment or fit of metal-mating actions such as the sliding of cores.

Yoke

In a large, single-cavity mold, the entire cavity and core plates usually form the mold cavity. In a smaller, multicavity mold, the core and the cavity blocks (inserts) are mounted on or in the various plates of the mold base. When various components are mounted in the plates, the plates are called yokes.

APPENDIX

Other tool steels		Comparison
0.9% carbon, 0.8% chromium		W5
0.9% carbon chrome		W5
09B		O1
1% carbon		W1
10% cobalt		T6
100 CD 6		1.2067
100 Cr 6		1.2067
100 MnCrW 4		1.2510
100 V 1		1.2833
1005	(SAE)	0.06C, 0.3Mn
1006	(SAE)	0.08C, 0.3Mn
1008	(SAE)	0.1C, 0.35Mn
100C6		1.2067
1010	(SAE)	0.1C, 0.45Mn
1011	(SAE)	0.1C, 0.75Mn
1012	(SAE)	0.12C, 0.45Mn
1013	(SAE)	0.13C, 0.65Mn
1015	(SAE)	0.15C, 0.45Mn
1016	(SAE)	0.15C, 0.75Mn
1017	(SAE)	0.17C, 0.45Mn
1018	(SAE)	0.17C, 0.75Mn
1019	(SAE)	0.17C, 0.85Mn
102 Cr 6 KU		1.2067

Table 17.24 Tool materials with near-matching chemical compositions

Other tool steels		Comparison
102 V 2 KU		1.2833
1020	(SAE)	0.2C, 0.45Mn
1021	(SAE)	0.2C, 0.75Mn
1022	(SAE)	0.2C, 0.85Mn
1023	(SAE)	0.2C, 0.45Mn
1024	(SAE)	0.22C, 1.5Mn
1025	(SAE)	0.25C, 0.45Mn
1026	(SAE)	0.25C, 0.75Mn
1027	(SAE)	0.26C, 1.35Mn
1029	(SAE)	0.28C, 0.75Mn
1030	(SAE)	0.31C, 0.75Mn
1035	(SAE)	0.35C, 0.75Mn
1036	(SAE)	0.34C, 1.35Mn
1037	(SAE)	0.35C, 0.85Mn
1038	(SAE)	0.38C, 0.75Mn
1039	(SAE)	0.4C, 0.85Mn
1040	(SAE)	0.4C, 0.75Mn
1041	(SAE)	0.4C, 1.5Mn
1042	(SAE)	0.43C, 0.75Mn
1043	(SAE)	0.43C, 0.85Mn
10-4-3-10		1.3207
1044	(SAE)	0.46C, 0.45Mn
1045	(SAE)	0.46C, 0.75Mn
1046	(SAE)	0.46C, 0.85Mn
1047	(SAE)	0.47C, 1.5Mn
1048	(SAE)	0.48C, 1.25Mn
1049	(SAE)	0.49C, 0.75Mn
105 C 6		1.2067
105 MV 01		1.2833
105 WC 13		1.2419
105 WCr 5		1.2419
105 WCr 6		1.2419
1050	(SAE)	0.51C, 0.75Mn
1051	(SAE)	0.5C, 1Mn
1052	(SAE)	0.51C, 1.3Mn
1053	(SAE)	0.51C, 0.85Mn
1055	(SAE)	0.55C, 0.75Mn
1060	(SAE)	0.6C, 0.75Mn
1061	(SAE)	0.6C, 0.9Mn
1064	(SAE)	0.65C, 0.65Mn
1065	(SAE)	0.65C, 0.75Mn
1066	(SAE)	0.65C, 0.1Mn
1069	(SAE)	0.7C, 0.55Mn
107 CrV 3 KU		1.2210

Table 17.24 Tool materials with near-matching chemical compositions *(continued)*

Other tool steels		Comparison
107 WCr 5 KU		1.2419
1070	(SAE)	0.7C, 0.75Mn
1072	(SAE)	0.7C, 1.15Mn
1074	(SAE)	0.75C, 065Mn
1075	(SAE)	0.75C, 0.55Mn
1078	(SAE)	0.78C, 0.45Mn
1080	(SAE)	0.81C, 0.75Mn
1084	(SAE)	0.86C, 0.75Mn
1085	(SAE)	0.86C, 0.85Mn
1086	(SAE)	0.86C, 0.4Mn
1090	(SAE)	0.91C, 0.75Mn
1095	(SAE)	0.96C, 0.4Mn
115 CrV 3		1.2210
12 Ch 13		1.4006
12 S		H21
120 C 2		1.2002
12-1-5-5		1.3202
12M		1.2C, 12Mn
1330 ~H	(SAE)	0.3C, 1.75Mn
1335 ~H	(SAE)	0.35C, 1.75Mn
1340 ~H	(SAE)	0.4C, 1.75Mn
1345 ~H	(SAE)	0.45C, 1.75Mn
14 Ch 17 N 2		1.4057
14% tungsten		0.65C, 4Cr, 14W, 0.5V
14/4/4		1.25C, 4Cr, 14W, 4V
145 Cr 6		1.2063
14HD		H24
1-5-1		A2
1513	(SAE)	0.13C, 1.25Mn
1518	(SAE)	0.18C, 1.25Mn
1522	(SAE)	0.21C, 1.25Mn
1524	(SAE)	0.22C, 1.5Mn
1525	(SAE)	0.26C, 0.95Mn
1525	(W. Nr)	1.1525
1526	(SAE)	0.25C, 1.25Mn
1527	(SAE)	0.25C, 1.35Mn
1536	(SAE)	0.33C, 1.35Mn
1541	(SAE)	0.4C, 1.5Mn
1545	(W. Nr)	1.1545
1547	(SAE)	0.47C, 1.5Mn
1548	(SAE)	0.48C, 1.25Mn
1551	(SAE)	0.5C, 1Mn
1552	(SAE)	0.51C, 1.35Mn
1561	(SAE)	0.6C, 0.9Mn

Table 17.24 Tool materials with near-matching chemical compositions *(continued)*

Other tool steels		Comparison
1566	(SAE)	0.65C, 1Mn
1572	(SAE)	0.7C, 1.15Mn
1625	(W. Nr)	1.1625
1645	(W. Nr)	1.1645
1663	(W. Nr)	1.1663
1673	(W. Nr)	1.1673
1730	(W. Nr)	1.1730
1740	(W. Nr)	1.1740
1750	(W. Nr)	1.1750
18 HD		H26
18% tungsten		T1
18-0-1		1.3355
18-0-2-10		1.3265
18-1-1-5		1.3255
1820	(W. Nr)	1.1820
1830	(W. Nr)	1.1830
1880	(SS)	1.1545
19F		O6
1C25		0.25C, 0.6Mn
1C35		0.35C, 0.65Mn
1C45		0.45C, 0.65Mn
1C55		0.55C, 0.75Mn
1C60		0.60C, 0.75Mn
1P		0.43C, 1Si, 1.4Cr
1U		1.9C, 12.5Cr, 0.75Mo, 0.25V
2P		S1
2% tungsten		S1
20 Ch 13		1.4021
20 Ch 17 N 2		1.4057
20 plus		P20
2002	(W. Nr)	1.2002
2008	(W. Nr)	1.2008
2057	(W. Nr)	1.2057
2067	(W. Nr)	1.2067
2080	(W. Nr)	1.2080
2082	(W. Nr)	1.2082
2083	(W. Nr)	1.2083
20S		H10
21 MnCr 5		1.2162
2-10-1-8		1.3247
2127	(W. Nr)	1.2127
2140		1.2510
2162	(W. Nr)	1.2162
22% tungsten		0.8C, 4.2Cr, 22W, 1.3V

Table 17.24 Tool materials with near-matching chemical compositions *(continued)*

Other tool steels		Comparison
2210	(W. Nr)	1.2210
2234	(SS)	1.2330
2241	(W. Nr)	1.2241
2242	(SS)	1.2344
2244	(SS)	1.2332
2260	(SS)	1.2363
227		H21
2302	(SS)	1.4006
2303	(SS)	1.4021
2303	(W. Nr)	1.2303
2304	(SS)	1.4028
2310	(SS)	1.2601
2311	(W. Nr)	1.2311
2312	(SIS)	1.2436
2312	(W. Nr)	1.2312
2316	(W. Nr)	1.2316
2321	(SS)	1.4057
2330	(W. Nr)	1.2330
2332	(W. Nr)	1.2332
2341	(W. Nr)	1.234
2343	(W. Nr)	1.2343
2344	(W. Nr)	1.2344
2361	(W. Nr)	1.2361
2361	(W. Nr)	1.2361
2363	(W. Nr)	1.2363
2365	(W. Nr)	1.2365
2367	(W. Nr)	1.2367
2378	(W. Nr)	1.2378
2379	(W. Nr)	1.2379
23S		D3
2419	(W. Nr)	1.2419
2436	(W. Nr)	1.2436
2442	(W. Nr)	1.2442
24S		0.24C, 2.5Ni, 3Cr 8.5W
2510	(W. Nr)	1.2510
2515	(W. Nr)	1.2515
2542	(W. Nr)	1.2542
2547	(W. Nr)	1.2547
2550	(W. Nr)	1.2550
2562	(W. Nr)	1.2562
2567	(W. Nr)	1.2567
2581	(W. Nr)	1.2581
25S		0.27C, 4.25Ni, 1.5Cr, 6.75W
2601	(W. Nr)	1.2601

Table 17.24 Tool materials with near-matching chemical compositions *(continued)*

Other tool steels		Comparison
2606	(W. Nr)	1.2606
2631	(W. Nr)	1.2631
2663	(W. Nr)	1.2663
2678	(W. Nr)	1.2678
2710	(SS)	1.2542
2711	(W. Nr)	1.2711
2713	(W. Nr)	1.2713
2714	(W. Nr)	1.2714
2718	(W. Nr)	1.2718
2721	(W. Nr)	1.2721
2722	(SS)	1.3343
2723	(SS)	1.3243
2735	(W. Nr)	1.2735
2744	(W. Nr)	1.2744
2762	(W. Nr)	1.2762
2764	(W. Nr)	1.2764
2766	(W. Nr)	1.2766
2767	(W. Nr)	1.2767
2770	(W. Nr)	1.2770
2782	(SS)	1.3348
2787	(W. Nr)	1.2787
2826	(W. Nr)	1.2826
2833	(W. Nr)	1.2833
2842	(W. Nr)	1.2842
2885	(W. Nr)	1.2885
28Mn6		0.28C, 1.5Mn
2-9-2		1.3348
2-9-2-8		1.3249
293		H21
2C25		0.25C, 0.6Mn
2C35		0.35C, 0.65Mn
2C45		0.45C, 0.65Mn
2C55		0.55C, 0.75Mn
2C60		0.60C, 0.75Mn
30Ch 13		1.4028
30CrMoV 12 27 KU		1.2365
30CrMoV 12		1.2365
30CrNiMo8		0.30C, 2Cr, 0.4Mo, 2Ni
32DCV 28		1.2365
32NCD 16		1.2766
3202	(W. Nr)	1.3202
3207	(W. Nr)	1.3207
3243	(W. Nr)	1.3243
3246	(W. Nr)	1.3246

Table 17.24 Tool materials with near-matching chemical compositions *(continued)*

Other tools steels		Comparison
3247	(W. Nr)	1.3247
3249	(W. Nr)	1.3249
3255	(W. Nr)	1.3255
3265	(W. Nr)	1.3265
32B		A2
32CrMo12		0.32C, 3Cr, 0.4Mo, 0.3Ni
32S		A2
3342	(W. Nr)	1.3342
3346	(W. Nr)	1.3346
3348	(W. Nr)	1.3348
3355	(W. Nr)	1.3355
34CD 4		1.2330
34Cr4		0.34C, 0.75Mn, 1Cr
34CrMo4		0.34C, 0.65Mn, 1Cr, 0.25Mo
34NiCrMo16		0.34C, 1.8Cr, 0.35Mo, 4Ni
35 B		A6
35 CrMo 4		1.2330
35 HM		1.2330
3505	(W. Nr)	1.3505
351		0.52C, 0.7Si, 1.1Cr, 1.9W, 0.3V
3551	(W. Nr)	1.3551
3554LW	(W. Nr)	1.33554LW
35CrNiMo6		0.35C, 0.65Mn, 1.5Cr, 0.25Mo, 1.5Ni
37Cr4		0.37C, 0.75Mn, 1Cr
38Cr2		0.38C, 0.65Mn, 0.5Cr
38Cr4		0.38C, 0.75Mn, 1Cr
39NiCrMo3		0.39C, 0.65Mn, 0.75Cr, 0.25Mo, 0.85Ni
3C25		0.25C, 0.6Mn
3C35		0.35C, 0.65Mn
3C45		0.45C, 0.65Mn
3C55		0.55C, 0.75Mn
3C60		0.60C, 0.75Mn
3Ch2W8F		1.2581
3Ch3M3F		1.2365
40 CDM 8		1.2311
40 CDM 8+S		1.2312
40 Ch 13		1.4034
40 CrMnMo 7		1.2311
40 CrMnMoS 8 6		1.2312
40 CrMo 4		1.2332
4005	(W. Nr)	1.4005
4006	(W. Nr)	1.4006
4012	(SAE)	0.11C, 0.85Mn, 0.2Mo
4014LW	(W. Nr)	1.4014LW

Table 17.24 Tool materials with near-matching chemical compositions *(continued)*

Other tool steels		Comparison
4021	(W. Nr)	1.4021
4023	(SAE)	0.22C, 0.8Mn, 0.25Mo
4024	(SAE)	0.22C, 0.8Mn, 0.25Mo
4027 ~H	(SAE)	0.27C, 0.8Mn, 0.25Mo
4028 ~H	(SAE)	0.27C, 0.8Mn, 0.25Mo
4028	(W. Nr)	1.4028
4032 ~H	(SAE)	0.32C, 0.8Mn, 0.25Mo
4034	(W. Nr)	1.4034
4037 ~H	(SAE)	0.37C, 0.8Mn, 0.25Mo
4042 ~H	(SAE)	0.42C, 0.8Mn, 0.25Mo
4047 ~H	(SAE)	0.47C, 0.8Mn, 0.25Mo
4057	(W. Nr)	1.4057
40Ch13		1.2083
40NiCrMo2		0.4C, 0.85Mn, 0.5Cr, 0.25Mo, 0.55Ni
40NiCrMo3		0.4C, 0.75Mn, 0.75Cr, 0.25Mo, 0.85Ni
410S21		1.4006
410S22		1.4006
4110	(W. Nr)	1.4110
4112	(W. Nr)	1.4112
4118 ~H	(SAE)	0.2C, 0.8Mn, 0.5Cr, 0.11Mo
4130 ~H	(SAE)	0.3C, 0.5Mn, 0.95Cr, 0.2Mo
4135 ~H	(SAE)	0.35C, 0.8Mn, 0.95Cr, 0.2Mo
4137 ~H	(SAE)	0.37C, 0.8Mn, 0.95Cr, 0.2Mo
4140 ~H	(SAE)	0.4C, 0.85Mn, 0.95Cr, 0.2Mo
4140	(W. Nr)	1.4140
4142 ~H	(SAE)	0.42C, 0.85Mn, 0.95Cr, 0.2Mo
4145 ~H	(SAE)	0.45C, 0.85Mn, 0.95Cr, 0.2Mo
4147 ~H	(SAE)	0.47C, 0.85Mn, 0.95Cr, 0.2Mo
4150 ~H	(SAE)	0.5C, 0.85Mn, 0.95Cr, 0.2Mo
416 Se		416
4161 ~H	(SAE)	0.6C, 0.85Mn, 0.8Cr, 0.3Mo
416S21		416
416S29		416
416S37		416
416S41		416
41Cr4		0.41C, 0.65Mn, 1Cr
41CrMo4		0.41C, 0.75Mn, 1Cr, 0.25Mo
42 CD 4		1.2332
420J2		1.2083
420S29		1.4021
420S37		420
420S45		1.4028
42CrMo4		0.42C, 0.65Mn, 1Cr, 0.25Mo
431 Se		431

Table 17.24 Tool materials with near-matching chemical compositions *(continued)*

Other tool steels		Comparison
4320 ~H	(SAE)	0.19C, 0.55Mn, 0.5Cr, 0.25Mo, 1.8Ni
4340 ~H	(SAE)	0.4C, 0.7Mn, 0.8Cr, 0.25Mo, 1.8Ni
4419 ~H	(SAE)	0.2C, 0.55Mn, 0.5Mo
441S49		431
4422	(SAE)	0.22C, 0.8Mn, 0.4Mo
4427	(SAE)	0.26C, 0.8Mn, 0.4Mo
45 NCD 17		1.2767
45 WCrSi 8		1.2542
45 WCrV 7		1.2542
45 WCrV 8KU		1.2542
4528	(W. Nr)	1.4528
45Cr2		0.45C, 0.65Mn, 0.5Cr
4615	(SAE)	0.15C, 0.55Mn, 0.25Mo, 1.8Ni
4617	(SAE)	0.17C, 0.55Mn, 0.25Mo, 1.8Ni
4620 ~H	(SAE)	0.19C, 0.55Mn, 0.25Mo, 1.8Ni
4621 ~H	(SAE)	0.2C, 0.8Mn, 0.25Mo, 1.8Ni
4626 ~H	(SAE~AIS)	0.26C, 0.55Mn, 0.2Mo, 0.85Ni
46Cr2		0.46C, 0.65Mn, 0.5Cr
47 CrMo 4		1.2332
47 Prima		W1
4718 ~H	(SAE)	0.18C, 0.8Mn, 0.45Cr, 0.35Mo, 1.05Ni
4720 ~H	(SAE)	0.19C, 0.6Mn, 0.45Cr, 0.2Mo, 1.05Ni
476 Special	(Mo Type)	D4
476 Specila	(W Type)	D6
476		D2
4815 ~H	(SAE)	0.15C, 0.5Mn, 0.25Mo, 3.5Ni
4817 ~H	(SAE)	0.17C, 0.5Mn, 0.25Mo, 3.5Ni
4820 ~H	(SAE)	0.2C, 0.6Mn, 0.25Mo, 3.5Ni
4CH5MF1S		1.2344
4CH5MFS		1.2343
4NHD		0.35C, 1.5Cr, 5.75W, 0.35V, 4Ni
4S28		1.5864
4WHD		0.35C, 1.1Si, 4W, 1.2Cr, 0.25V
5 CTDS		H12
5% cobalt		T4
50 CMV 4		1.2241
50 MCD 5		0.5C, 1.2Mn, 0.65Cr, 0.2Mo
50 NiCr 13		1.2721
50 WCV 22		1.2547
50100		1.05C, 0.35Mn, 0.5Cr
5015	(SAE)	0.14C, 0.4Mn, 0.4Cr
503A37		0.37C, 0.7Mn, 0.85Ni
503A42		0.42C, 0.7Mn, 0.85Ni
503H37		0.37C, 0.85Mn, 0.85Ni

Table 17.24 Tool materials with near-matching chemical compositions *(continued)*

Other tool steels		Comparison
503H42		0.42C, 0.85Mn, 0.85Ni
503M40		0.4C, 0.85Mn, 0.85Ni
5046 -H	(SAE)	0.45C, 0.85Mn, 0.27Cr
5060 -H	(SAE)	0.6C, 0.85Mn, 0.5Cr
50B40 -H		0.4C, 0.85Mn, 0.5Cr
50B44 -H		0.45C, 0.85Mn, 0.5Cr
50B46 -H		0.46C, 0.85Mn, 0.25Cr
50B50 -H		0.5C, 0.85Mn, 0.5Cr
50B60 -H		0.6C, 0.85Mn, 0.5Cr
50CrV4		0.5C, 0.9Mn, 1Cr, 0.15V
51100		1.05C, 0.35Mn, 1Cr
5115 -H	(SAE)	0.15C, 0.8Mn, 0.8Cr
5120 -H	(SAE)	0.19C, 0.8Mn, 0.8Cr
5130 -H	(SAE)	0.3C, 0.8Mn, 0.95Cr
5132 -H	(SAE)	0.32C, 0.7Mn, 0.85Cr
5135 -H	(SAE)	0.35C, 0.7Mn, 0.9Cr
5140 -H	(SAE)	0.4C, 0.8Mn, 0.8Cr
51410		410
51416Se		416
51420		420
51420F		420
51420FSe		420
51431		431
51440A		440A
51440B		440B
51440C		440C
51440F		1.1C, 1.2Mn, 17Cr, 0.7Mo
51440FSe		1.1C, 1.2Mn, 17Cr, 0.7Mo
5145 -H	(SAE)	0.45C, 0.8Mn, 0.8Cr
5147 -H	(SAE)	0.48C, 0.85Mn, 1Cr
5150 -H	(SAE)	0.5C, 0.8Mn, 0.8Cr
5155 -H	(SAE)	0.55C, 0.8Mn, 0.8Cr
5160 -H	(SAE)	0.6C, 0.85Mn, 0.8Cr
51B60 -H		0.6C, 0.85Mn, 0.8Cr
52100		1.05C, 0.35Mn, 1.45Cr
526		M2
526M60		0.6C, 0.65Mn, 0.65Cr
530A30		0.3C, 0.7Mn, 1Cr
530A32		0.32C, 0.7Mn, 1Cr
530A36		0.36C, 0.7Mn, 1Cr
530A40		0.4C, 0.7Mn, 1Cr
530H30		0.3C, 0.7Mn, 1Cr
530H32		0.32C, 0.7Mn, 1Cr
530H36		0.36C, 0.7Mn, 1Cr

Table 17.24 Tool materials with near-matching chemical compositions *(continued)*

Other tool steels		Comparison
530H40		0.4C, 0.7Mn, 1Cr
530M40		0.4C, 0.75Mn, 1Cr
534A99		1C, 1.4Cr
535A99		1C, 0.6Mn, 1.4Cr
53S		H13
54 NiCrMo V6		1.2711
55 NCD 13		1.2721
55 NCDV 6		1.2711
55 NCDV 7		1.2713
55 NiCr 10		1.2718
55 NiCrMoV 6		1.2713
55 WC 20		1.2542
55 WCrV 8KU		1.2550
56 NiCrMoV 7		1.2714
57 NiCrMoV 7 7		1.2744
58 S		H12
58 WCr9KU		1.2550
5864	(W. Nr)	1.5864
5CC		A2
5ChNM		1.2713
6/5/2		M2
60 CrSi 8		1.2550
60MnSiCr 4		1.2826
60 WCrV 7		1.2550
60 WCS 20		1.2550
605A32		0.32C, 1.5Mn, 0.27Mo
605A37		0.37C, 1.5Mn, 0.27Mo
605H32		0.32C, 1.5Mn, 0.27Mo
605H37		0.37C, 1.5Mn, 0.27Mo
605M30		0.3C, 1.5Mn, 0.27Mo
605M36		0.36C, 1.5Mn, 0.27Mo
606M36		0.36C, 1.5Mn, 0.27Mo
608H37		0.37C, 1.5Mn, 0.5Mo
608M38		0.38C, 1.5Mn, 0.5Mo
6118 ~H	(SAE)	0.18C, 0.6Mn, 0.6Cr, 0.12V
6150 ~H	(SAE)	0.5C, 0.8Mn, 0.95Cr, 0.15V
640A35		0.35C, 0.75Mn, 0.65Cr, 1.3Ni
640H35		0.35C, 0.75Mn, 0.65Cr, 1.25Ni
640M40		0.4C, 0.75Mn, 0.65Cr, 1.3Ni
65 NCD 6		1.2714
6-5-2		1.3343
6-5-2-5		1.3243
653M31		0.31C, 0.6Mn, 1Cr, 3Ni
6542 Molybdenum (Cook)		M2

Table 17.24 Tool materials with near-matching chemical compositions *(continued)*

Other tool steels		Comparison
6582	(W. Nr)	1.6582
6747	(W. Nr)	1.6747
67S		F3
69S		D2
6ChV25		1.2550
7 PCR		W4
70 MDC 8		A6
708A37		0.37C, 0.85Mn, 1Cr, 0.2Mo
708A42		0.42C, 0.85Mn, 1Cr, 0.2Mo
708H37		0.37C, 0.85Mn, 1Cr, 0.2Mo
708H42		0.42C, 0.85Mn, 1Cr, 0.2Mo
708M40		0.4C, 0.85Mn, 1Cr, 0.2Mo
709M40		0.4C, 0.85Mn, 1Cr, 0.3Mo
722M24		0.24C, 0.6Mn, 3.25Cr, 0.55Mo
7-4-2-5		1.3246
74S		0.3C, 2.3Cr, 0.3Mo, 4.25W, 0.6V
75 CrMoNiW 6 7		1.2762
75		1.1750
785M19		0.19C, 1.6Mn, 0.25Mo, 0.55Ni
80		1.1625
8115	(SAE)	0.15C, 0.8Mn, 0.4Cr, 0.11Mo, 0.3Ni
816M40		0.4C, 0.6Mn, 1.2Cr, 0.15Mo, 1.5Ni
817M40		0.4C, 0.6Mn, 1.2Cr, 0.3Mo, 1.5Ni
81B45 ~H		0.45C, 0.85Mn, 0.45Cr, 0.11Mo, 0.3Ni
823M30		0.3C, 2Cr, 0.4Mo, 2Ni
826M31		0.31C, 0.65Cr, 0.55Mo, 2.5Ni
826M40		0.4C, 0.65Cr, 0.55Mo, 2.5Ni
82S		0.33C, 1Si, 3Cr, 2.8Co, 1W, 2.8Mo, 0.9V
830M31		0.31C, 1Cr, 0.3Mo, 3Ni
835M30		0.3C, 1.25Cr, 0.3Mo, 4.1Ni
85 NiV4		0.85C, 0.8Ni, 0.1V
8615	(SAE)	0.15C, 0.8Mn, 0.5Cr, 0.2Mo, 0.55Ni
8617 ~H	(SAE)	0.17C, 0.8Mn, 0.5Cr, 0.2Mo, 0.55Ni
8620 ~H	(SAE)	0.2C, 0.8Mn, 0.5Cr, 0.2Mo, 0.55Ni
8622 ~H	(SAE)	0.22C, 0.8Mn, 0.5Cr, 0.2Mo, 0.55Ni
8625 ~H	(SAE)	0.25C, 0.8Mn, 0.5Cr, 0.2Mo, 0.55Ni
8627 ~H	(SAE)	0.27C, 0.8Mn, 0.5Cr, 0.2Mo, 0.55Ni
8630 ~H	(SAE)	0.3C, 0.8Mn, 0.5Cr, 0.2Mo, 0.55Ni
8637 ~H	(SAE)	0.37C, 0.85Mn, 0.5Cr, 0.2Mo, 0.55Ni
8640 ~H	(SAE)	0.4C, 0.85Mn, 0.5Cr, 0.2Mo, 0.55Ni
8642 ~H	(SAE)	0.42C, 0.85Mn, 0.5Cr, 0.2Mo, 0.55Ni
8645 ~H	(SAE)	0.45C, 0.85Mn, 0.5Cr, 0.2Mo, 0.55Ni
8650 ~H	(SAE)	0.5C, 0.85Mn, 0.5Cr, 0.2Mo, 0.55Ni
8655 ~H	(SAE)	0.55C, 0.85Mn, 0.5Cr, 0.2Mo, 0.55Ni

Table 17.24 Tool materials with near-matching chemical compositions *(continued)*

Other tool steels		Comparison
8660 ~H	(SAE)	0.6C, 0.85Mn, 0.5Cr, 0.2Mo, 0.55Ni
86B30H		0.3C, 0.75Mn, 0.5Cr, 0.2Mo, 0.55Ni
86B45 ~H		0.45C, 0.85Mn, 0.5Cr, 0.2Mo, 0.55Ni
8720 ~H	(SAE)	0.2C, 0.8Mn, 0.5Cr, 0.25Mo, 0.55Ni
8740 ~H	(SAE)	0.4C, 0.85Mn, 0.5Cr, 0.25Mo, 0.55Ni
8822 ~H	(SAE)	0.22C, 0.85Mn, 0.5Cr, 0.35Mo, 0.55Ni
88HP		T1
897M39		0.39C, 3.25Cr, 1Mo, 0.2V
90 MCV 8		1.2842
90 MnCrV 8		1.2842
90 MnVCr 8 KU		1.2842
90 MV 8		1.2842
90 MWCV 5		1.2510
905M31		0.31C, 1.6Cr, 0.2Mo, 1.1Al
905M39		0.39C, 1.6Cr, 0.2Mo, 1.1Al
9254	(SAE)	0.55C, 1.4Si, 0.7Mn, 0.7Cr
9255	(SAE)	0.55C, 2Si, 0.8Mn
9257	(Cook)	W2
9260 ~H	(SAE)	0.6C, 2Si, 0.85Mn
9310 ~H	(SAE)	0.1C, 0.55Mn, 1.2Cr, 0.11Mo, 3.25Ni
945A40		0.4C, 1.4Mn, 0.5Cr, 0.2Mo, 0.75Ni
945M38		0.38C, 1.4Mn, 0.5Cr, 0.2Mo, 0.75Ni
94B15 ~H		0.15C, 0.85Mn, 0.4Cr, 0.11Mo, 0.45Ni
94B17 ~H		0.17C, 0.85Mn, 0.4Cr, 0.11Mo, 0.45Ni
94B30 ~H		0.3C, 0.85Mn, 0.4Cr, 0.11Mo, 0.45Ni
95 MCWV 5		1.2510
95 MnCrW 5		1.2510
95 MnWCr 5 KU		1.2510
9Ch1		1.2067
9Ch5VF		1.2363
9ChVG		1.2510
A 100		0.32C, 0.6Mn, 0.65Cr, 2.5Ni, 0.55Mo
A 13		0.4C, 1.15Cr, 1.5Ni, 0.3Mo, 0.6Mn
A11		D3
A25CrMo4		0.25C, 0.65Mn, 1Cr, 0.25Mo
A6		D2
AB 213		0.33C, 1Cr, 0.3Ni, 1.75W
AB 75		S4
ADIC		H13
ADS		H11
AGS		L1
AH chrome die		A2
AHA		1.2542
AHK		1.2550

Table 17.24 Tool materials with near-matching chemical compositions *(continued)*

Other tool steels		Comparison
Alloy C WPS		D3
Alloy mold		1.2766
Alum die mold		H11
ALZ		H13
AM1		0.35C, 1Si, 1.35W, 5Cr, 1.5Mo, 0.45V
AM3		0.4C, 1Si, 5Cr, 1.35Mo, 1V
Amutit S		1.2510
Amutit		1.2419
Ardho No 2		0.47C, 1.1Ni, 0.6Cr
ARH Medium		420
ARH Tough		420
Ark Superior Triumph Superb		T1
Ark Superlative Triumph Superb 1000		T4
ARL.GB		431
Arne		1.2510
ARW		410
ARWS.GB		416
ASP 23		1.27C, 0.5Si, 0.3Mn, 4.2Cr, 6.4W, 5Mo,
3.1V		
Astra		0.55, 1Cr, 2W
AT		0.45C, 1.2Si, 0.7Mn, 3Ni, 0.5Cr
Austenite		0.65C, 14W, 4Cr, 0.5V
AW		0.73C, 14W, 4Cr, 1V
AZ		H21
B 20V		1.1545
B10		1.1730
B25CrMo4		0.25C, 0.65Mn, 1Cr, 0.25Mo
BA 2		1.2363
BA 500		0.33C, 1.15Cr, 0.25Mo, 4.15Ni
BA 6		0.7C, 2Mn, 1Cr, 1.4Mo
BCC	(Carr)	0.5C, 1Si, 1.2Cr, 2.25W, 0.25V
BCC	(Huntsman)	D5
BCD	(Carr)	S1
BCD	(Huntsman)	D3
BCD37		0.37C, 1.6Cr, 0.3V, 2W
BCHV		1.8C, 12.5Cr, 0.25V, 0.8Mo
BCRS		D1
BCW		D6
BD 2A		1.75C, 12.5Cr, 0.8Mo, 0.6V
BD 2		1.2379
BD 3		1.2080
BD2A		D2A
Benum		0.3C, 1.25Cr, 4.1Ni

Table 17.24 Tool materials with near-matching chemical compositions *(continued)*

Other tool steels	Comparison
Best Warranted Cast Steel	W1
BF 1	1.25C, 0.4Cr, 0.3V, 1.45W
BH 10	1.2365
BH 10A	H 10A
BH 11	1.2343
BH 12	1.2606
BH 13	1.2344
BH 19	0.4C, 4.25Co, 4.25Cr, 0.45Mo, 2.2V, 4.25W
BH 21 A	0.25C, 2.75Cr, 0.6Mo, 2.25Ni, 0.5V, 9.25W
BH 21	1.2581
BH 26	0.55C, 0.6Co, 4.1Cr, 0.6Mo, 1.25V, 18W
BKV	D3
BL 3	1.2067
Black Label Steels	W1
Blue Band Cast Steel	W1
Blue Label Steels	W1
Blue Label	W2
BM 1	1.3346
BM 15	1.50C, 5Co, 4.75Cr, 3Mo, 5V, 6.6W
BM 2	1.3343
BM 34	1.3249
BM 4	1.3C, 0.6Co, 4.1Cr, 4.75Mo, 4V, 6.1W
BM 42	1.3247
BO 1	1.2510
BO 2	1.2842
BR 1	L5
BS 1	1.2542
BS 2	0.5C, 1Si, 0.45Mo, 0.2V
BS 224	1.2711
BS 5	0.55C, 1.85Si, 0.7Mn, 0.45Mo, 0.2V
BS50 Extra	0.54C, 1.1Cr, 4Ni, 0.3Mo, 0.5W
BS50	0.5C, 1.1Cr, 3.3Ni
BSS	M1
BST	1.2767
BT 1	1.3355
BT 15	1.3202
BT 2	0.8C, 0.6Co, 4.1Cr, 0.7Mo, 1.9V, 18W
BT 20	0.8C, 0.6Co, 4.6Cr, 1Mo, 1.5V, 21.75W
BT 21	0.65C, 0.6Co, 3.8Cr, 0.7Mo, 0.5V, 14W
BT 4	1.3255
BT 42	1.3207
BT 5	1.3265
BT 6	0.8C, 11.75Co, 4.1Cr, 1Mo, 1.5V, 20.5W

Table 17.24 Tool materials with near-matching chemical compositions *(continued)*

Other tool steels	Comparison
BV	T6
BW 1A	1.1750
BW 1B	1.1625
BW 1C	1.20C, 0.15Cr, 0.1Mo, 0.2Ni
BW 2	1.2833
C 100 KU	1.1545
C 102	1.1645
C 105 W 1	1.1545
C 105 W 2	1.1645
C 120 KU	1.1663
C 120	1.1663
C 1215	D2
C 1220	D3
C 125 W	1.1663
C 135 W	1.1673
C 140 KU	1.1673
C 45 W	1.1730
C 60 W	1.1740
C 75 W	1.1750
C 80 KU	1.1525
C 80 W 1	1.1525
C 80 W 2	1.1625
C 80	1.1625
C 85 W	1.1830
C.M.V.	1.2344
C2–C6	W1
C36	0.36C, 0.65Mn
C46	0.46C, 0.65Mn
C53	0.53C, 0.6Mn
CA 510	A2
CA1220	2.1C, 1.2Cr, 0.7W
CA71215	1.65C, 12Cr, 0.5W, 0.7Mo
Calmax	0.6C, 0.8Mn, 4.5Cr, 0.5Mo, 0.2V
Capital 305	M1
Capital 398	M35
Capital 405	M42
Capital 50	M50
Capital 562	M2
Cast Steel	W1
CCM 10	T6
CCMP 15	0.81C, 22W, 4.75Cr, 1.6V, 15Co, 0.5Mo
CCR 350	0.75C, 3.5Cr
CCR 50	W5
CCR130	L1

Table 17.24 Tool materials with near-matching chemical compositions *(continued)*

Other tool steels	Comparison
CCR2	0.6C, 0.6Cr
CCV	0.4C, 0.65Mn, 1.25Cr, 0.15V
CCW	S1
CCZ	H10A
CD	D3
CD3	F2
CDD	1.9C, 0.6Cr, 6W
CDS2	H13
CDV 4	D5
CDV	H13
CDV1	D4
CDV2	D2
Celfor	W1
CGH	0.42C, 2.4Cr, 0.15V, 0.25Mo, 0.8Ni
Ch12	1.2080
Ch12F1	1.2379
Ch12M	1.2601
CHD 7	W1
CHD	H12
CHD2	H14
CHD3	H12
CHD4	0.65C, 3.75Cr, 0.45V, 0.6Mo
Chrom Special	1.2080
Chromo Triple Three	0.32C, 3Cr, 0.9V, 3Co, 2.8Mo, 1W
ChWG	1.2419
CJM	P20
CKK	L1
CL 15	0.4C, 3.25Ni
CL 222	0.35C, 0.8Cr, 0.5W, 0.7Mo
CL 225	0.25C, 1.25Ni, 0.5Cr, 0.25Mo
CL 244	0.55C, 1.5Ni, 0.75Cr, 0.3Mo
CL 40T	S4
CL 40X	S5
CL 444	H12
CL 45	L2
CL 562	M2
CL 60	0.6C, 0.7Mn, 0.6Cr
CL 99	0.4C, 1.2Cr, 0.25Mo, 4.2Ni
CM 1255	T15
CM Extra 150	0.55C, 0.65Mn, 0.65Cr, 0.3Mo, 1.5Ni
CM Extra 300	0.31C, 0.6Cr, 0.6Mo, 2.6Ni
CM Extra 450	0.41C, 0.6Cr, 0.6Mo, 2.55Ni
CM	0.4C, 1.15Cr, 0.27Mo, 1.55Ni
CM5	T4

Table 17.24 Tool materials with near-matching chemical compositions *(continued)*

Other tool steels		Comparison
CMC	(Cook)	O2
CMC	(Thomas Turton)	M42
CMI		0.07C, 5Cr, 0.3V, 1Mo
CMV		H13
CMVM		0.5C, 0.65Mn, 0.95Cr, 0.2V
CMW		H12
Co 500		T4
Co 512		T8
Cobalt Extra Special		T6
Cobaltcrom		D5
Cobra		0.29C, 7W, 2.13Cr, 0.25V, 0.75Mo, 4.75Co
Common Hardening		0.55C, 0.7Mn
Conqueror 14%		T7
Conqueror LC		H24
Conqueror Tempers 1–6		W1
Constant Special		1.2842
Covas		T15
CPM 10V		2.45C, 5.25Cr, 10V, 1.3Mo
CPM 3V		0.8C, 7.5Cr, 1.3Mo, 2.75V
CPM 420V		2.2C, 13Cr, 1Mo, 9V
CPM 9V		1.8C, 0.5Mn, 0.9Si, 5.25Mo, 1.3Mo, 9V
CPM M2 HSHC		M2
CPM REX 20		M62
CPM REX 25		M61
CPM REX 76		M48
CPM REX M35S		M35
CPM REX M3HS		M3
CPM REX M42		M42
CPM REX M45		1.3C, 4Cr, 5Mo, 3V, 6.25W, 8Co
CPM REX M4HC HS		M4
CPM REX T15		T15
CPM T440V		2.15C, 0.4Mn, 17Cr, 0.4Mo, 5.5V
CR 80		D1
CRD Special		1.2601
CRD Supra		1.2379
CRD		1.2080
CRLS		1.2363
CRM2		0.6C, 0.6Cr, 0.3Mo
CRMI		A2
Cromax		0.45C, 1.30Mn, 0.7Cr, 0.2Mo
Cromodie HC		A2
Cromodie HCV		A7
Cromodie W		H12
Cromodie		H13

Table 17.24 Tool materials with near-matching chemical compositions *(continued)*

Other tool steels		Comparison
Crovaco 14		1.12C, 13.5W, 4.35Cr, 4.1V, 0.65Mo, 4.8Co
CRP		O1
CRU-Die 2		0.55C, 1.2Cr, 1.65Ni, 0.5Mo, 0.2V
CRV 14		0.68C, 14.5W, 3.75Cr, 0.65V
CRV 1444		1.2C, 13.75W, 4.4Cr, 3.75V, 0.5Mo
CS 13M Extra		D2
CS 13M		D3
CSM 420		1.2083
CSM2		P20
CSP		O1
CSV 4		0.38C, 1.5Cr, 0.1V, 1.5Si
CSV 5		0.45C, 1.5Cr, 0.1V, 1.5Si
CSV 6		0.61C, 1.2Cr, 0.1V, 0.9Si
CT 150		H12
CT		H20
CTU		0.37C, 1.1Si, 5Cr, 1.75W, 1V, 1.30Mo
CV	(Barworth)	0.45C, 2.5Cr, 0.2V
CV	(Osborne)	W2
CV	(Stone)	L2
CV 18		T1
CV 22		0.79C, 22W, 4.5Cr, 1.4V
CV Punch and Die		L2
CV1842		T2
CVHD		0.6C, 0.7Cr, 0.2V
CVM		H11
CVM2		H12
CVM3		H13
CVM4		0.58C, 0.9Si, 5Cr, 1.1V, 1.35Mo
CVM6		0.38C, 0.9Si, 2W, 5Cr, 0.95V, 1.75Mo
Cyclone 4V		M4
Cyclone 56		M2
Cyclone 92		M1
Cyclone 92CW		M1
Cyclone DG		T4
Cyclone MC33		M33
Cyclone MC42		M42
Cyclone MC46		M46
Cyclone MC50		M50
Cyclone MC6		M35
Cyclone MC7		M7
D 421		S1
DAC		1.2344
DBC		1.2606

Table 17.24 Tool materials with near-matching chemical compositions *(continued)*

Other tool steels	Comparison
DBS	1.2721
DC 1	1.2080
DC 11	1.2379
DC 3	1.2379
DC12	1.2363
DCM	A2
DDW	1.2567
DE-CP10V	1.2080
DE-CPPK	1.2601
DE-CPPU	1.2379
DE-CPR	1.2C, 12Cr, 1.2Mo
DE-CPV2	1.2378
DE-CPW	1.2436
DE-DMo5	1.3343
DE-EMo12	1.3207
DE-G42	1.1525
DE-G43	1.1545
DE-NHF	1.2542
DE-NHW	1.2550
DE-PS	1.2063
DE-VNC4M	1.2767
DE-WP7V	0.5C, 1Si, 0.8Mn, 8.3Cr, 0.3Ni
DE-Z1b	1.2842
DE-Z2T	1.2419
DH21	1.2344
DH2F	1.2344
DH4	1.2567
DH5	1.2581
DH6	1.2343
DH62	1.2606
DH72	1.2365
DHA1	1.2344
DHS	1.3343
Diehard HCD	D4
Diehard LC	0.8C, 12Cr, 0.5Mo, 0.5V
Diehard Standard	D2
DJS	1.3346
DJS	M1
DJSY	M7
DOH	0.9C, 1.7Mn
DOM 5	A2
DOM MW	D2A
DOM VM	D2
Dominant Extra Superb	T5

Table 17.24 Tool materials with near-matching chemical compositions *(continued)*

Other tool steels	Comparison
Dominant Hierom	D3
Dominant OH	0.95C, 1Cr, 1Mn, 0.5W
Dominant SA	0.4C, 1.75W, 1.75Cr, 0.3V
Dominant Special	0.68C, 14W, 3.5Cr, 0.5V
Dominant Superb	T4
Dominant SX	D4
Dominator	D3
Double Conqueror Vanadium	W2
Double Extra	W1
Double Griffin	0.35C, 1.75Cr, 3.5Ni
Double Horseman	T4
Double Rapid	T4
Double Seven	D5
Double Shear Temper	W1
Double Six	D3
Dreadnought 30BW	0.3C, 1.5Cr, 3.75, 0.3V, 5.75W
Dreadnought 5/6/2	M2
Dreadnought CTVM	M15
Dreadnought FP4T	0.3C, 2.3Cr, 0.3Mo, 0.6V, 4.3W
Dreadnought FPHD(N)	H21
Dreadnought FPHD	H21
Dreadnought M42	M42
Dreadnought Select	T1
Dreadnought Superior	T4
Dreadnought Supreme	T6
DS 122	D3
DS 133	D4
DS 144	D2
DS 200	O1
DS 400	L4
DS 600	O2
DUX 4	S1
DVS	M3
E	0.65C, 14.25W, 3.75Cr, 0.5V
E4340 ~H	0.4C, 0.75Mn, 0.8Cr, 0.25Mo, 1.8Ni
E51100	1.05C, 0.35Mn, 1Cr
E52100	1.05C, 0.35Mn, 1.45Cr
EE	H24
EF	O2
Electem	0.55C, 0.7Mn, 0.75Cr, 0.28Mo, 1.5Ni
EMS 45	1.1730
EMS 60	1.1740
EMS 85	1.1830
EN 100	0.38C, 1.4Mn, 0.5Cr, 0.2Mo, 0.75Ni

Table 17.24 Tool materials with near-matching chemical compositions *(continued)*

Other tool steels	Comparison
EN 100C	0.4C, 1.4Mn, 0.5Cr, 0.2Mo, 0.75Ni
EN 11	0.6C, 0.65Mn, 0.65Cr
EN 110	0.4C, 0.6Mn, 1.2Cr, 0.15Mo, 1.5Ni
EN 111	0.4C, 0.75Mn, 0.65Cr, 1.3Ni
EN 111A	0.35C, 0.75Mn, 0.65Cr, 1.3Ni
EN 12	0.4C, 0.85Mn, 0.85Ni
EN 12B	0.37C, 0.7Mn, 0.85Ni
EN 12C	0.42C, 0.7Mn, 0.85Ni
EN 13	0.19C, 1.6Mn, 0.25Mo, 0.55Ni
EN 16	0.36C, 1.5Mn, 0.27Mo
EN 16B	0.32C, 1.5Mn, 0.27Mo
EN 16C	0.37C, 1.5Mn, 0.27Mo
EN 16D	0.3C, 1.5Mn, 0.27Mo
EN 16M	0.36C, 1.5Mn, 0.27Mo
EN 17	0.38C, 1.5Mn, 0.5Mo
EN 18	0.4C, 0.75Mn, 1Cr
EN 18A	0.3C, 0.7Mn, 1Cr
EN 18B	0.32C, 0.7Mn, 1Cr
EN 18C	0.36C, 0.7Mn, 1Cr
EN 18D	0.4C, 0.7Mn, 1Cr
EN 19	0.4C, 0.85Mn, 1Cr, 0.3Mo
EN 19A	0.4C, 0.85Mn, 1Cr, 0.2Mo
EN 19B	0.37C, 0.85Mn, 1Cr, 0.2Mo
EN 19C	0.42C, 0.85Mn, 1Cr, 0.2Mo
EN 23	0.31C, 0.6Mn, 1Cr, 3Ni
EN 24	0.4C, 0.6Mn, 1.2Cr, 0.3Mo, 1.5Ni
EN 25	0.31C, 0.65Cr, 0.55Mo, 2.5Ni
EN 26	0.4C, 0.65Cr, 0.55Mo, 2.5Ni
EN 27	0.31C, 1Cr, 0.3Mo, 3Ni
EN 30B	0.3C, 1.25Cr, 0.3Mo, 4.1Ni
EN 31	1C, 1.4Cr
EN 40B	0.24C, 0.6Mn, 3.25Cr, 0.55Mo
EN 40C	0.39C, 3.25Cr, 1Mo, 0.2V
EN 41A	0.31C, 1.6Cr, 0.2Mo, 1.1Al
EN 41B	0.39C, 1.6Cr, 0.2Mo, 1.1Al
En 56A	410
En 56AM	416
En 56B	420
En 56BM	416
En 56C	420
En 56CM	416
En 56D	420
EN 57 Se	431
EN 9	0.55C, 0.7Mn

Table 17.24 Tool materials with near-matching chemical compositions *(continued)*

Other tool steels	Comparison
EN24	0.24C, 1Cr, 0.3Mo, 1.5Ni
ERCO 3	0.32C, 3Cr, 2.8Mo, 0.5V, 3Co
ETA	0.9C, 0.5W, 0.15V
ETH	W110
EW 15 Special	0.19C, 1.3Cr, 0.2Mo, 4Ni
EW 52 H	0.2C, 1.2Cr, 0.25Mo, 1Mn
EWPX	1.2581
EWX 40	0.07C, 4Cr, 0.5Mo
EXD 5	H23
EXD1	0.33C, 1.5Cr, 5.5W, 3.75Ni
Extra Double Horseman	T6
Extra Extra Special M	0.77C, 22W, 4.25Cr, 1.5V
Extra Quality Hard Temper Chrome	W5
Extra Quality Tough and Hard Temper	W2
Extra Tough	W1
Extra Triple Conqueror	1.55C, 0.55Cr, 5.5W
Extra Triple Griffin	1.5C, 0.25Cr, 6W
Extra Zah Special	1.1525
Extra Zah	1.1525
Extra Zahhart Special	1.1545
Extra Zahhart	1.1545
E-Z 85 WCDV 6	1.3554LW
F 2	0.58C, 0.8Mn, 0.75Cr, 0.2V
F 543	0.09C, 4.8Ni, 3.9Cr, 3Mo
F 5553	1.3207
F.5107	1.1625
F.5117	1.1645
F.5123	1.1663
F.5211	1.2601
F.5212	1.2080
F.5213	1.2436
F.5220	1.2510
F.5227	1.2363
F.5230	1.2067
F.5233	1.2419
F.5241	1.2542
F.5242	1.2550
F.5267	1.2316
F.5307	1.2711
F.5313	1.2365
F.5317	1.2343
F.5318	1.2344
F.5323	1.2581

Table 17.24 Tool materials with near-matching chemical compositions *(continued)*

Other tool steels	Comparison
F.5520	1.3355
F.5530	1.3255
F.5540	1.3265
F.5563	1.3202
F.5603	1.3343
F.5607	1.3348
F.5611	1.3249
F.5613	1.3243
F.5615	1.3246
F.5617	1.3247
Favorit	1.2842
FCW5	1.35C, 5W
FDAC	1.2343
Firthob	P3
Flashut	0.95C, 3.75Cr, 0.18V, 0.23Mo
FMP 035	W2
FMP 1850	0.55C, 18W, 4.1Cr, 0.7V, 1Mo
FMP 200	O1
FMP 328	H11
FMP 329	H13
FMP 336	D2
FMP 338	D3
FMP 348	0.42C, 1.55Cr, 0.3Mo, 4Ni
FMP 379	A2
FMP 399	S1
FMP 455	T4
FMP 470	0.8C, 22W, 4.5Cr, 1.5V, 1Mo
FMP 501	M1
FMP 504	M4
FMP 505	H21
FMP 507	0.3C, 9W, 3Cr, 0.3V, 0.5Mo, 2.5Ni
FMP 513	H12
FMP 526	M41
FMP 530	M30
FMP 536	M15
FMP 542	M42
FMP 555C	T15
FMP 563	M3
FMP 599	0.7C, 14W, 4.25Cr, 0.8V, 0.75Mo
FMP 622	T1
FMP 644	1.25C, 13.5W, 4.75Cr, 4V
FMP 682	H43
FMP 808	T6
FMP 828	T5

Table 17.24 Tool materials with near-matching chemical compositions *(continued)*

Other tool steels	Comparison
FMP 842	T2
FMP 922	M7
FMP 928	M34
FMP 929	M43
FMP 933	1.3C, 9.25W, 4.25Cr, 3.5V, 3.75Mo, 10Co
FMP 948	M10
FT 125	0.4C, 1.5Ni, 1.2Cr, 0.3Mo
FT 95V	W2
G 1 Special	0.32C, 1.3Cr, 0.3Mo, 4.1Ni
G 14C	0.44C, 1.5Cr, 0.4Mo, 3.5Ni
Geordie	W1
GFS	O1
Gigant 50	1.3355
Gigant M5	1.3343
Gigant M5Co	1.3243
Gigant M9	1.3346
GN	D2
GO 31	1.2419
GO3	1.2419
GOA	1.2510
Grandios Extra 655	1.3243
Grane	1.2721
Green Lable	W1
Ground Flat Stock	O1
GS	O1
GS3	1.2721
Guss. Stahl 3	1.1830
Guss. Stahl 5 H	1.1730
Guss. Stahl 4 W	1.1740
GZ	T4
H2	L1
H33	A2
H41	1.3346
H42	D2
H50	H13
H61	H10
Hammer & Zange 45	1.1730
Hammer & Zange 60	1.1740
Hardenite (CHD)	W1
HD 3MX	0.33C, 0.9Si, 1W, 3Cr, 0.9V, 2.8Mo, 3Co
HD10	H21
HD3	H21
HDB1	W1
HDB2	0.6C, 0.65Mn, 1.25Ni

Table 17.24 Tool materials with near-matching chemical compositions *(continued)*

Other tool steels	Comparison
HDB3	0.55C, 0.65Mn, 0.65Cr, 1.5Ni
HDB5	0.55C, 0.65Mn, 0.6Cr, 0.2Mo, 1.65Ni
HDC	1.2581
HDS	H13
HDZ	0.3C, 3.4Cr, 0.34V, 8.4W
Hecla 105	0.5C, 0.7Mn, 1Cr
Hecla 135	0.6C, 2Cr, 2Ni, 0.5Mo
Hecla 139	0.55C, 0.7Cr, 0.25Mo, 1.75Ni
Hecla 149C	H21
Hecla 15	O2
Hecla 150	0.4C, 0.7Mn, 1Cr, 0.2Mo
Hecla 159	D2
Hecla 174	H11
Hecla 175	A2
Hecla 177	H12
Hecla 18	W1
Hecla 28	W2
Hecla 67	0.3C, 1.3Cr, 4.25Ni
Hecla 67B	0.3C, 1.3Cr, 4.25Ni, 0.3Mo
Hecla D17	0.6C, 0.8Mn
HJS 202	0.11C, 0.5Cr, 1.25Ni
HJS 555	0.16, 1.15Cr, 0.3Mo, 4.25Ni
HJS 5-6-2	M2
HJS 626	T4
HJS M1	M1
HM3	H21
HMI	O2
Holdax	0.4C, 1.5Mn, 1.9Cr, 0.2Mo
Horseman Brand (14% W)	0.7C, 14W, 3.75Cr, 0.5V
Horseman Brand (22% W)	0.78C, 22W, 4.5Cr, 1V
Horseman	T1
HOV	H12
HP	H21
HRO 1243	1.2721
HRS	L1
HS 10-4-3-10	1.3207
HS 18-0-1	1.3355
HS 18-0-1-10	1.3265
HS 1-8-1	1.3346
HS 18-1-1-5	1.3255
HS 2-9-1-8	1.3247
HS 2-9-2	1.3348
HS 6-5-2	1.3343
HS 6-5-2-5	1.3243

Table 17.24 Tool materials with near-matching chemical compositions *(continued)*

Other tool steels	Comparison
HS 7-4-2-5	1.3246
HSB 1	0.85C, 0.6Mn
HSB 4 Special	0.8C, 0.5Cr, 0.3V
HSC 6-5-3	1.3342
HSM/W9A	H21
HV5	T15
HW 5	H12
HW2N	0.28C, 2.25Ni, 2.1W, 0.85Cr, 0.3V, 0.5Mo
HW4	0.3C, 4.5W, 2.5Cr, 0.55V, 0.3Mo
HWX	0.36C, 0.7Mn, 5.8Ni, 2.8W, 13.5Cr, 0.65V
Hyblade	S5
Hyform	W1
IAS	O7
IBD	0.32C, 1.25Cr, 0.35Mo, 4.25Ni
ICN	0.2C, 1Si, 1Mn, 14Cr, 1Ni
ICS	W1
ICW	D3
IDI	O1
Impax Supreme	0.37C, 1.4Mn, 2Cr, 1Ni, 0.2Mo
Inmanite	T4
Intra	F1
Intrinsic Special	H21
Invincible 18%	T1
Invincible 22%	0.75C, 4Cr, 22W, 1.25V
IR	D3
Iron Duke	0.4C, 0.3Cr, 3.25Ni
IU	1.9C, 12.5Cr, 0.75Mo, 0.3V
J24	M2
J28	H19
J34	M2
J35	M1
J36	T15
J37	1.3C, 9.5W, 4.5Cr, 3.5V, 4Mo, 10Co
J4 V	W2
J42	M42
JA	A2
JC 20	H20
Je	H13
JEM	W1
Jethete	0.14C, 0.7Mn, 12Cr, 0.35V, 1.8Mo, 2.4Ni
JG	0.5C, 1Mn, 1Cr, 0.3Mo
K 4 Special	O1
K 9	O1
K.E. 1006	1.2833

Table 17.24 Tool materials with near-matching chemical compositions *(continued)*

Other tool steels	Comparison
K.E. 1036	0.9C, 13Cr
K.E. 15	410
K.E. 160	1.05C, 0.5Cr
K.E. 169	0.14C, 0.85Cr, 3.4Ni, 0.15Mo
K.E. 200	1.4C, 13Cr, 0.6Mo, 3.5Co
K.E. 226	1.05C, 0.5Cr, 2.1W
K.E. 227	0.5C, 1.2Mn, 0.6Cr, 0.2Mo
K.E. 25	1.4021
K.E. 275	H21
K.E. 339	0.28C, 2.25Ni, 2.5Cr, 9.5W, 0.15V
K.E. 35	420
K.E. 355	1.2766
K.E. 396	1.2714
K.E. 40A	416
K.E. 43	431
K.E. 595	1.25C, 0.85Mn, 1.2Cr, 1.3W
K.E. 621	W5
K.E. 637	O1
K.E. 672	1.2510
K.E. 805	1.6582
K.E. 839	1.2067
K.E. 896	1.2241
K.E. 897	1.5864
K.E. 960	1.2547
K.E. 961	1.55C, 12.5Cr, 0.55W
K.E. 965	0.4C, 1.6Si, 13Cr, 2.75W, 0.55Mo
K.E. 970	1.2080
K.E. A203	440B
K.E. A207	440C
K.E. A28	0.43C, 13.25Cr, 1Ni
K.E. A505	431
K.E. A508	420
K.E. Diamond No. 10	1.5C, 0.6Cr, 5.75W
K.E.A. 108	1C, 0.55Mn
K.E.A. 138	H23
K.E.A. 145	1.2344
K.E.A. 162	1.2363
K.E.A. 172	A6
K.E.A. 180	1.2601
K.E.A. 205	1.4C, 3.45V
K.E.A. 220	1.2330
K.E.A. 222	H19
K.E.A. 227	0.5C, 1.2Mn, 0.65Cr, 0.2Mo
K.E.A. 275	H21

Table 17.24 Tool materials with near-matching chemical compositions *(continued)*

Other tool steels	Comparison
K.E.A. 28	0.45C, 13.25Cr, 1Ni
K.E.A. 476	1.2601
K.E.A. 505	431
K100	1.2080
K105	1.2601
K110	1.2379
K18	H12
K190	2.3C, 12.5Cr, 4V, 1.1Mo
K200	1.2067
K305	1.2363
K306	0.51C, 1Si, 5Cr, 1.4V, 1.4Mo
K4 Special KSA	O2
K455	1.2550
K460	1.2510
K5	T4
K5M	1.3243
K600	1.2767
K605	1.2721
K630	1.2770
K720	1.2842
K980	W1
K990	W110
KAO	F2
KAOC	F3
KAOK	1.3C, 5W
KCNM	1.2721
Kelock 1014	0.75C, 6Cr, 14.5W, 0.8Mo, 1.4V
Kelock 1021	T5
Kelock 237	1.3355
Kelock 795	0.7C, 4.1Cr, 14.25W, 0.6V
Kelock 873	T4
Kelock A157	1.3343
Kelock A182	1.3346
Kelock A229	M42
KK	M2
KKK	2.1C, 0.65Mn, 0.85Ni, 13Cr, 0.65Mo
KLAH	S1
KLD	1.2550
KM	1.2631
KMV	D2
KN90	1.2770
KNL	D2A
Komalp 3Herz	1.3355
Komalp MO	1.3346

Table 17.24 Tool materials with near-matching chemical compositions *(continued)*

Other tool steels	Comparison
Komalp WM	1.3343
Kova 57	1.25C, 4Cr, 10W, 3.5V, 3Mo, 9Co
KW10	410
KW30	420
KWB	431
L.T.A.H.	A6
LCHD	H21
LE	0.4C, 3.25Ni
LTAH	A6
LTTS	O7
M Brand	0.6C, 14.5W, 3.6Cr, 0.3V
M200	1.2312
M201	1.2311
M238	1.2378
M300	1.2316
M310	1.2083
M314	0.3C, 0.7Si, 1.1Mn, 16.8Cr, 0.15Mo
M390PM	1.9C, 20Cr, 1Mo, 4V, 0.6W
Malloy	0.6C, 1.1Si, 1.1Cr, 0.25Mo
Mammut Special	1.3355
Maxmith	0.4C, 0.6Mn, 3.25Ni
MC	M1
MCMO	A6
MCT	O1
MCV	0.5C, 1.25Mn, 0.55Cr, 0.25Mo, 0.55Ni
MIC 4	O1
MIC 8	O6
MIC	O2
Minerva HC	0.53C, 1.8Cr, 1.9W, 0.2V
Minerva LC	0.43C, 1.8Cr, 1.9W, 0.2V
MJW CR	1.25C, 0.5–3Cr
MJW LT	1.25C, 2.5W
MJW OH	O2
MJW	W1
MJWCV	W2
MKZ	H21
Mo 500	M2
Mo 53	M3
Mo 550H	M41
Mo 550	M35
Mo 900	M1
Mo 92	M7
Mo 980H	M42
MOG 111	0.45C, 1.5Cr, 0.8V, 0.5W, 0.5Mo

Table 17.24 Tool materials with near-matching chemical compositions *(continued)*

Other tool steels	Comparison
MOG 330	0.32C, 3Cr, 0.6V, 3Mo
MoG 510V	H13
MoG 510	H11
MoG 511	H12
MOHD	0.36C, 6Mo, 1W, 3.5Cr, 0.75V
Molycut 562	M2
Monarch (HCR)	D6
Monarch (OHB)	O1
Monarch (SSM-H)	0.75C, 22W, 7.5Cr, 1.30V
Monarch 652	M2
Monarch BLA	H13
Monarch DW8	0.3C, 7.75W, 2.75Cr, 0.45V
Monarch General Utility	0.6C, 0.7Mn, 0.7Cr
Monarch NCG	0.35C, 1.5Cr, 0.3Mo, 4Ni
Monarch PCS	0.8C, 1.25Cr, 0.5Ni, 1.9W
Monarch Special TAN	0.30C, 0.65Mn, 2.5Ni, 0.6Mo, 0.75Cr
Monarch TAN	L6
Morapid Extra 500	1.3243
Morapid Extra 9	1.3346
More 397	M7
More 500	M35
More 9S	M1
More V30	M3
Motor Magnus	M2
Motor Maximum	T1
Motor Special	0.68C, 4Cr, 14W, 0.25V
Movan Special	1.25C, 9.5W, 4.25Cr, 3.2V, 3.25Mo, 10Co
MOW 562	M2
MP	1.2721
MS	1.2419
MS1	1.2510
MST	1.2842
MT2	S1
MTDS	H21
Multitherm	H12
MY	0.55C, 0.95Cr, 0.95Si
Myextra	1.2542
N100	410
N316	416
N320	1.4021
N324	420
N350	1.4057
N530	1.4028
N540	1.4034

Table 17.24 Tool materials with near-matching chemical compositions *(continued)*

Other tool steels	Comparison
N555	0.6C, 14.1Cr, 0.7Mo, 0.1V
N685	1.2361
N690	1.4528
NAT	H11
NBS	1.2721
NC1510	0.4C, 1Cr, 1.5Ni
NC41M	0.3C, 1.25Cr, 0.3Mo, 4.25Ni
NCM	0.52C, 0.6Cr, 0.25Mo, 1.5Ni
NCM1	0.35C, 1.1Cr, 0.2Mo, 1.55Ni
NCM2	0.35C, 0.6Cr, 2.5Ni
NCM3	0.35C, 1.2Cr, 3.5Ni
NCM4	0.35C, 1.3Cr, 0.3Mo, 4.1Ni
ND	T1
New Capital	0.7C, 3.75Cr, 14.5W, 0.75V
Newhall	O1
NF	0.38C, 1.65Ni, 1.2Cr, 0.25Mo
NG	0.55C, 0.7Cr, 0.1V, 0.2Mo, 1.7Ni
NG2 Supra	0.55C, 1.1Cr, 0.1V, 0.5Mo, 1.7Ni
NH	0.31C, 4.1Ni, 1.3Cr, 0.3Mo
NHP	0.26C, 9W, 3Cr, 0.3V, 0.5Mo, 2.5Ni
Nita	0.53C, 1.5Cr, 0.21Mo, 1.1Al
No 0 Hardenite	W1
No 1 Hardenite	W1
No 1 Monarch	0.65C, 4Cr, 14W, 0.6V
No 1 Steel	0.6C, 0.7Mn
No 10 Hardenite	1.2C, 3Cr, 1W
No 2 Hardenite	W1
No 3 Hardenite	W1
No 4 Hardenite Van	1.4C, 0.4Cr, 3.6V, 0.4Mo
No 5 Hardenite	L1
No 5	H21
No 6 Hardenite	1C, 1Mn, 1.35Cr
No 7 Hardenite (70H)	O1
No 8 Hardenite	0.85C, 1W, 1.35Cr, 0.2V
Nonvar	O2
Novo 6/5/2	M2
Novo 9/2	M1
Novo C	0.75C, 13.5W, 6.25Cr, 1.25V, 0.75Mo
Novo Enormous	T6
Novo Max	T4
Novo Superb	1.2C, 10.5W, 4.25Cr, 3.25V, 3.75Mo, 10Co
Novo Superior SS	0.75C, 21W, 4.5Cr, 1.25V
Novo Superior	T1
Novo TCV	1.5C, 11W, 5Cr, 5V

Table 17.24 Tool materials with near-matching chemical compositions *(continued)*

Other tool steels		Comparison
Novo V		T15
Novo VHC		M42
Novo		0.7C, 14W, 0.4Cr, 0.5V
NRM		D2
NRW		D3
NS 12		W5
NSC		1C, 1.1Mn, 1.4Cr
NSCD		O2
NSCM1 (HC)		0.6C, 0.95Cr, 0.3Mo, 1.4Ni
NSCT		L1
NSS 3		O2
NTC	(Barworth)	0.5C, 1Mn, 1Cr, 0.35Mo
NTC	(Huntsman)	0.4C, 0.5W, 0.4Cr
Nu-Die Xtra		H13
NV		H12
Nyblade		0.47C, 1Si, 3Ni, 0.6Cr, 0.2Mo
O		0.35C, 1.5Mn
O6S		W2
O9B		O1
OHD		1C, 1.6Cr, 0.5W
OK Crown		S1
OO		0.6C, 0.7Si, 1.1Cr, 1.9W, 0.3V
Optimax		0.38C, 0.9Si, 0.5Mn, 13.6Cr, 0.3V
Orvar Supreme		0.37C, 1Si, 0.4Mn, 5.3Cr, 1.4Mo, 0.9V
P 704		O2
P 720		L1
P1008		1.2083
P1009		1.2316
P256		0.55C, 0.7Mn
P280		0.6C, 0.6Cr
P552		0.3C, 4.25Ni, 1.25Cr, 0.3Mo
P553		0.4C, 1.5Ni, 1.1Cr, 0.3Mo
P558		0.3C, 2.5Ni, 0.7Cr, 0.5Mo
P564		0.3C, 3Ni, 0.74Cr
P576		1.2767
P602		0.4C, 0.65Mn, 1.1Cr, 0.3Mo
P609		0.4C, 0.8Mn, 1Cr
P618		0.4C, 0.3Ni, 3Cr, 1Mo, 0.25V
P973		1.2770
PAC		H19
Pax No. 2		1.2547
Pax Non-break No 2		S1
PCS		H12
PCSK		S1

Table 17.24 Tool materials with near-matching chemical compositions *(continued)*

Other tool steels	Comparison
PDS1	1.1740
PDX	0.28C, 0.85Cr, 2.25W, 0.5Mo, 0.3V, 2.25Ni
PE655	1.3243
PGT	O7
Pitho	O1
Plain Carbon Steels	W1
Plasmould	0.35C, 4.3Ni, 1.3Cr, 0.3V
PLMB/1	0.55C, 0.65Mn, 1.63Ni, 0.7Cr, 0.28Mo
PLMC/1	0.42C, 2.6Ni, 0.65Cr, 0.6Mo
PLMC/2	0.4C, 4.25Ni, 1.4Cr, 0.25Mo
Pluto Paramount	0.78C, 4.5Cr, 0.7Mo, 18.75W, 1.25V, 10Co
Pluto Perfectum	T4
Pluto Plus	M42
Pluto Premier	M15
Plutocrat	0.78C, 4.25Cr, 0.5Mo, 22W, 1Co
Plutog	1.3355
PN1	S5
Pneumo	0.45C, 0.75Si, 1.25Cr, 2W
Pnusnap OH	0.43C, 1Si, 1.8W, 1Cr
Pnusnap WH	0.35C, 1Cr, 1.8W
Premo	0.5C, 4Cr, 0.5Mo
Prima Mittel	1.1645
Prima Zah	1.1625
Prima Zahhart	1.1645
PRN2.DCCM	1.05C, 1.3Mn, 1.3Cr
PWE 893	0.25C, 8.25W, 2.4Cr, 0.18V, 0.4Mo, 2.35Ni
Q	0.4C, 0.8Mn
R 9030	0.35C, 0.75Cr, 0.5Mo, 2.75Ni
R18	1.3355
R18K5F2	1.3255
R6AM5	1.3343
R6M5	1.3554LW
R6M5K5	1.3243
RAB 420ESR	0.37C, 0.5Mn, 0.75Si, 13.5Cr, 0.3V
RAB1	0.4C, 4.1Ni, 1.3Cr, 0.3Mo
RAB20	0.4C, 1.5Mn, 2Cr, 0.2Mo
Ramax S	0.33C, 1.4Mn, 16.7Cr
RB 10	1.1645
RBD	H21
RCC Spezial	1.2601
RCC Supra	1.2379
RCC	1.2080
RCW2H	1.2581
RDC1	1.2606

Table 17.24 Tool materials with near-matching chemical compositions *(continued)*

Other tool steels	Comparison
RDC2	1.2343
RDC2V	1.2344
Red Band Cast Steel	W1
Red Lable	W1
REGIN 3	1.2542
Regor	A2
Remount	W1
RHB-E	0.22C, 2.25Ni, 10W, 2.25Cr, 0.45Mo, 0.20V
Rigor	A2
RKCM	1.2363
ROP 57	1.2601
ROP19	1.2344
ROP21	1.2363
RPG3	1.2365
RSD 13	A2
RT 10	1.1545
RT 1733	1.2510
RT 8	1.1525
RTO912	1.2542
RTW2H	1.2542
RTWK	1.2550
RUS	1.2842
RUS3	1.2510
RUS4	1.2419
RV	L3
RWA	1.2567
RWS	1.2510
S 10-4-3-10	1.3207
S 12-1-4-5	1.3202
S 18-0-1	1.3355
S 18-1-2-10	1.3265
S 18-1-2-5	1.3255
S 2-10-1-8	1.3247
S 2-9-1	1.3346
S 2-9-2	1.3348
S 2-9-2-8	1.3249
S 6-5-2	1.3343
S 6-5-2-5	1.3243
S 6-5-3	1.3344
S 7-4-2-5	1.3246
S.K. Silver Steel	1.2002
S.R.E.	T1
S200	1.3355

Table 17.24 Tool materials with near-matching chemical compositions *(continued)*

Other tool steels	Comparison
S305	1.3255
S400	M7
S401	1.3346
S402	M1 Special
S500	1.3247
S600	1.3343
S605	M3
S690	M4
S705	1.3243
SA1	1.2419
Saban Extra	T1
Saben 6-5-2	1.3343
Saben 652	M2
Saben HC	1.25C, 13.5W, 4.6Cr, 3.75V, 0.3Mo
Saben Kerau	0.75C, 22W, 4.25Cr, 1.5V, 0.25Mo
Saben Tenco	T5
Saben Wunda	T4
Sabex	0.3C, 9.5W, 2.25Cr, 0.15V, 1.75Ni
SBM	0.58C, 1Cr, 0.25V
SBR	1.1830
SC 13 NEOR	D3
SC 25	D2
SC 26	D3
SC 38	1.5C, 12Cr, 1.1V, 0.75Mo
SC 40	0.4C, 12Cr, 0.5Ni
SC 45 168	0.75C, 17Cr
SC 6-5-2	1.3342
SC	S4
SC90	1.5C, 1.1Mn, 0.9Cr
SCD	1.2363
SCH 3	1.4028
SCR1	0.47C, 1Si, 1.4Cr
SCS 2	1.4021
SCV 5	A2
SCV	L2
SD 20	0.35C, 1.2Cr, 1.75Ni
Senator A	H11
Senator B	H14
Senator C	H21
SG	H21
SGHV	1.2581
SGT	1.2419
SHC1	0.44C, 1Si, 0.65Mn, 0.65Cr, 3Ni
SICV	1.15C, 0.7Cr, 0.1V

Table 17.24 Tool materials with near-matching chemical compositions *(continued)*

Other tool steels		Comparison
Silver Steel		1.2002
SIW	(Balfour)	0.67C, 1.15Si, 0.7Mn, 0.6Cr, 1.3W
SIW	(Fortuna)	1.20C, 0.1V, 1W
SK 1		1.1673
SK 2		1.1663
SK 3		1.1645
SK 4		0.95C
SK 5		1.1625
SK 53		1.2510
SK 6		1.1625
SK 7		1.1740
SKC 11		1.2057
SKC 24		0.38C, 0.7Mn, 0.5Cr, 0.3Mo, 3Ni
SKC 3		1.1625
SKC 31		0.2C, 0.9Mn, 1.5Cr, 0.55Mo, 3Ni
SKD 1		1.2080
SKD 11		1.5C, 12Cr, 1Mo, 0.35V
SKD 12		1.2363
SKD 4		1.2567
SKD 5		1.2581
SKD 6		1.2343
SKD 61		1.2344
SKD 62		1.2606
SKD 7		0.33C, 3Cr, 2.75Mo, 0.55V
SKD 8		0.4C, 4.15Co, 4.35Cr, 0.4Mo, 2V, 4.15W
SKH 10		1.5C, 4.7Co, 4.1Cr, 4.7V, 12.5W
SKH 2		1.3355
SKH 3		1.3255
SKH 4		1.3265
SKH 4A		1.3265
SKH 51		1.3343
SKH 52		1.3344
SKH 53		1.3344
SKH 54		1.3C, 4.1Cr, 4.2V, 5.9W
SKH 55		1.3243
SKH 56		0.9C, 8Co, 4.1Cr, 5Mo, 1.95V, 6.2W
SKH 57		1.3207
SKH 58		1C, 4Cr, 8.7Mo, 1.95V, 1.8W
SKH 59		1.1C, 8Co, 4Cr, 9.5Mo, 1.2V, 1.5W
SKS 11		1.2562
SKS 2		1.2419
SKS 21		1.2515
SKS 3		1.2419
SKS 31		1.2419

Table 17.24 Tool materials with near-matching chemical compositions *(continued)*

Other tool steels	Comparison
SKS 4	0.5C, 0.75Cr, 0.75W
SKS 41	0.4C, 1.25Cr, 3W
SKS 43	1.2833
SKS 44	0.85C, 0.2V
SKS 5	0.8C, 0.35Cr, 1Ni
SKS 51	0.8C, 0.35Cr, 1.65Ni
SKS 7	1.2442
SKS 8	1.2008
SKS 93	1C, 0.95Mn, 0.4Cr
SKS 94	0.95C, 0.95Mn, 0.4Cr
SKS 95	0.85C, 0.95Mn, 0.4Cr
SKT 3	0.4C, 0.8Mn, 1Cr, 0.4Mo, 0.4Ni, 0.2V
SKT 4	1.2713
SLD 2	1.2379
SLD	1.2379
SLZ	0.36C, 1.75Si, 13Cr, 2.75W, 0.65V, 5.75Ni
SMN	S4
SMO	S5
SMV 200	O2
SND	W2
SNK	1.2770
SNSC	O1
Somcold	D2
Somdie	0.48C, 1Si, 0.5Mo, 1.20Cr, 0.07V, 1Ni
Somtuf	?C, 2Ni, 2W
SP3	1.1830
SP5W	1.1740
SP6W	1.1730
SPCR	O1
Spear 5/6/2	M2
Spear 50	0.55C, 0.6Mn
Spear 75	S1
Spear 9/2/1	M1
Spear AHC	S1
Spear B1	0.32C, 1.25Cr, 4.25Ni
Spear B2	0.20C, 13Cr
Spear D1	L4
Spear D12	D3
Spear D13	D4
Spear D14	D5
Spear D15	A2
Spear D16	D2
Spear D17	D2
Spear D5	H13

Table 17.24 Tool materials with near-matching chemical compositions *(continued)*

Other tool steels	Comparison
Spear D7	H12
Spear D9	H21
Spear DX	0.28C, 2.2W, 2.25Ni, 0.85Cr
Spear Leapfrog	0.65C, 14W
Spear M42	M42
Spear Mermaid	T1
Spear No 2 Vanadium	W2
Spear No 2	W1
Spear NS	O1
Spear PS	L2
Spear Superior	T6
Spear Triple Mermaid	T4
Special Ardho	0.35C, 3.2Ni, 0.78Cr
Special BB/HDS	H21
Special Cold Pressing Vanadium Steel	W2
Special Conqueror Vanadium	W2
Special HW	H21
Special K	D3
Special M	T1
Speedicut Leda	T4
Speedicut Maximum 18	T1
Speedicut Superleda	T5
Speedicut Vanleda	T15
Spenard	O2
Spezial K	1.2080
Spezial K5	1.2363
Spezial K8	1.2631
Spezial KMV	1.2379
Spezial KNL	1.2601
SPG Extra V	1.2344
SPG Extra	1.2343
SPG Special W	1.2606
SPM	O1
SPS	0.5C, 0.75Si, 0.2V, 0.7Mn, 2W, 0.9Cr
Spur 6/5/2	M2
Spur 9/2/1	M1
SPW	0.25C, 13Cr
SRE 500	T4
Sremo	M2
SS652	1.3343
SS921	1.3346
SSC	0.58C, 1.05Cr, 0.3V
ST	0.7C

Table 17.24 Tool materials with near-matching chemical compositions *(continued)*

Other tool steels	Comparison
Stag Extra Special	T4
Stag Major	T6
Stag Mo 562	M2
Stag V55	T15
Stag Vanco	1.52C, 17.5W, 5Cr, 8.75Co, 0.5Mo
Stage Special	T1
Stagmold	0.32C, 4.1Ni, 1.3Cr, 0.3Mo
Stamold H.T.	0.36C, 1Si, 1Mn, 16Cr, 1.2Mo
Statos CRD Spezial	1.2601
Statos CRD Supra	1.2379
Statos CRLS	1.2363
Statos Extra	1.2842
Statos Spezial	1.2080
Statos Superior	1.2419
Stavax ESR	0.38C, 0.9Si, 0.5Mn, 13.6Cr, 0.3V
STNS	O1
Stora 16	O1
Stora 18	S1
Stora 214	W2
Stora 25	T1
Stora 27	T4
Stora 29	M2
Stora 30	1.2C, 6.5W, 4Cr, 3.4V, 5Mo, 10Co
Stora 323	0.3C, 5.3W, 1.5Cr, 0.1V, 0.6Mo, 4.8Co
Stora 364	D2
Stora 424	M35
Stora 431	M1
Stora 433	M7
Stora 62 G	D3
Stora 62	D6
Stora 65	A2
Stora 67	H13
Stora 85	0.55C, 1Cr, 3Ni, 0.3Mo
SUJ 1	1.2057
SUJ 2	1.3505
SUJ 3	1.2127
SUJ 4	1.3505
SUJ 5	1C, 1Mn, 1Cr, 0.2Mo
Super Austenite	T1
Super AW23	0.3C, 4.1Ni, 1.25Cr, 0.25Mo
Super C12	D2
Super Capital	1.3C, 4.25Cr, 9W, 3.5V, 8.5Co, 3Mo
Super Chromium	D3
Super CS	D5

Table 17.24 Tool materials with near-matching chemical compositions *(continued)*

Other tool steels	Comparison
Super Dominant	T1
Super Inmanite	0.8C, 22W, 4.5Cr, 1.5V, 1Mo, 10Co
Super Invincible 5% Cobalt	T4
Super Invincible Advance 10% Cobalt	T5
Super Invincible TB	T8
Super Maxel	0.5C, 1.25Mn, 0.65Cr, 0.18Mo
Super Monarch	T1
Super Rapid Extra Mo	1.3343
Super Rapid	T1
Super Super Monarch (E1)	T4
Super TKL	S1
Superapid Extra	1.3355
Superior Spur	T1
Supermax 98	M42
Superwear	A7
SUS 403	1.4006
SUS 410	1.4006
SUS 420 J 1	1.4021
SUS 420 J 2	1.4028
SUS 431	1.4057
SUS 440 B	440 B
Sverker 21	D2
Sverker 3	D6
SW 111	1.05C, 1Cr, 1Mn, 1.2W
SW 55	O1
SW1	1.2419
SWS	1.2510
T 1040	W4
T Quality	W1
T4 (Rochling)	1.1730
T5 (Rochling)	1.1740
T6 (Rochling)	1.1740
TA	W1
TB	W1
TD	W1
TDC	F1
Tenax C1	L1
Tenax C2	0.6C, 0.7Mn, 0.7Cr
Tenax C9/5	W5
Tenax CTS	S1
Tenax CV	0.5C, 0.7Mn, 1Cr, 0.25V
Tenax DCB	0.3C, 4.25Ni, 1.25Cr, 0.3Mo
Tenax FPC 5	H11

Table 17.24 Tool materials with near-matching chemical compositions *(continued)*

Other tool steels	Comparison
Tenax FPCT	H12
Tenax FPDD	F2
Tenax K 250	D3
Tenax K250 CN	D5
Tenax K250MV	D2
Tenax PNS	O1
Tenitkl Special	1.2550
Tenitw Special	1.2542
Thermodie	0.55C, 0.9Cr, 0.75Mo, 2.15Ni
Thremalloy	0.27C, 8.5Cr, 0.55V, 9.5Co, 4Mo
Thyrapid 3243	1.3243
Thyrapid 3343	1.3343
Thyrapid 3346	1.3346
Thyrapid 3355	1.3355
Thyrodur 1525	1.1525
Thyrodur 1545	1.1545
Thyrodur 1625	1.1625
Thyrodur 1645	1.1645
Thyrodur 1730	1.1730
Thyrodur 1740	1.1740
Thyrodur 1830	1.1830
Thyrodur 2067	1.2067
Thyrodur 2080	1.2080
Thyrodur 2363	1.2363
Thyrodur 2379	1.2379
Thyrodur 2419	1.2419
Thyrodur 2436	1.2436
Thyrodur 2510	1.2510
Thyrodur 2542	1.2542
Thyrodur 2550	1.2550
Thyrodur 2601	1.2601
Thyrodur 2606	1.2606
Thyrodur 2718	1.2718
Thyrodur 2721	1.2721
Thyrodur 2762	1.2762
Thyrodur 2767	1.2767
Thyrodur 2826	1.2826
Thyrodur 2833	1.2833
Thyrodur 2842	1.2842
Thyroplast 2083	1.2083
Thyroplast 2162	1.2162
Thyroplast 2311	1.2311
Thyroplast 2312	1.2312
Thyroplast 2316	1.2316

Table 17.24 Tool materials with near-matching chemical compositions *(continued)*

Other tool steels	Comparison
Thyroplast 2341	1.2341
Thyroplast 2711	1.2711
Thyroplast 2764	1.2764
Thyrotherm 2343	1.2343
Thyrotherm 2344	1.2344
Thyrotherm 2365	1.2365
Thyrotherm 2367	1.2367
Thyrotherm 2567	1.2567
Thyrotherm 2581	1.2581
Thyrotherm 2713	1.2713
Thyrotherm 2714	1.2714
Thyrotherm 2744	1.2744
Thyrotherm 2885	1.2885
TI	T1
TKL	S1
TM	W1
TMS	O2
Toba	0.4C, 0.7Mn, 1Cr, 0.2V, 0.5Ni
TOH	O1
Toughard	0.4C, 0.45Cr, 0.5W
TPM	O1
Treble Extra Cast Steel	W1
Treble Super Monarch (E4)	T6
TRG	W1
Triple 5 Monarch	T15
Triple Conqueror	F3
Triple Crescent	1.25C, 1.25Cr, 4.3W, 0.3V
Triple Griffin	1.35C, 0.25Cr, 2.75W
Triple Spur	T6
Triple Velos	T4
Triumphator 5	1.2363
Triumphator MW	1.2601
Triumphator VM	1.2379
Triumphator	1.2080
TTQ Triumph Superb Double Thousand	T6
Tufdie	?C, 13Cr, 3W, ?Ni, ?V
Tungsten Diamond	F3
Two Spur	T4
Tyrann Extra V	1.2542
U10	1.1645
U10A	1.1545
U13	1.1663
U3	1.2550

Table 17.24 Tool materials with near-matching chemical compositions *(continued)*

Other tool steels	Comparison
U8	1.1625
U8A	1.1525
UHB 11	0.5C
UHB 16	1.1525
UHB 19Va	W2
UHB 20	1.1545
UHB 29	1.3343
UHB 424	1.3243
UHB 431	1.3346
UHB 9	1.1730
UHB Special	1.2606
UHB20	W1
Ultra Capital 22	0.76C, 4.25Cr, 22W, 3.5V, 8.5Co, 3Mo
Ultra Capital 395	M4
Ultra Capital Plus 1	T4
Ultra Capital Plus 2	T6
Ultra Capital	T1
US Ultra 2	1.2344
US Ultra 4	1.2606
US Ultra	1.2343
V 175	T2
V6N	1.2770
VALAND1	1.2581
Vamox	0.3C, 2.25W, 0.8Cr, 0.3V, 0.5Mo, 2.25Ni
VAP	T1
VC 12	T6
Velos 42	M42
Velos UR	T1
Velos	M2
VF	H19
VG	M15
Viaduct 15	0.45C, 0.7Si, 1.1Cr, 1.9W, 0.3V
Vigilant	0.55C, 0.7Mn
VMC (H)	0.33C, 3Cr, 0.13V, 0.5Mo
VMC	H13
VWMC	H12
W 10 Extra	1C
W 10 Prima	1C
W 11 Prima	1.1C
W 1230	H23
W 182	T2
W 63 K	0.65C, 1Mn
W 63	0.6C
W 85 K	0.8C, 1Mn

Table 17.24 Tool materials with near-matching chemical compositions *(continued)*

Other tool steels	Comparison
W 93 K	0.9C, 1Mn
W100	1.2581
W108	1.2678
W10V	1C, 0.1V
W14 MO	O6
W18	T1
W2N	1.2770
W300	1.2343
W302	1.2344
W304	H12
W320	1.2365
W321	1.2885
W324	H15
W43	0.45C
W500	1.2711
W8N	0.85C, 0.1V, 0.8Ni
WA 235	0.35C, 1Cr, 0.2V, 2W, 1Si
WA 245	0.45C, 1Cr, 0.2V, 2W, 1Si
WA 250	S1
WA 255	0.6C, 1Cr, 0.2V, 2W
WA 530	0.30C, 2.5Cr, 0.6V, 4.5W
WA 930	H20
WAMV	1.2567
Warranted Crucible Cast Steel	W1
WATCO	0.33, 0.9Cr, 0.7Mn, 0.12V
WCD	1.2606
WCD2	1.2343
WCDV	1.2344
WCM Co	H10A
WCM	1.2365
WCO	H19
WCPS	0.48C, 0.7Si, 1.2W, 1.5Cr, 0.15V, 0.25Mo, 0.7Ni
WEL	L1
WF 8	0.62C, 1Mn, 0.6Cr, 1Si
WGKL	1C, 1.5Cr
WH2	1.3355
WHC	0.58C, 0.5Mn
White Band Cast Steel	W1
White Label	W2
Wing 111 A	1C, 1Cr, 1Mn
Wing 111 B	1C, 0.5Cr, 1.5Mn, 0.25V
Wing's Double Shear Temper	W1
WK5K	1.2567

Table 17.24 Tool materials with near-matching chemical compositions *(continued)*

Other tool steels	Comparison
WKM33	1.2365
WKZ	1.2581
WKZ50	1.2567
WL	A7
WM 13	0.4C, 13Cr
WMD Extra	H10A
WMD	1.2365
WMEV	1.2344
WMN	0.28C, 3Cr, 9.5W, 2Ni, 0.15V
WMO	1.2365
WMS	1.2606
WMWH	1.2581
WO 3	0.6C
Wolfram	S1
WS	1.07C, 1.2Mn, 1.7W, 1.1Cr
WSB Extra	1.15C, 2W
WSMA	1.2343
WVC	H19
WZ	0.4C, 3Cr, 0.25V, 0.4Mo, 11W
X 10 Cr 13	1.4006
X 100 CrMoV 5 1 KU	1.2363
X 100 CrMoV 5 1	1.2363
X 100 CrMoV 5	1.2363
X 105 CrCoMo 18 2	1.4528
X 12 Cr 13	1.4006
X 155 CrVMo 12 1 KU	1.2379
X 155 CrVMo 12 1	1.2379
X 16 CrNi 16	1.4057
X 160 CrMoV 12	1.2601
X 165 CrMo	1.2601
X 165 CrMoV 12	1.2601
X 165 CrMoW 12 KU	1.2601
X 19 NiCrMo 4	1.2764
X 20 Cr 13	1.4021
X 205 Cr 12 KU	1.2080
X 21 Cr 13 KU	1.4021
X 210 Cr 12	1.2080
X 210 CrW 12	1.2436
X 215 CrW 12 1 KU	1.2436
X 22 CrNi 17	1.4057
X 220 CrVMo 12 2	1.2378
X 30 Cr 13	1.4028
X 30 NiCrMo 16 6	1.6747
X 30 WCrV 5 3 KU	1.2567

Table 17.24 Tool materials with near-matching chemical compositions *(continued)*

Other tool steels	Comparison
X 30WCrV 5 3	1.2567
X 30WCrV 9 3KU	1.2581
X 30WCrV 9 3	1.2581
X 30WCrV 9	1.2581
X 31Cr 13KU	1.4028
X 32CrMoCoV 3 3 3	1.2885
X 32CrMoV 3 3	1.2365
X 36CrMo 17	1.2316
X 37CrMoV 5 1KU	1.2343
X 37CrMoV 5	1.2343
X 37CrMoW 5 1	1.2606
X 38CrMo 16 1	1.2316
X 38CrMo 16	1.2316
X 38CrMoV 5 1	1.2343
X 40Cr 13	1.4034
X 40Cr 14	1.4034
X 40CrMoV 5 1 1KU	1.2344
X 40CrMoV 5 1	1.2344
X 40CrMoV 5 3	1.2367
X 40CrMoV 5	1.2344
X 41Cr 13KU	1.2083
X 42Cr 13	1.2083
X 45Cr 13	1.4034
X 45NiCrMo 4	1.2767
X 50CrMoW 9 11	1.2631
X 55CrMo 14	1.4110
X 6CrMo 4	1.2341
X 82WMoV 6 5	1.3343
X 91CrMoV 18	1.2361
XC 100	1.1545
Y 100C 6	1.2067
Y_1 105V	1.2833
Y_1 105	1.1545
Y_1 80	1.1525
Y_1 90	1.1525
Y100C6	1.2067
Y1120	1.2631
Y_2 120	1.1663
Y_2 140	1.1673
Y_3 45	1.1730
Y_3 60	1.1740
YC3	1.1545
YC5	1.1830
YCK2	1.2631

Table 17.24 Tool materials with near-matching chemical compositions *(continued)*

Other tool steels	Comparison
YDC	1.2567
Yellow Label Steels	W1
Yellow Label	W1
YEM	1.2365
YHX2	1.3355
YK3	1.1545
YK5	1.1830
YK50	1.2842
YXM1	1.3343
YXM4	1.3243
YXMT	1.3346
Z 10 C 13	1.4006
Z 100 CDV 5	1.2363
Z 100 DCWV 09-04-02-02	1.3348
Z 110 DKCWV 09-08-04-02-01	1.3247
Z 110 WKCDV 07-05-04-04-02	1.3246
Z 12 C 13	1.4006
Z 120 WDCV 06-05-04-03	1.3344
Z 130 WDCV 06-05-04-04	1.3344
Z 130 WKCDV 10-10-04-04-03	1.3207
Z 15 CN 16-02	1.4057
Z 155 CDV 12	1.2601
Z 155 CVD 12-1	1.2379
Z 160 CDV 12	1.3279
Z 20 C 13	1.4021
Z 200 C 12	1.2080
Z 200 C 13	1.2080
Z 210 CV 13	1.2080
Z 210 CW 12	1.2436
Z 25 WCNV 9	0.28C, 2.25Ni, 2.5Cr, 9.5W, 0.15V
Z 30 C 13	1.4028
Z 30 CDV 12-28	1.2365
Z 30 WCV 9	1.2581
Z 30 WCV 9-3	1.2581
Z 32 WCV 5	1.2567
Z 35 CDV 5	1.2344
Z 35 CWDV 5	1.2606
Z 38 CDV 5	1.2343
Z 38 CDV 5-3	1.2367
Z 40 C 14	1.4034
Z 40 CDV 5	1.2344
Z 40 CDV 5-1	1.2344
Z 75 WV 18-01	1.3355
Z 8 WDCV 6	1.3343

Table 17.24 Tool materials with near-matching chemical compositions *(continued)*

Other tool steels	Comparison
Z 80 WCV 18-04-01	1.3355
Z 80 WDV 06 05	1.3343
Z 80 WKCV 18-05-04-01	1.3255
Z 85 DCWV 08-04-02-01	1.3346
Z 85 WDCV 06-05-04-02	1.3343
Z 85 WDKCV 06-05-05-04-02	1.3243
Z 90 WDCV 06-05-04-02	1.3342
Z 90 WDKCV 06-05-05-04-02	1.3243
ZN	0.25C, 8.5W, 3Cr, 0.25V, 2.25Ni

~H May include the letter 'H'.

Table 17.24 Tool materials with near-matching chemical compositions *(continued)*